GAS

TECHNICAL KNOWLEDGE FOR GAS OPERATIVES

Published by Pearson Education Limited, Edinburgh Gate, Harlow, Essex, CM20 2JE.

www.pearsonschoolsandfecolleges.co.uk

Text © JTL, 2013
Typeset by Tek-Art, Crawley Down, West Sussex
Original illustrations © Pearson Education Ltd, 2013
Illustrated by Tek-Art, Crawley Down, West Sussex

The right[s] of JTL to be identified as author of this work have been asserted by them in accordance with the Copyright, Designs and Patents Act 1988.

First published 2013

16 15 14 13
10 9 8 7 6 5 4 3 2 1

British Library Cataloguing in Publication Data
A catalogue record for this book is available from the British Library.

ISBN 978 1 447 93553 7

Printed by L.E.G.O. S.p.A. in Italy

Acknowledgements
The authors and publisher would like to thank the following individuals and organisations for permission to reproduce photographs:

(Key: b-bottom; c-centre; l-left; r-right; t-top)

Alamy Images: Art Directors & TRIP 137t, Justin Kase 35t, 467, Photofusion Picture Library 137c, Stephen French 39c, StockImages 408, Trevor Smithers ARPS 409t; **DK Images:** Steve Gorton 126b, 254cl; **Imagemore Co., Ltd:** 90, 255c; **Imagestate Media:** John Foxx Collection 74; **Pearson Education Ltd:** Clark Wiseman / Studio 8 27t, 27c, 45, 61, 253tc, 253cl, 253br, 255br (Cable locator), Gareth Boden 3, 7, 9b, 12, 13b, 14, 25t, 26, 35c, 35b, 36, 37, 39t, 44tl, 44b, 48, 57, 60, 71, 73c, 77, 81, 92c, 92b, 108, 115b, 116, 117, 124l, 124r, 125, 126c, 129, 137b, 151, 152, 153, 154, 176c, 176b, 178, 179, 189, 226, 238, 240, 242tc, 244, 245tr, 246br, 253tl, 253tr, 254tc, 254tr, 254bc, 254br, 255, 255br (Bits), 256tl, 256cl, 256bl (Circular saw), 256bl (Grinder), 256bl (Rotary hammer), 258cr, 259, 264t, 297tr, 297br, 298tr, 298bl, 302, 311, 313, 333, 374, 381, 383tr, 383br, 385, 402, 403, 404, 406, 422, 424, 426, 429, 430, 434, 453, 463, 464, 472, 475cl, 475bl, 479, HL Studios 23, 73t, 256br, Ian Wedgewood 258tr (Mixer), Jules Selmes 255cl, 255cr, 255br (Chalk line), 258tr (Plastering trowel), 258br (Site store), Naki Photography 9t, 13t, 25b, 147, 150t, 150c, 150b, 236, Trevor Clifford 253cr, 254bl, 255tr, Tsz-shan Kwok 264b, Coleman Yuen 86; **PhotoDisc:** 15, 253c, 254tl; **Photos.com:** jakatics 24; **Science Photo Library Ltd:** Martyn F. Chillmaid 94, Ria Novosti 461; **Shutterstock.com:** Africa Rising 455, Karin Hildebrand Lau 321, Mary Rice 165, Minerva Studio 1, Monkey Business Images 29, Nomad Soul 307, ronfromyork 83, Stripped Pixel 409b, Uroš Medved 251, vrvalerian 121, WimL Page Design (Grey pipe); **Sozaijiten:** 92t, 107, 115c; **SuperStock:** Cultura Limited 53; **Veer/Corbis:** Carolina Smith 371, iofoto 270tr, Ivonne 450, Kayros Studio Be Happy Page Design (Splashes), majaphoto 270tl, Tomas Skopal 285, tuja66 258br (Tool box), Zastavkin 191

Cover images: *Front:* **Shutterstock.com:** Shchipkova Elena

All other images © Pearson Education

The authors and publisher would also like to thank the following for their kind permission to reproduce material in this book:

Page 407 – Fault finding guide reproduced with permission of Baxi Heating UK Ltd

Page 376, 377, 384, 387 (Figure 12.23), 389, 392, 393 – artworks reproduced with permission of Worcester Bosch Group

Every effort has been made to contact copyright holders of material reproduced in this book. Any omissions will be rectified in subsequent printings if notice is given to the publishers.

Websites
Pearson Education Limited is not responsible for the content of any external internet sites. It is essential for tutors to preview each website before using it in class so as to ensure that the URL is still accurate, relevant and appropriate. We suggest that tutors bookmark useful websites and consider enabling students to access them through the school/college intranet.

This book is produced in partnership with JTL, the leading training provider for plumbing and electrical installation. JTL works with over 3,500 businesses and supports over 9,000 apprentices across all sectors.

JTL has a network of highly experienced training officers throughout England and Wales who give training, guidance and advice to thousands of people working through their apprenticeship.

This book uses the expertise of JTL's trainers to bring the essential knowledge for the Diploma to life in an accessible and highly visual style.

Contents

Introduction

This book has been developed with you in mind. It has a dual purpose: to support your learning and to be a useful reference book after you have gained your qualification as a gas operative.

The book supports students who are undertaking the Level 3 NVQ Diploma in Domestic Plumbing and Heating (Gas Options). This pathway in the diploma has been prepared by working with the awarding organisations to provide a qualification for those seeking a career in the gas sector.

These awards are provided by three awarding organisations:

- City and Guilds
- BPEC Certification Ltd
- EMTA Awards Ltd (EAL).

All of these organisations have been approved by the sector skills councils for both plumbing and gas. These are SummitSkills for Building Services Engineering and Energy and Utility Skills (EU Skills) for the gas, power, waste management and water industries.

WHO THE QUALIFICATION IS AIMED AT

The new Diploma qualification is aimed both at new entrants to the profession, such as apprentices or adults changing career, and at members of the existing workforce who are looking to update or improve their skills. It is intended to train and assess candidates so that they can be recognised as occupationally competent by the plumbing industry.

Learners will:

- gain the skills to work as a competent plumber at this high level
- achieve a qualification recognised by the Joint Industry Board (JIB) for professional grading to the industry
- complete an essential part of the SummitSkills Advanced Apprenticeship.

This pathway is part of the Level 3 qualification. It is one of two optional pathways you can choose to take following completion of the Level 3 Diploma in Plumbing and Heating.

To complete the level 3 (Gas option) routes you must complete plumbing and gas assessments in both the workplace and the training centre. On completion of the award you will be placed on the Gas Safe Register, where your employer can then apply for your competency card.

ABOUT THIS BOOK

This book is designed to support the following qualifications:

- 600/1134/8 City and Guilds Level NVQ Diploma in Domestic Plumbing and Heating (Gas Fired Water and Central Heating Appliances)
- 600/1124/5 City and Guilds Level 3 NVQ Diploma in Domestic Plumbing and Heating (Gas Fired Warm Air Appliances)

- 600/1451/9 EAL Level 3 NVQ Diploma in Domestic Plumbing and Heating (Gas Fired Warm Air Appliances)
- 600/1657/7 EAL Level 3 NVQ Diploma in Domestic Plumbing and Heating (Gas Fired Water and Central Heating Appliances)
- 600/6285/X BPEC Level 3 Diploma in Domestic Plumbing and Heating (Gas Fired Warm Air Appliances)
- 600/6284/8 BPEC Level 3 Diploma in Domestic Plumbing and Heating (Gas Fired Water and Central Heating Appliances)

These qualifications are approved on the Qualifications and Credit Framework (QCF). The QCF is a new government framework which regulates all vocational qualifications to ensure they are structured and titled consistently and quality assured. While awarding organisations will design their own assessment content, they will all follow the same unit structure and assessment strategy.

This book is mapped against the unit *Understand core gas safety principles for natural gas within domestic building services engineering* with each chapter taking a thematic approach to cover a particular area of knowledge and skill in the gas industry. References are also made to regulations for further detailed reading such as British Standards, Gas Regulations and Building Regulations.

There are progress checks and knowledge checks throughout the book which will enable you to assess your own level of knowledge and understanding at various stages of the course.

This book has been written by vocational lecturers with many years of experience in the plumbing and heating trade, as well as in further education, where they currently teach plumbing and heating qualifications to wide and diverse groups of students. The development of this resource has been fully supported by JTL, the leading training provider in building services.

USING THIS BOOK

Although this book is aimed at the current QCF awards, it has also been written with the intention that it can be used as a handbook for operatives undertaking an initial ACS assessment or re-taking their ACS assessments.

The book is not intended as a handbook to an actual installation. Nor is it a code of practice or a guidance note. British standards, manufacturer's data and HSE documents are among the materials to be used and referred to for actual installation work.

FEATURES OF THIS BOOK

This book has been fully illustrated with artworks and photographs, some of which are from training centres and not real life situations. They will help to give you more information about certain concepts and procedures, as well as helping you to follow step-by-step processes or identify particular tools or materials.

This book also contains a number of different features to help your learning and development.

Safe working 🚫

Red safety tips remind you of things you **should not** do.

Safe working ❗

Blue safety tips remind you of things you **should** do.

Safe working ⚠️

Yellow safety tips indicate warnings of **hazards** or danger.

Safe working ➕

Green safety tips indicate first aid or health issues.

Progress check

These features contain a series of short questions and usually appear at the end of each learning outcome. They give you the opportunity to check and revise your knowledge.

Case study

These features highlight real life events or situations which are relevant to the plumbing and heating industry.

Knowledge check

This is a series of multiple choice questions at the end of each unit. Answers to the questions are supplied on the training resource disk.

Chapter 2 of this book covers the Gas Safety (Installation and Use) Regulations 1998 in depth. In this chapter, quotes from the Regulations are interspersed with commentary, case studies and real life examples to place these in context.

The text containing information about the regulations is contained in a shaded box. It will give direct quotes from the Regulations.

The commentary on the Regulation quoted will be found in a lighter shaded box. This text will explore and develop the ideas in the Regulations, and give real life examples to place these theories into context.

ACKNOWLEDGEMENTS

Pearson and JTL would like to thank Barry Spick and Shaun O'Malley for their hard work and dedication in writing the content of this book. Thanks also go to Gary Nutton for his development of Chapter 13 of this book.

Pearson would like to thank all those who have contributed to the development of this book making sure that standards and quality remained high through to the final product.

Thanks to Philip Brewis of JTL Training for carrying out a comprehensive review of the books contents and Ron Denton of Leicester College for his invaluable advice and assistance with Chapter 13 of this book.

Thanks also to Philip Bunce and Stuart Farrimond of Worcester, Bosch Group for their help and kind permission with reproducing artworks for Chapter 12.

Pearson particularly wish to thank Jim Dawson and all the staff of the Gateshead Skills Academy for their help, enthusiasm and dedication in setting up the photo shoot for this book and for their technical advice and support. Thanks also to Sam Rowland, Steve Rowland, Michael Wilson and Neil Wilson for appearing as models in many photos and Gareth Boden for his inspiration and hard-work as photographer.

Shaun O'Malley would like to thank his wife and family for their patience and Mum (Jean) and Dad (Jim) for their assistance and support. He would also like to thank Riverside College for their assistance and help and continued support.

Chapter 1

Health
and Safety

THIS UNIT COVERS

- health and safety at work legislation
- safety signs
- safe handling and storage of tools and equipment
- safe working in occupied premises
- safe working in confined spaces
- safe working with electricity
- safe working with heating equipment
- safe working with asbestos
- safe working at height
- safe manual handling
- first aid provision at work
- protection from loud noise at work

Key term

So far as is reasonably practicable
The degree of risk in a particular situation can be balanced against the time, trouble, cost and physical difficulty of taking measures to avoid the risk.

INTRODUCTION

Health and safety is essential for the safe installation and use of gas systems, to both the user and installer. This chapter will focus on the safety aspects of gas systems and appliance installation.

HEALTH AND SAFETY AT WORK LEGISLATION

Health and Safety at Work Act 1974 (HASAWA)

The HASAWA applies to everyone concerned with work activities, ranging from employers, employees and the self-employed, to manufacturers, suppliers and importers of materials for use at work, and people responsible for work premises. It also includes provisions to protect members of the public. The duties apply both to individuals and to public and private bodies, including corporations, partnerships, local authorities and nationalised industries. The duties are expressed in general terms, so that they apply to all types of work activity and work situations. Every employer has a duty to ensure, **so far as is reasonably practicable**, the health, safety and welfare at work of their employees. The principles of safety responsibility and safe working are expressed in sections 2–9 of the Act. Employers and the self-employed are required to carry out their undertakings in such a way as to ensure, so far as is reasonably practicable, that they do not expose people who are not their employees to risks to their health and safety (sections 3(1) and 3(2)). In some areas, the general duties of the Act have been supplemented by specific requirements, set out in regulations made under the Act. Specific legal requirements are included in earlier legislation which is still in force. Failure to comply with the general requirements of the Act or specific requirements found elsewhere in the legislation may result in legal proceedings.

There are numerous regulations within HASAWA with which you have to comply:

- The Workplace (Health, Safety and Welfare) Regulations (1992)
- Fire Precautions Act (1971)
- The Electricity at Work Regulations (1989)
- Health and Safety (First Aid) Regulations
- Provision & Use of Work equipment Regulations (PUWER)
- Personal Protective Equipment at Work Regulations (2002)
  - Management of Health & Safety at Work Regulations Hazards and Risk assessments
  - Working at Height Regulations (2005)
  - Manual Handling Operations Regulations (1992)
  - Safety Signs & Signals Regulations (1996)
  - Control of Asbestos at Work regulations

— Control of Lead at Work Regulations

— Control of Substances Hazardous to Health Regulations (COSHH)

— The Confined Spaces Regulations (1997)

▪ Construction (Design and Management) Regulations 2007 (CDM)

▪ Reporting of Injuries, Diseases and Dangerous Occurrences Regulations 1995 (RIDDOR).

COSHH

The regulations were first introduced in 1988 and provide a legal framework for the control of substances hazardous to health in all types of business, including factories, farms, offices, shops, plumbing, gas and many other occupations and businesses. So you are likely to use a substance governed by the regulations in your daily work.

You can acquire information about the substance from either data sheets which are available from wholesalers or the company themselves. Information regarding the substance can normally be found on the packaging and are identified by the use of signs (see Figure 1.01).

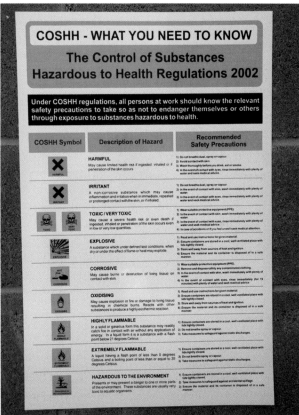

Figure 1.01 Examples of common COSHH labelling

RIDDOR

On 6 April 2012, RIDDOR was amended, in particular with regard to the reporting of 'over three-day' injuries. The time period has now been increased from over three days to over seven days' **incapacitation**; however, this does not count the day on which the accident happened.

Key term

Incapacitation
This means that the worker is absent or is unable to work when they would reasonably be expected to, as part of their normal work.

Health and safety records and reports

Under RIDDOR, employers must still keep a record of all 'over three-day' injuries; these can be recorded in the accident book which has been approved by the Health and Safety Executive (HSE).

These records must include the following information:

- date, time and place of the incident
- name and job of the injured or ill person
- details of the injury/illness and what first aid was given
- what happened to the person immediately afterwards (e.g. if they went back to work, went home, went to hospital)
- name and signature of the first-aider or person dealing with the incident.

See Chapter 11 for more details of RIDDOR and gas incidents.

The deadline for reporting the injury has also been increased from 10 to 15 days from the day of the accident. Injuries can be reported online using form F2508 – Report of an Injury by going to www.hse.gov.uk/riddor/report.htm where all reporting forms can be accessed.

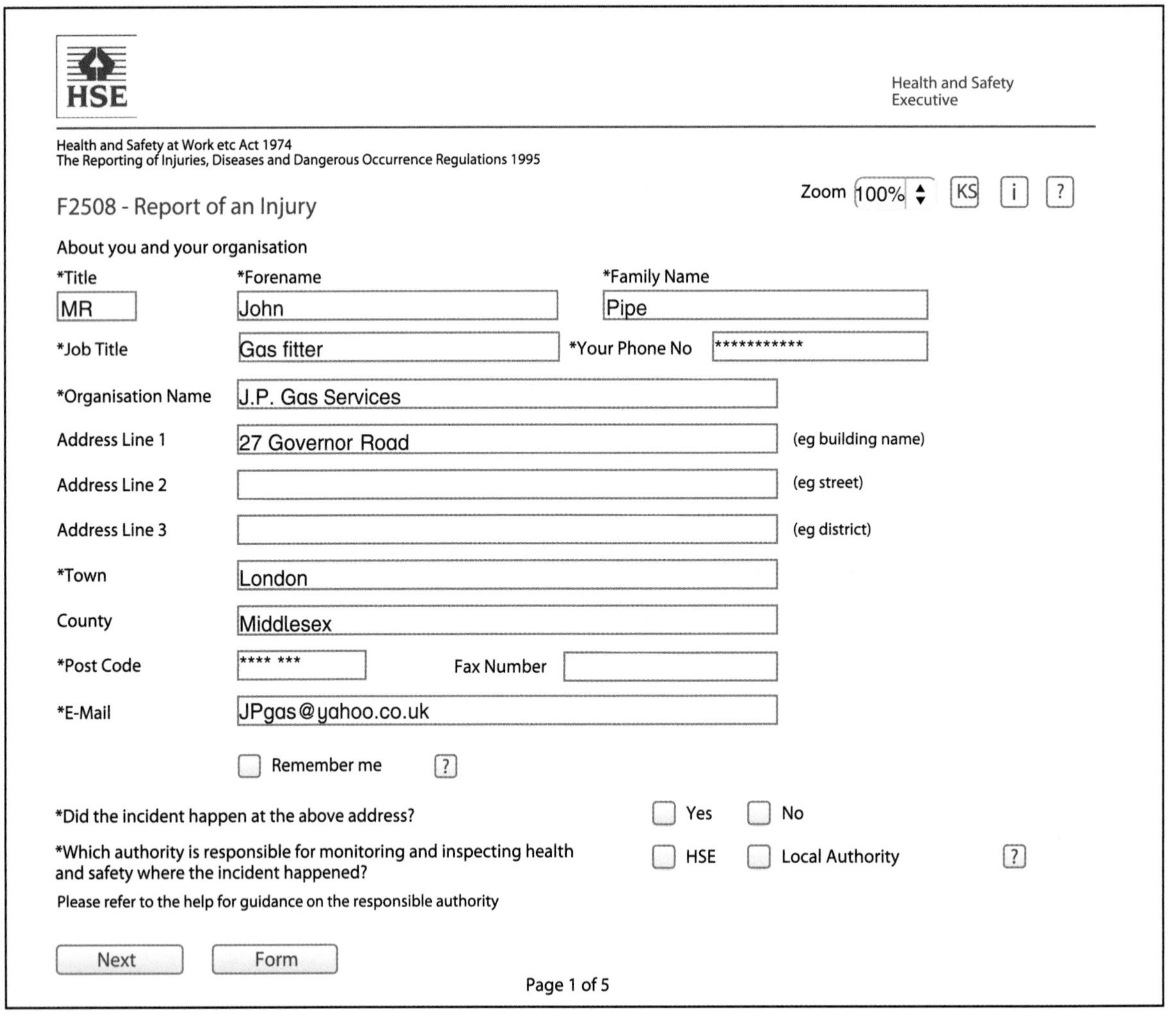

HSE

Health and Safety Executive

Health and Safety at Work etc Act 1974
The Reporting of Injuries, Diseases and Dangerous Occurrence Regulations 1995

F2508 - Report of an Injury

Zoom 100% ▲▼ KS i ?

About you and your organisation

*Title: MR

*Forename: John

*Family Name: Pipe

*Job Title: Gas fitter

*Your Phone No: ***********

*Organisation Name: J.P. Gas Services

Address Line 1: 27 Governor Road   (eg building name)

Address Line 2:   (eg street)

Address Line 3:   (eg district)

*Town: London

County: Middlesex

*Post Code: **** ***   Fax Number:

*E-Mail: JPgas@yahoo.co.uk

☐ Remember me   ?

*Did the incident happen at the above address?   ☐ Yes   ☐ No

*Which authority is responsible for monitoring and inspecting health and safety where the incident happened?   ☐ HSE   ☐ Local Authority   ?

Please refer to the help for guidance on the responsible authority

[Next]   [Form]

Page 1 of 5

Figure 1.02 A sample HSE online reporting form

Other forms used for reporting are:

- F2508 – Report of a Dangerous Occurrence
- OIR9B – Report of an Injury Offshore
- OIR9B – Report of a Dangerous Occurrence Offshore
- F2508a – Report of a Disease.

There are two versions of form F2508 that are specifically for reporting gas incidents:

- F2508G1 – Report of a Flammable Gas Incident
- F2508G2 – Report of a Dangerous Gas Fitting.

The reporting of immediately dangerous (ID) situations is mainly concerned with instances such as gas escapes, inadequate ventilation or flues, and poor workmanship. These can lead to a death or a serious injury, for example carbon monoxide poisoning, which can be fatal.

Examples of reportable dangerous gas fittings

- Serious gas leaks, leaving open ends on pipework, caused by poor workmanship and the use of materials deemed not fit for purpose.
- A defective flue – this is where combustibles are not being cleared and could cause carbon monoxide poisoning.
- A flued gas appliance installed without a flue.
- An appliance connected to the wrong kind of gas, for example a natural gas boiler connected to an LPG system.
- A gas appliance fitted with inadequate ventilation.
- Safety devices on a faulty appliance being rendered deliberately inoperative so that the appliance will still work, for example the flame supervision device (FSD) being wedged open.
- An appliance connected to the gas supply with a flexible connection hose which is not fit for the purpose, for example using a washing machine hose to connect a cooker to the system.
- Open flued appliances installed in a bathroom or shower room after 24 November 1984.
- Any open flued appliance, installed before 24 November 1984, which has serious ventilation and flueing defects.

Examples of reportable flammable gas incidents

These reportable incidents can be caused when a person (e.g. an employee, employer or member of the public) is using gas:

- personal injuries
- fire
- asphyxiation
- carbon monoxide poisoning
- gas explosion.

SAFETY SIGNS

It is important that you understand the different types of safety signs that may be displayed on site or in many other places. The types of signs are:

- prohibition – red circle on a white background with a red line across
- mandatory – blue circle with a white image identifying mandatory use of safety equipment
- hazard – yellow triangle with a black border and black image
- information or safe condition signs – green square with a white image.

Figure 1.03 Safety signs

There are also fire signs and combination signs, which are a combination of the symbol, writing and, sometimes, a direction arrow.

SAFE HANDLING AND STORAGE OF TOOLS AND EQUIPMENT

As a gas operative, you use a wide variety of tools and equipment, both electrical and manual. It is important that these tools are handled and stored correctly.

Hand tools

Hand tools, as with all tools, need to be looked after to make them last longer and continue to perform reliably. This is particularly important for cutting tools, such as hacksaws, wood saws and chisels. Hacksaw blades should be changed regularly and at the first sign of damage. Both wood and cold chisels should be sharpened and protectors placed on the sharp ends when being stored. When dealing with pipe-cutters, change the cutting wheels when blunt or damaged.

Portable electrical tools and equipment

Care should be taken with electrical tools such as drills, transformers, battery drills and combustion performance analysis equipment. All electrical tools, when used in the construction industry, must hold a current PAT certificate and be tested every three months.

Appliances, pipes and equipment

Bear in mind the following safety advice when handling appliances, pipes and equipment:

- When lifting a heavy object, be sure the load is not too heavy for one person. It may be a two-person job.
- Ensure the weight lifted is evenly balanced to avoid injury and accident.
- Follow the Manual Handling Operations Regulations (1992).
- Stack objects at a manageable height.
- If the manual handling activity cannot be avoided, the employer must take the appropriate measures to remove or reduce the level of risk involved.

The manual handling regulations do not specify a maximum load that can be lifted. They place duties on employers to manage or control risk; measures to take to meet this duty will vary depending on the circumstances of the task.

SAFE WORKING IN OCCUPIED PREMISES

The Construction (Design and Management) Regulations, 2007 (CDM) stipulate how the Acts can be implemented in a sensible way and cover the subjects of:

- working environment
- safety
- facilities.

A contractor working in an occupied building has the legal responsibility to safeguard the regular users of the building and meet the requirements of the Acts, for example:

- When working in a domestic property, you become responsible for the occupants' welfare, as it becomes your place of work.
- Machinery or equipment that has been provided to ensure everyone's health and safety should not be tampered with.

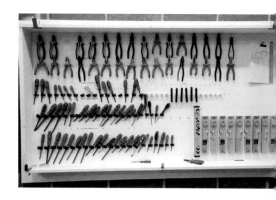

Figure 1.04 It is important to keep hand tools in good condition

- It is the employer's responsibility to ensure that equipment is kept in good working order.
- As an employee, you have a responsibility to keep the workplace tidy, and therefore reduce the risk of incident.

When working in premises, whether they are domestic or commercial, you have a responsibility to ensure your own safety and the safety of other people who are within your working environment.

SAFE WORKING IN CONFINED SPACES

A confined space is a place which is substantially enclosed (though not always entirely), and where serious injury can occur from hazardous substances or other conditions within the space or nearby (e.g. lack of oxygen).

As a gas operative, you may find yourself working in the following confined spaces:

- ductwork
- unventilated or poorly ventilated rooms
- under-stair cupboards
- under floorboards
- in roof spaces.

The law states that a suitable and sufficient risk assessment for all the work to be carried out in a confined space must be undertaken along with any necessary safety measures.

Where work in confined spaces is unavoidable, you should make sure you have a safe system of work in place for working in such an environment. You can find more information at www.hse.gov.uk/confinedspace/index.htm or download the information sheet INDG258 Safe Work in Confined Spaces.

SAFE WORKING WITH ELECTRICITY

The Electricity at Work Regulations 1989 (SI 1989/635) (as amended) came into force on 1 April 1990. The purpose of the regulations is to help ensure precautions are taken against the risk of death or personal injury from electricity in work activities. The regulations impose specific duties on employers to put into place measures to protect their employees against death or personal injury from the use of electricity at work.

Electricity can kill or severely injure people and cause damage to property. Each year around 20 people die from electric shock or electric burns at work and about 30 die from electrical accidents in the home (HSE, *Electricity at work: Safe working practices* (2003)). Most of these accidents are preventable by following the relevant safety procedures. Many electrical accidents occur because people are working on or near equipment that is:

- thought to be dead but is actually live
- known to be live but those involved do not have adequate training or the appropriate equipment, or they have not taken adequate precautions.

Safe isolation

As a gas operative, you will work on many appliances and components which are supplied by electricity. It is important to know and understand how to safely isolate the electrical supply by following the safe isolation procedure. Before working on any electrical supply, you must make sure that it is completely dead and cannot be switched on accidentally without you knowing. This is a requirement of the Electricity at Work Regulations 1989 and it is *essential* for your personal safety and for those around you. The proper way to test if a circuit is live is to use an approved voltage-indicating device (see Figure 1.05).

Voltage-indicating device

Voltage-indicating devices may use either an illuminated lamp or a meter scale to indicate the presence of a voltage. Test lamps are normally fitted with a 15 W lamp and must be constructed in such a way that they are not dangerous if the lamp is broken. They must also be fitted with protection against excess current, either by a fuse not exceeding 500 mA or a current-limiting resistor and a fuse. The test leads should be secured and sealed into the body of the voltage detector. The maximum voltage that may be tested by the test lamps and voltage indicators should be clearly marked on the device, as should the maximum voltage the device can withstand.

Testing the electrical supply

Most fatal accidents involving electricity occur at the isolation stage. This is when you must take great care and fully focus on what you are doing, as you may have no idea of the type of supply you are confronted with. Do not take any risks and, if you are not sure, seek assistance.

To safely test the electrical supply, follow these steps:

- identify all sources of supply
- isolate the supply
- secure isolation
- test that the equipment or system is dead
- begin work.

Test equipment and checking system is dead

Any circuit you work on *must* be tested to ensure it is dead. Test equipment must be regularly checked to make sure it is in good and safe working order. Your test equipment must have a current calibration certificate, indicating that the instrument is working properly and providing accurate readings. If it is not calibrated, test results could be inaccurate. Before starting work:

- check the equipment for any damage – look to see if the case is cracked or broken which could result in false readings
- check that the insulation on the leads and the probes is not damaged, and that it is complete and secure.

Test the voltage indicator on a proven supply before you start; this will confirm that the device is working. The best piece of equipment for doing this is a proving unit (see Figure 1.06).

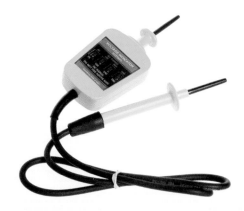

Figure 1.05 Voltage-indicating device

Safe working

You should *never* test a live supply. The use of a homemade test lamp or neon screwdriver is *not acceptable*, as this will not detect a low-voltage supply.

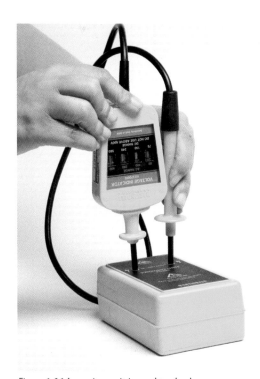

Figure 1.06 A proving unit is used to check that a voltage indicator is working correctly

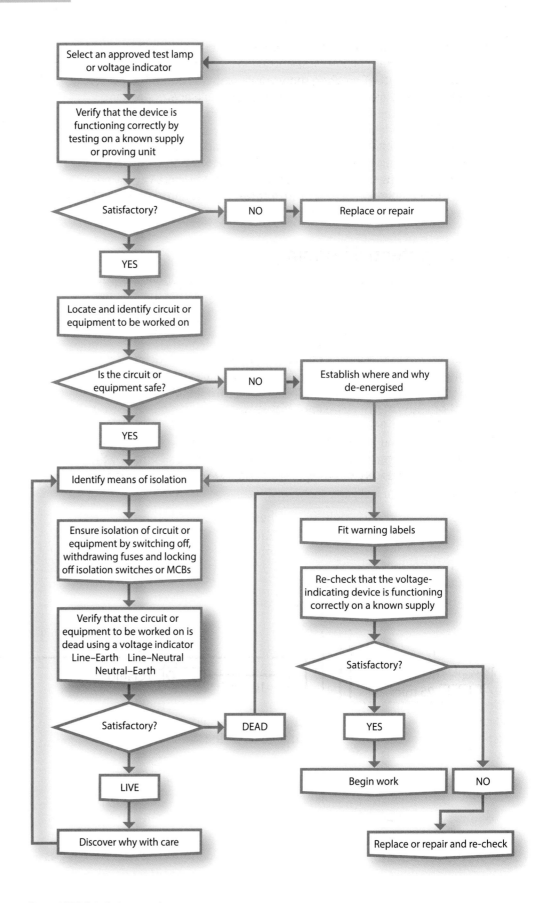

Figure 1.07 Safe isolation procedure

Standardised procedures have been drawn up by the Electrical Contracting Industry and these are used as the standard for safe working in the plumbing and gas industries. By following the flow chart shown opposite in Figure 1.07 you should be able to safely isolate an appliance or component from the electrical supply.

Secure isolation

To prevent the supply being turned on accidentally by someone else while you are working, the fuse or circuit breaker should be removed and kept in your pocket, or the isolator locked off with a special locking-off device or padlock. As an extra precaution, a sign saying 'work in progress and system switched off' must be displayed at the consumer unit or the area in which you are working.

Main equipotential bonding

Bonding of pipework is an essential aspect of electrical systems, most systems that are installed require earth bonding and you need to be aware if the bonding has been installed correctly. Copper pipework can provide a route for stray electrical currents and can cause corrosion of the pipework.

The method shown in Figure 1.08 covers the requirements of the electrical wiring regulations (BS 7671). Note the size of the bonding cables, which are to be 10 mm^2; the connections to the pipework have to be as close as possible to the point of entry into the dwelling; and the gas has to be within 600 mm of the meter. They must also be labelled: 'SAFETY ELECTRICAL CONNECTION, DO NOT REMOVE'. If you have to break the earthing of the pipework then you must ensure that continuity of the earthing circuit is maintained by the use of temporary continuity bonding. See Chapter 5 for more information on equipotential bonding.

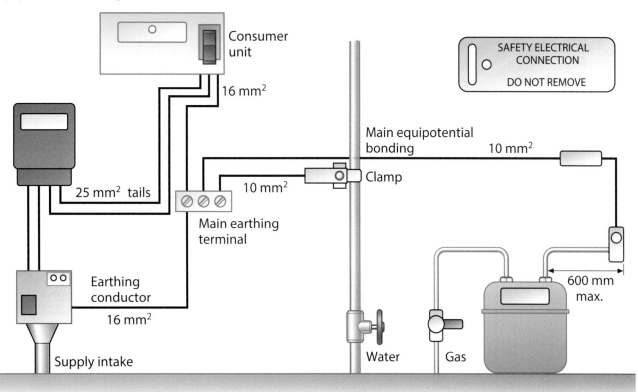

Figure 1.08 Main equipotential bonding

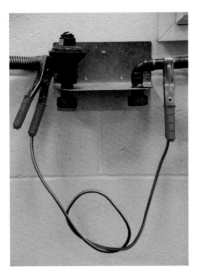

Figure 1.09 Temporary bonding when a meter is removed

Temporary continuity bonding

When working on gas pipework it is essential that you check the pipework is bonded correctly. When removing sections of pipework or meters it is important that the continuity of the earth is maintained. To do this, use temporary bonding clips as shown in Figure 1.09.

To check that continuity is being maintained use a low-reading ohmmeter or multimeter set to Ω (ohms). Place one probe on one section of pipework away from the bonding clips and then place the second probe on the opposite section (see Figure 1.10 below). The reading should be at least 1.00. If it is less than 1.00 you will not have full continuity and should check that the clips are secured correctly and re-check this before continuing work. See Chapter 5 for more information on temporary continuity bonding.

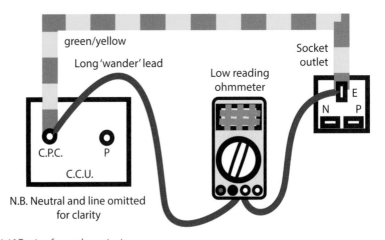

Figure 1.10 Testing for earth continuity

SAFE WORKING WITH HEATING EQUIPMENT

Fire prevention

Figure 1.11 Fire triangle

Combustion (fire or burning) is the rapid combination of a fuel with oxygen (air) at high temperature. A fire can reach temperatures of up to 1,000 °C within a few minutes of starting. For a fire to start there are three requirements: combustible substance (the fuel), oxygen (usually as air) and a source of heat (spark, friction, match). When these three items come together in the correct combination, a fire occurs. Fires can spread rapidly. Once established, even a small fire can generate sufficient heat energy to spread and accelerate the fire to surrounding combustible materials.

Fire prevention is largely a matter of 'good housekeeping'. The workplace should be kept clean and tidy. Plumbing is a particularly risky job in terms of potential fire hazards. The use of blow torches and welding equipment – which are often deployed near combustible materials, and sometimes in tight positions or difficult to access areas – means you are particularly vulnerable.

Because 'hot working' is such a risky job, your employer is required to provide strict working methods. These procedures constitute basic common sense, but are also a requirement of insurance companies. They usually include:

- providing a fire extinguisher in the immediate working area
- completing work with a blow or welding torch a minimum of an hour before leaving a site.

Building sites can be dirty places. Timber shavings and other combustible materials find their way under floors, baths etc. Make sure the area you are working in is clean before you start and after you finish to ensure such combustibles are no longer posing a fire threat. Another major cause of fires is electrical faults. All alterations and repairs in electrical installations must only be carried out by a qualified person, and must be to the standards laid down in the Institute of Electrical Engineers' Regulations.

You may sometimes work in an occupied building, such as an office block. You must be aware of the building's fire safety procedures, and find out about normal and alternative escape routes. Know where your assembly point is located and report your presence to your supervisor.

Fire extinguishers

Fires are commonly classified into four groups, according to fuel type and how the fires are extinguished:

- **Class A** – fires involving solid materials, extinguished by water
- **Class B** – fires involving flammable liquids, extinguished by foam or carbon dioxide
- **Class C** – fires involving flammable gases, extinguished by dry powder
- **Class D** – fires involving flammable metals, extinguished by dry powder.

If a fire is small, it may be possible to put it out quickly and safely. However, if your efforts to contain a fire prove fruitless and it starts to get out of control, when you decide to make your escape be aware that you could have difficulty finding your way through the smoke and the fumes. Smoke and fumes are just as lethal as the fire itself so always make your way to an exit as soon as possible.

Fire-fighting equipment, including extinguishers, buckets of sand or water and fire-resistant blankets, should be readily available in all public buildings. In larger premises, you will find automatic sprinklers, hose reels and hydrant systems. The table below shows each type of fire extinguisher and their main uses.

Type of extinguisher	Colour code	Main use
Water	Red	Wood, paper or fabrics
Foam	Cream	Petrol, oil, fats and paints
Carbon dioxide (CO_2)	Black	Electrical equipment (do *not* use in enclosed spaces)
Dry powder	Blue	Liquids, gasses and electrical equipment

Table 1.01 Fire extinguishers by type, colour and use

Heating equipment

As a gas operative you will be using heat-producing equipment (e.g. blow lamps) on a daily basis. See Figure 1.12.

When using this type of equipment, you need to be aware of what is around your work space as there could be flammable materials such as:

- cardboard
- newspapers

Figure 1.12 Blow lamps for domestic (top) and larger (bottom) pipework

- curtains
- carpets
- lagging (also known as hair-felt)
- wooden joists and floorboards
- plastics (can cause toxic fumes when burnt).

When using flame-producing equipment in industrial premises, you would normally need to obtain a Hot Work Permit before any of this type of equipment can be used. When working in domestic premises, it may also be advisable to get the permission of the householder before using the equipment.

Propane torches

Before using propane torches, you should:

- check that the sealing cap is on when the propane bottle is not in use
- visually inspect the bottle connection for any damage before fitting the regulator, after removing the plastic cap
- check the connection of the hose to the handle and make sure metal crimping rings are used not jubilee clips
- visually inspect the handle for damage to valves and make sure the control knob is free
- check that the needle valve is clean and does not leak
- check that the valve connection thread to the bottle is not damaged and that the 'O' ring is not damaged where one is used to seal the regulator to the bottle
- inspect the nozzle for blockages
- examine the bottle for any damage or corrosion.

When storing propane torches, you should:

- remove the hoses, handles and regulator
- replace the plastic cap to protect the cylinder needle valve connection.
- keep the propane bottles away from sources of heat.

Be aware:

- Propane is heavier than air so propane vapours will sink to the bottom if there is a leak.
- After lighting the torch, do not leave unattended, switch off when not in use.
- Take care when working in confined spaces, roof areas, below floors.
- Check there is nothing that can be damaged or catch fire within the vicinity.
- Under no circumstances should propane torches be used in areas where there could be a build up of gases.
- Ensure that you protect the surrounding area, e.g. decorations (wallpaper), from the naked flame; to prevent scorching use a heat resistant mat.

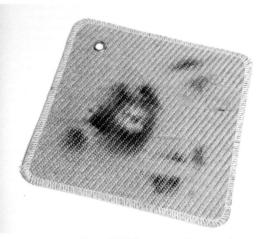

Figure 1.13 A heat mat will protect the building fabric

What to do if your equipment is leaking

During a work task there is a possibility, if a leak occurs, that the equipment might catch fire. If this occurs then proceed as follows:

- Turn off the gas supply to the blow lamp, extinguishing the flame if lit.
- Using an approved or leak detector fluid only, check the suspected parts for leaks first.
- After checking suspected parts, inspect other parts of the equipment.
- Repair or replace if necessary any of the defective parts.
- Re-energise the gas supply and test with soap solution.
- If all is ok then re-light torch and proceed to use.

SAFE WORKING WITH ASBESTOS

The Control of Asbestos Regulations 2012 came into force on 6 April 2012. It updated previous asbestos regulations to take account of the European Commission's view that the UK had not fully implemented the EU Directive on exposure to asbestos (Directive 2009/148/EC).

In practice the changes are fairly limited. They mean that some types of non-licensed work with asbestos now have additional requirements, i.e. notification of work, medical surveillance and record-keeping. *All other requirements remain unchanged.*

Progress check

1 Which one of the following fire extinguisher types carries a cream panel?
   a Carbon dioxide (CO_2)
   b Foam
   c Water
   d Dry powder
2 Which three elements make up the fire triangle?
3 List the steps to follow when safely isolating electrical appliances/components.
4 When a reportable accident occurs, within how many days should the accident be reported?
   a 5 days
   b 2 weeks
   c 15 days
   d 3 weeks
5 Which piece of legislation covers health and safety for people in the workplace?

Figure 1.14 You must always take precautions when working in areas where asbestos is present

What's new in the 2012 regulations?

The 2012 regulations contain the following new stipulations:

- From 6 April 2012, some non-licensed work needs to be notified to the relevant enforcing authority.
- Brief written records should be kept of non-licensed work and this has to be notified, for example a copy of the notification with a list of workers on the job and the level of likely exposure of those workers to asbestos. This does not require air monitoring on every job provided an estimate of degree of exposure can be made based on experience of similar past tasks or published guidance. More advice for reportable non-licensed work is available to download from the HSE website.
- By April 2015, all workers/self-employed doing notifiable non-licensed work with asbestos must be under health surveillance by a doctor. Workers who are already under health surveillance for licensed work need not have another medical examination for non-licensed work. *However*, medicals for notifiable non-licensed work are not acceptable for those doing licensed work.
- Some updates to the terminology and other changes have been made to reflect other legislation, for example the prohibition section has been removed because the prohibition of supply and use of asbestos is now covered by Registration, Evaluation, Authorisation and Restriction of Chemicals Regulations 2006 (REACH).

The following points from the 2012 regulations remain unchanged:

- If existing asbestos-containing materials are in good condition and are not likely to be damaged, they may be left in place, for example flue pipes from a boiler.
- If you want to do any building or maintenance work in premises, or on plant or equipment that might contain asbestos, you need to identify where it is and its type and condition; assess the risks, and manage and control these risks.
- The requirements for licensed work remain the same: in the majority of cases, work with asbestos needs to be done by a licensed contractor. This work includes most asbestos removal, all work with sprayed asbestos coatings and asbestos lagging, and most work with asbestos insulation and asbestos insulating board (AIB).
- The control limit for asbestos is 0.1 asbestos fibres per cubic centimetre of air (0.1 f/cm3). The control limit is not a 'safe' level and exposure from work activities involving asbestos must be reduced to as far below the control limit as possible.
- Training is mandatory for anyone liable to be exposed to asbestos fibres at work. This includes maintenance workers and others who may come into contact with or disturb asbestos (e.g. cable installers), as well as those involved in asbestos removal work.

As a gas operative, you will be maintaining or removing many types of older boilers and gas-fired appliances. In many of these, asbestos would have been used as gaskets on flueways and access covers. In order to work safely with these, you need to use the following procedure:

How to work with asbestos

Checklist

PPE	Tools and equipment	Source information
• Disposable overalls fitted with a hood • Boots without laces (laced boots are hard to decontaminate) • Respiratory protective equipment	• 500-gauge polythene sheeting and duct tape • Warning tape and notices • Class H vacuum cleaner (BS 8520) to collect adhering gasket residues • Scraper • Garden-type sprayer containing wetting agent (water) • Bucket of water and rags • Asbestos waste container, e.g. labelled polythene sack • Clear polythene sack	• HSE publication a11 asbestos essentials – removing asbestos cement (ac) debris • This information and more essential information on asbestos is available to download from the HSE website

Prepare the work area

- Ensure there is safe access to the work area.
- Restrict the access to a minimum number of operatives.
- Close doors, and use tape and notices to warn others of the work in progress.
- Ensure there is adequate lighting in the work area.

Procedure

- Ensure the system has been made safe (pipework emptied, electrical supply isolated, etc.).
- Protect nearby surfaces from contamination by covering with 500-gauge polythene sheeting and fix with duct tape to non-asbestos surfaces.
- Protect vulnerable components with polythene sheeting.

Removal

- Unbolt or dismantle the equipment.
- Once accessible, dampen the asbestos using the sprayer and continue dampening as it is exposed.
- Ease the gasket or rope seal away with the scraper, and place in the waste container.
- Keep the surface damp and ease away asbestos residues.
- Gently scrape off residues using e.g. 'shadow vacuuming' (i.e. hold the vacuum cleaner nozzle close to the task area (e.g. screw removal); the vacuum will remove local dust at the cutting point; enclose the tool (e.g. drill bit) with a cowl and attach the nozzle).

Cleaning and disposal

- Clean the equipment and the area with the Class H vacuum cleaner and/or damp rags.
- Put used rags, polythene sheeting and other waste in the asbestos waste container and tape it closed.
- Put the asbestos waste container in a clear polythene sack and tape it closed.

Disposal

- Double-wrap and label asbestos waste – standard practice is to use a red inner bag with asbestos warnings printed on and a clear outer bag with a Carriage of Dangerous Goods sign on it.
- Do not overfill bags and beware of sharp objects as they can pierce the plastic bags.
- Dispose of contaminated waste safely at a licensed disposal site.
- Complete a Waste Consignment Note and keep copies of these documents for 3 years.
- You can download more information on working with and disposing of a variety of asbestos cement materials from www.hse.gov.uk/asbestos/essentials/index.htm

SAFE WORKING AT HEIGHT

Working as a gas operative can require you to work at height and it is imperative that you understand the dangers associated with the nature of this workplace.

The types of jobs that may require you working at height include:

- installing pipework in ceiling voids
- installing pipework in trenches
- fitting boilers at high level
- installing flue liners
- installing boiler flues
- penetrating roofs to terminate flues.

The Work at Height Regulations 2005 (as amended)

The Work at Height Regulations 2005 (as amended) govern working at height practices and were written as during 2005/06 falls from height accounted for 46 fatal accidents at work and approximately 3350 major injuries. These still remain the biggest cause of workplace deaths and one of the main causes of injury.

'Work at height' is defined in regulation 2 as '[a] place is at height if a person could be injured falling from it, even if it is at or below ground level'. And 'work' includes moving around at a place of work, but not travel to and from a place of work.

In order to comply with the Work at Height Regulations, your employer or any person who controls the work of others (e.g. your supervisor or manager) must:

- adhere to the duties of employers/dutyholders as laid out in the regulations
- respect the hierarchy for managing and selecting equipment for working at height
- undertake a risk assessment prior to work being carried out.

As an employee, you must report any safety hazards to those in charge and use the equipment supplied (including safety devices) properly, following any training and instructions given (unless you think that would be unsafe).

Access equipment

Access equipment refers to items such as:

- ladders
- roof ladders
- trestles, steps
- independent scaffolds
- putlog scaffolds
- tower scaffolds.

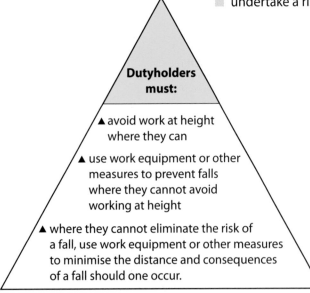

Dutyholders must:

- ▲ avoid work at height where they can
- ▲ use work equipment or other measures to prevent falls where they cannot avoid working at height
- ▲ where they cannot eliminate the risk of a fall, use work equipment or other measures to minimise the distance and consequences of a fall should one occur.

Figure 1.15 Dutyholders' responsibilities

The list also affects working in excavations because of the need to use ladders and/or scaffolds to access below ground level installations. Working in excavations also comes under the working at height regulations.

Access equipment should be kept in good order. Regular checks of the equipment should be carried out by a competent person and recorded appropriately.

Ladders

Types of ladder

There are three classes of ladder in use:

- Class 1 – Industrial (Heavy duty) BS2037 Class 1
- EN131 – replaced Class 2 (Light trades)
- BS2037 Class 3 (Domestic use).

Types of ladder include:

- pole ladders
  — used mainly for accessing scaffolds
  — are a Class 1 ladder
  — are available in various lengths
- extension ladders:
  — wooden – used when working with electricity
  — aluminium – used for heavier mobile jobs
  — can be a Class 1 or Class two ladder (you should always use a Class 1 ladder when working as a gas operative)
- roof ladders – used when fitting boiler flues, flue liners and flashings, e.g. slate for a roof penetration

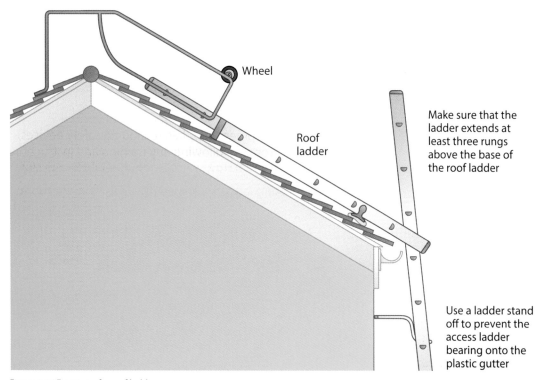

Wheel

Roof ladder

Make sure that the ladder extends at least three rungs above the base of the roof ladder

Use a ladder stand off to prevent the access ladder bearing onto the plastic gutter

Figure 1.16 Erection of a roof ladder

- stepladders
  - wooden steps – used where there could be a hazard with sparking e.g. petro-chemical works and when working with electricity
  - aluminium steps – used for general heavy duty work; not to be used when working with electricity
  - fibreglass steps – used for general work; can also be used when working with electricity (good insulator).

Stepladders are often used by plumbers. When in use, make sure the ground is level and firm and that all four legs rest firmly and squarely on the ground. Never use the top of the steps unless it is constructed as a working platform. On wooden steps check the hinges and the ropes are of equal length.

When using **roof ladders**, check the stiles are straight and sound. The rungs should also be sound and the hook must be firmly over the ridge. Make sure the wheels are firmly fixed and free running, and the pressure plates (i.e. the parts that rest on the roof surface) are sound.

Checking ladders and steps

When checking your ladders and steps, look out for:

- cracks on stiles – could split when weight is put on the ladder or step
- splits on the rungs/steps – could break when weight is put on the rungs/steps

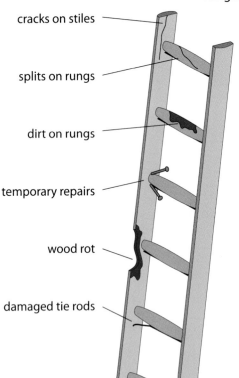

cracks on stiles

splits on rungs

dirt on rungs

temporary repairs

wood rot

damaged tie rods

warping

Figure 1.17 Ladder defects

- mud/dirt on the rungs/steps – when wet can be extremely slippery and could cause a fall
- temporary repairs – may not have been done correctly; you should never use a ladder/steps that's been repaired
- wood rot – ladder/steps with rot should not be used, can cause breakages in the ladder
- damaged tie rods – the stiles could come apart and the rungs/steps would come out of their recesses
- warping or bent stiles – would mean not being able to position the ladder/steps squarely, would cause the ladder/step to be unstable.

You will be expected to undertake these checks whenever you have any access equipment to use.

Before using ladders, consider the following questions:

- Is a ladder the most suitable means of access?
- Does the risk assessment justify using one?
- Is the chosen ladder suitable for the job?
- Are there enough people to handle the ladder?
- Has the ladder any loose or damaged parts?

Short ladders can be carried by one person on the shoulder in either the horizontal or vertical position. Long ladders should be carried by two people horizontally on the shoulders, one at either end, holding the upper stile. When carrying ladders you should take care when rounding corners and when passing between and under obstacles.

After ladders are erected, check that:

- both ladder feet are on firm and level ground
- the ladder is set at an angle of 75 ° (to get the ladder to 75 °, use the ratio 4:1, for example, if your ladder is 4 m high, then set it 1m out at the base)
- the ladder is secured top or bottom
- the ladder projects at least 1 m (five rungs) above the landing place
- you have enough clear space at each rung for your feet at the working platform
- you are not going to overreach when working
- your hands are free to climb when lifting tools
- you still have three points of contact with the ladder.

If any one of these cannot be achieved then you should not be working from a ladder.

Figure 1.18 A trestle

Trestles

If a job cannot be carried out safely using stepladders, trestle scaffolds should be used instead.

To set out a trestle scaffold:

- erect it on firm, level ground, fully opened
- ensure the platform is at least 450 mm or two boards wide
- ensure the platform is no more than two-thirds of the way up the trestle
- ensure the scaffold boards are of equal length and do not overhang by more than four times their own thickness
- the maximum span is 1.3 m for 40 mm board and 2.5 m for 50 mm board
- toe boards and guard rails are used if the platform is more than 2 m off the ground.

Trestles **must not be used** where anyone can fall from a height of more than **4.5 m**.

Scaffolding

A lot of jobs working at height involve scaffolding.

Types of scaffolding

Types if scaffolding include:

- independent – totally independent of the structure which it is against
- putlog – built into the walls as the building is being built
- mobile – mainly made from aluminium, moves about with wheels to enable greater flexibility when working.

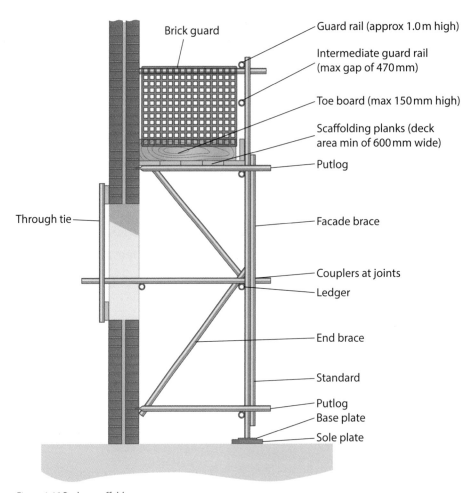

Brick guard

Guard rail (approx 1.0 m high)

Intermediate guard rail (max gap of 470 mm)

Toe board (max 150 mm high)

Scaffolding planks (deck area min of 600 mm wide)

Putlog

Through tie

Facade brace

Couplers at joints

Ledger

End brace

Standard

Putlog

Base plate

Sole plate

Figure 1.19 Putlog scaffold

Putlog scaffolding must be tied into a building. The guard rail height should be between 0.91 m and 1.00 m. The flattened end of putlog should be inserted into the wall by at least 75 mm.

Independent tied scaffolding uses ledger braces installed alternately to improve its stability. The putlog ends are not inserted into the brickwork; the scaffold is tied to the building through openings such as windows. This prevents the scaffold from falling away from the building.

There are many types of **mobile scaffolding**. It is made out of light alloy tube and can come with wheels for ease of movement or it can be fitted with base plates to create a static tower. The structure is built by slotting the components together until the desired height is reached. It should be erected following a safe method of work and the manufacturer's instructions, as it can become unstable at great height. PASMA guidelines for methods of erection are often used and they also run training courses throughout the country.

When used in exposed or external conditions the height of the tower must be taken into consideration to ensure safety. The height of the working platform must be no more than three times the minimum base dimension. For external work with a 2 m base, the maximum height of the scaffold would therefore be 2 m x 3 = 6 m. For internal work this can be increased to 3.5 times the base, so for a 2 m base, the maximum height of the scaffold would be 2 m x 3.5 = 7 m.

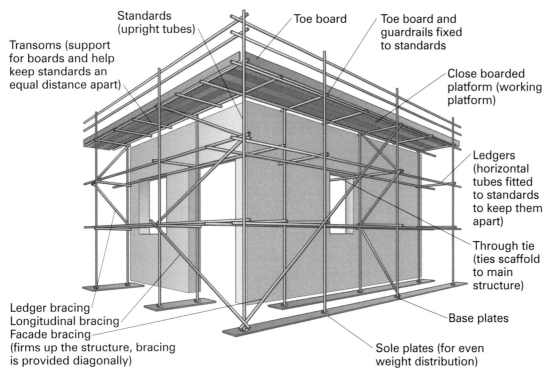

Standards (upright tubes)

Transoms (support for boards and help keep standards an equal distance apart)

Toe board

Toe board and guardrails fixed to standards

Close boarded platform (working platform)

Ledgers (horizontal tubes fitted to standards to keep them apart)

Through tie (ties scaffold to main structure)

Ledger bracing
Longitudinal bracing
Facade bracing
(firms up the structure, bracing is provided diagonally)

Base plates

Sole plates (for even weight distribution)

Figure 1.20 Independent scaffold

When working on platforms over 2 m from the ground, mobile scaffolding should:

- be fitted with guardrails – at least 950 mm high from working platform
- be fitted with rung spacings of between 230 mm and 300 mm
- be fitted with an intermediate guardrail so the unprotected gap does not exceed 470 mm
- have all four wheels locked
- never be moved without first removing tools, materials and people
- always be accessed using the internal ladder secured to the tower
- be secured to the building if over 9 m
- not to be built higher than 12 m.

When using mobile scaffolding on a pavement, check with your local council as you may need a pavement licence. Before work begins *always* inspect the tower to check if any braces or clips have been removed. When working on these types of scaffold ensure that you wear a safety harness and fall arrestor.

Outriggers are used to increase the stability of a mobile scaffold. Stability depends on the height ratio and can be increased by increasing the area of the base. Outriggers should be marked with hazard warning tape to prevent people walking into them.

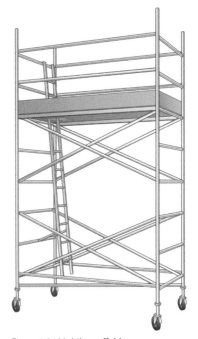

Figure 1.21 Mobile scaffold tower

Operators have died when trapped in the mobile elevated working platform (MEWP) basket or when the machine has overturned. Great care must be taken to select the most appropriate MEWP and ensure that use of the machine is properly planned and managed. It is essential that operators are properly trained and instructed.

Figure 1.22 Mobile elevated working platform (MEWP)

Progress check

1 List the checks that should be made before using ladders and/or steps.

2 What PPE should you use when working with asbestos?

3 At what angle should a ladder be erected?

  a 45°

  b 55°

  c 65°

  d 75°

4 List the checks that should be made before using a scaffold.

MEWP operators have to attend a recognised operator training course and will receive a certificate, card or licence, listing the categories of MEWP the bearer is trained to operate.

The expiry date of the training licence or card should be checked.

In addition to formal training for the type of MEWP, operators should have familiarisation training on the controls and operation of the specific make and model of MEWP they are using.

Using a ladder to access a scaffold

The ladder used to access a scaffold should be tied back to the scaffold. You should ensure the ladder is set at the correct angle using the 4:1 ratio outlined on page 21. The ladder must extend past the working platform by 1 m.

Checking scaffold

Scaffolds must be checked each day before use by a competent person using the following checklist:

- Does it look safe?
- Are the scaffold tubes plumb and level?
- Is there a sufficient number of braces and scaffold boards?
- Are adequate toe boards and guardrails fitted?
- Is it free from excessive loads?
- Is there proper access via a ladder?
- Are there no gaps in the boards?

Only use the scaffold when authorised to do so.

SAFE MANUAL HANDLING

The manual handling or lifting of objects is the cause of more injuries on worksites than any other accident. Back strains and associated injuries are the main source of lost hours in the building services industries. Manual handling can involve: pushing, pulling, and the lifting and lowering of loads (tools, cylinders boilers, radiators, etc.). The movement of loads requires careful

1 The task – does it involve:	2 The load – is it:
• stooping? • twisting? • excessive lifting or lowering distances? • excessive carrying distances? • excessive pushing or pulling distances? • frequent or prolonged physical effort? • the sudden risk of the load moving?	• heavy? • bulky or unwieldy? • difficult to grasp? • unstable or with contents that are likely to shift? • sharp, hot or otherwise potentially damaging?
3 The working environment – does it have:	4 The individual – does he or she have:
• space constraints? • slippery or unstable floors? • trip hazards? • variation in levels? • poor lighting? • hot/cold/humid conditions?	• any restriction on their physical capability? • the knowledge and training for manual handling?

Table 1.02 Safe manual handling questions

Figure 1.23 There should be information giving guidance on manual handling in your workplace

planning in order to identify potential hazards before they cause injuries. You should follow the safety precautions and codes of practice at all times.

Before handling a heavy item by hand, ask yourself whether the load can be moved another way with less risk of personal injury, for example using a sack trolley or a cart. If you decide to move the heavy load yourself, first consider the questions in Table 1.02 (shown opposite).

These are factors that must be taken into account when carrying out manual handling operations. Take a few moments to consider the information in each of the four lists and the impact each factor could have on the safe movement of tools and materials. Extreme care must be taken when lifting or moving heavy or awkward objects manually. The maximum load a fit man is likely to safely lift is 20 kg.

It is the responsibility of the employer to carry out risk assessments for the safe lifting and carrying of loads.

More information for manual handling is available at www.hse.gov.uk.

FIRST AID PROVISION AT WORK

The Health and Safety (First-Aid) Regulations 1981 states that:

'The employer should make an assessment of first-aid needs appropriate to the circumstances of each workplace.'

Many gas operatives work for small companies and many jobs are one-off jobs mainly in domestic dwellings. How much first-aid provision the employer has to make depends upon each of the workplace circumstances. There is no need for the assessment of first-aid needs to be formal or written down but employers need to justify their level of first-aid provision. Under the Management of Health and Safety at Work Regulations 1999 this would

Figure 1.24 A first aid kit is a requirement for all premises

require that the employer makes an assessment of the risks to the health and safety of employees at work and to identify what measures would need to be undertaken.

Companies should provide a first-aid kit in each of the company vehicles. There is no mandatory list of items to be included for first-aid kits, but general guidance is given in the regulations as comprising:

- a leaflet giving general guidance on first aid (for example, HSE's leaflet *Basic advice on first aid at work*)
- 20 individually wrapped sterile plasters (assorted sizes), appropriate to the type of work (hypoallergenic plasters can be provided, if necessary)
- two sterile eye pads
- four individually wrapped triangular bandages, preferably sterile
- six safety pins
- two large sterile individually wrapped unmedicated wound dressings
- six medium-sized individually wrapped unmedicated wound dressings
- a pair of disposable gloves (see HSE's leaflet *Latex and you*).

Your workplace should store its health and safety equipment in a first-aid station. This is likely to contain a first aid-kit and information as well as other features such as an eye wash station. Remember, if it is a serious incident you should contact the emergency services.

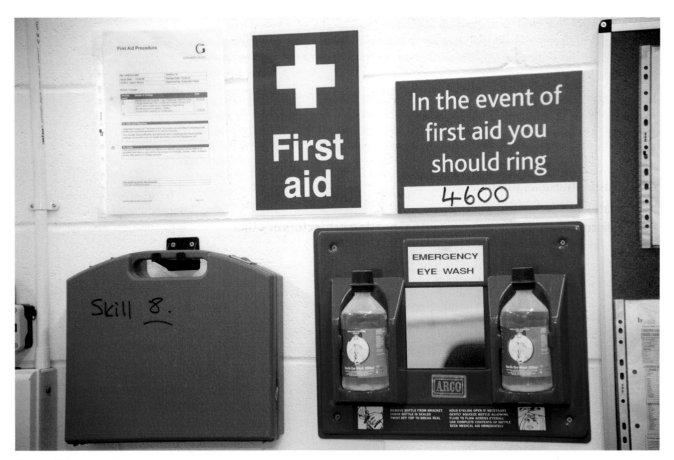

Figure 1.25 First-aid station

PROTECTION FROM LOUD NOISE AT WORK

The regulation which this guidance is taken from is The Control of Noise at Work Regulations 2005.

Hearing damage can be caused by exposure to excessive levels of noise when working. This hearing loss can be permanent and is in many cases incurable. Research has shown that over two million people are exposed to noise levels above those which may be harmful.

Hearing loss is usually gradual when you have prolonged exposure to excessive noise levels. Hearing damage can also be caused immediately by a sudden exposure to extremely loud noises. Exposure to noise at work can also cause 'tinnitus' which gives the sensation of ringing or buzzing in the ear.

These effects are preventable if:

- employers take action to reduce exposure to noise, and provide personal hearing protection and health surveillance to employees
- manufacturers design tools and machinery to operate more quietly
- employees make use of the personal hearing protection or other control measures supplied.

Figure 1.26 Ear defenders

Figure 1.27 Ear plugs

Regulation 7 (1) of the noise at work regulations 2005 states:

The employer shall make personal hearing protectors available upon request to any employee who is so exposed.

Regulation 7(2) states:

If an employer is unable by other means to reduce the levels of noise to which an employee is likely to be exposed to below an upper exposure action value, he shall provide personal hearing protectors to any employee who is so exposed.

As an employee you also have a duty under the regulations. Regulation 8 (2) states that every employee shall:

make full and proper use of personal hearing protectors provided to them by their employer in compliance with Regulation 7(2) and of any other control measures provided by his employer in compliance with his duties under these Regulations.

It also states:

if they discover any defect in any personal hearing protectors or other control measures they should report it to their employer as soon as is practicable.

Your employer must make a valid decision about whether an employee is at risk from exposure to noise and the actions that may have to be taken to prevent or adequately control the exposure to the noise.

Safe working

Exposure action levels for noise:

- Lower action level = 80 decibels
- Upper action level = 85 decibels

Where employees are likely to be exposed at or above the upper exposure action values action must be taken to reduce noise exposure with a planned programme of noise control.

Knowledge check

1 Which type of sign is described by a circular shape with a blue background and a white symbol?

a Mandatory sign
b Prohibition sign
c Safe condition sign
d Warning sign

2 Which of the following would be the subject of a prohibition sign?

a Safety helmets
b No smoking
c Emergency exit
d Eye protection area

3 Which of the following parties is responsible for enforcing safety legislation?

a The Department of Trade and Industry
b The Health and Safety Commission
c The Local Authority
d The Health and Safety Executive

4 What type of fuel is present in a Class A fire?

a Solid fuel
b Flammable gases
c Flammable liquids
d Flammable metals

5 How far above the stepping-off point should a ladder extend?

a A minimum of 1 m
b A maximum of 1 m
c A minimum of 2 m
d A maximum of 2 m

6 Which one of the following fire extinguisher types carries a black panel?

a Carbon dioxide (CO_2)
b Foam
c Water
d Dry powder

7 If the height of the ladder leaned against the wall is 4 m, what is the correct distance horizontally from the base of the wall to the foot of the ladder?

a 1,000 mm
b 900 mm
c 800 mm
d 1,200 mm

8 Which of the following would most likely require a site-specific work permit?

a Soldering copper pipework
b Decommissioning a central heating system
c The use of a hydraulic bending machine
d Maintaining discharge pipe work

9 When did the current Control of Asbestos at Work Regulations come into force

a 6 April 2010
b 6 May 2012
c 6 April 2012
d 6 June 2011

10 What is the recognised industry procedure for checking that a circuit is dead?

a Safety procedure
b Safe isolation procedure
c Safety compliance procedure
d Safe insulation procedure

Gas legislation and regulation

Chapter 2

THIS UNIT COVERS

- Gas Safety (Installation and Use) Regulations 1994
- standards
- the Health and Safety Executive
- Gas Safety (Installation and Use) Regulations 1998
- other related regulations

INTRODUCTION

Gas work was regulated by law for the first time in 1994. The regulations were introduced to reduce the number of deaths that occurred due to poor gas installations and maintenance. The regulations are not retrospective – they only apply to new installation practice from the time they were introduced. This means you have to be up-to-date with all legislation and regulations. This chapter sets out the key aspects of the relevant regulations and explains what they mean in practical terms and how they apply to a gas operative.

GAS SAFETY (INSTALLATION AND USE) REGULATIONS 1994

The Gas Safety (Installation and Use) Regulations (GSIUR) 1994 were introduced by Parliament in response to growing concerns about the dangers posed to the public from unqualified people being allowed to carry out gas work unsupervised.

For the first time the regulations required all people carrying out gas work to be **competent** and legal entities (people or businesses). Anyone charging money for gas work had to be Gas Safe registered.

The Gas Safe Register only accepts companies for membership who can prove their gas operatives are competent. This meant that passing exams became compulsory for all gas fitters.

However, the regulations were drafted by lawyers not by expert gas engineers (although they were of course consulted) and there are many ambiguities and contradictions in the law. The interpretation of the law is in the hands of the Health and Safety Executive (HSE), the Gas Safe Register and, ultimately, the courts. The 1994 regulations were tightened and updated in 1998, but few of the ambiguities were addressed.

The HSE has published its own interpretation of the regulations in a book called *Safety in the Installation and Use of Gas Systems and Appliances (Guide)*. This is widely relied upon in the gas industry but ultimately the court still has the last say.

Key term

Competent
A competent person is someone who has sufficient skills and knowledge to be able to carry out a specific task to a high standard.

STANDARDS

Any reference to an 'appropriate standard' is a reference to any of the following which are current at the time of the work activity:

- a British Standard (see Appendix 4 of the HSE's *Safety in the Installation and Use of Gas Systems and Appliances (Guide)* for a full list)
- a relevant standard or code of practice of a national standards body of any member state of the European Union (EU)
- a relevant technical specification acknowledged for use as a standard by a public authority of any member state of the EU.

Case study

In August last year, Mr. Jones was undertaking general building work to a house. As part of this work he installed a gas fire despite not being a Gas Safe registered engineer and having had no experience of this type of work. He was seen by the householders who asked him about his qualifications. He told them a registered engineer had put the gas pipe in and that he was just assembling the fire.

A few months after installation the residents noticed the pilot light wasn't working and asked a Gas Safe registered engineer to service the fire. The engineer classed it **immediately dangerous** and disconnected the gas supply. The fire had not been installed in accordance with the manufacturers' instructions and the flue had not been connected to the fire, risking carbon monoxide poisoning and potentially fatal consequences.

Mr. Jones was taken to court. He had breached section 3(2) of the HASAWA 1974 and regulations 3(3) and 27(1) of the GSIUR 1998. He was fined £7,000 and ordered to pay £1,753 in costs for the regulation 3(3) breach. In addition, Mr. Jones was ordered to pay the householders £1,644 in compensation. No penalties were imposed for the other charges.

Anyone fitting a gas appliance needs to be registered with the appropriate approved body to work legally. Mr. Jones defied this requirement and endangered lives with an installation that he was not competent or legally allowed to install. He failed, amongst other things, to connect the flue to the fire and, as a result, he placed the householders and their guests at risk from carbon monoxide poisoning, which can be fatal.

Every year around 15 people die from carbon monoxide poisoning caused by gas appliances and flues that have not been properly installed, maintained or ventilated. Badly fitted and poorly serviced gas appliances can cause fires, explosions, gas leaks and carbon monoxide poisoning.

Homeowners should always request ID from engineers and check their engineer is properly accredited.

Key term

Immediately dangerous
If an installation is deemed immediately dangerous, if used or left connected to the gas supply, it is considered to be an immediate danger to life or property and therefore puts people's health and safety at risk.

THE HEALTH AND SAFETY EXECUTIVE

The Health and Safety Executive (HSE) is Britain's national regulator for workplace health and safety. It aims to reduce work-related death, injury and ill health. It does this through:

- research, information and advice
- promoting training
- new or revised regulations and codes of practice
- working with local authority partners by inspection, investigation and enforcement.

The HSE website (www.hse.gov.uk) contains a lot of information about the powers and actions of the HSE.

GAS SAFETY (INSTALLATION AND USE) REGULATIONS 1998

There are 41 gas safety regulations within the GSIUR 1998 and they are split into groups as follows:

- Part A – General
- Part B – Gas fittings – general provisions

- Part C – Meters and regulators
- Part D – Installation pipework
- Part E – Gas appliances
- Part F – Maintenance
- Part G – Miscellaneous.

It is not the intention here to cover every regulation, but to interpret only the most important parts of the regulations and clarify exactly what they mean. The full regulations and guidance document should always be referred to.

Part A – General

Regulation 1: Citation and commencement

This regulation simply states that the gas safety and use regulations came into force in October 1998.

Regulation 2: General interpretation and application

This part of the regulations defines the terms used within the regulations. Some of these definitions are given below, followed by brief explanations.

2(1) 'Emergency control' means a valve for shutting off the supply of gas in an emergency being a valve intended for use by a consumer of gas.

The emergency valve is fitted near the meter so that in an emergency – such as a gas escape – the whole domestic system can be turned off by the consumer of gas (normally the customer or householder).

'The responsible person' means the occupier of the premises or, where there is no occupier or the occupier is away, the owner of the premises or any person with authority for the time being to take appropriate action in relation to any gas fitting therein.

Someone has to be responsible. If, for example, a gas leak occurred on the property, who is authorised to have it repaired? The responsible person is often the occupant of the house and, if they are not available, the owner of the house. If the owner is not available, anyone in authority is responsible, such as the public gas transporter.

Case study

A person walking down a street smells gas coming from a terraced house and, being concerned, knocks on the door of the house but gets no reply. A neighbour comes out of her house. She is also concerned and says she has the landlord's phone number. She gives him a ring, but the landlord is not available.

Concern grows. The neighbour rings the gas transporter who comes down and has the right to enter the house. He discovers the gas cooker has been left on so immediately turns it off. Without this chain of events and action being possible, an explosion could have occurred resulting in possible death or injury. By identifying the responsible person in any scenario, it is easier to establish responsibility so that the necessary action can be taken.

'Work' in relation to a gas fitting includes any of the following activities carried out by any person whether an employee or not, that is to say –

(a) installing or re-connecting the fitting;

(b) maintaining, servicing, permanently adjusting, disconnecting, repairing, altering or renewing the fitting or purging it of gas or air;

(c) where the fitting is not readily movable, changing its position; and

(d) removing the fitting.

But the expression does not include the connection or disconnection of a bayonet fitting or other self-sealing connector.

It is allowable for an unqualified person to disconnect a gas cooker from a bayonet in order to remove the appliance. However, if someone decides to do any 'work' on the gas system, they must be competent.

2(2) For the purpose of these regulations:

(a) any reference to installing a gas fitting include a reference to converting any pipe, fitting, meter, appliance to gas use; and

(b) a person to whom gas is supplied and who provides that gas for use in a flat or part of premises let by him shall not in so doing be deemed supplying gas.

This section clarifies that a landlord does not supply the gas to their buildings themselves, i.e. they are not the supplier.

2(3) (3) Subject to paragraphs (4) and (5) below, these regulations shall apply to or in relation to gas fittings used in connection with:

(a) gas which has been conveyed to premises through a distribution main; or

(b) gas conveyed from a gas storage vessel.

The term 'gas fitting' only applies in these regulations in situations where gas is supplied via a distribution main or from a gas storage vessel outside the property.

2(4) Save for regulations 37, 38 and 41 and subject to regulation 3(8), these regulations shall not apply in relation to the supply of gas to, or anything done in respect of a gas fitting at, the following premises, that is to say:

(a) a mine or quarry within the meaning of the Mines and Quarries Act 1954[3] or any place deemed to form part of a mine or quarry for the purposes of that Act;

(b) a factory within the meaning of the Factories Act 1961[4] or any place to which any provisions of the said Act apply by virtue of sections 123 to 126 of that Act; but they shall apply in relation to such premises or part thereof used for domestic or residential purposes or as sleeping accommodation.

The act is clear that mines and factories are not included in the regulations but if a change of use or part of it is used as a domestic residence than they shall apply.

Example

Disconnecting cooker
A householder needs to move his cooker to clean behind it. To do so, he needs to disconnect it from the bayonet fitting to safely pull the cooker clear. This fitting is designed so that it can be disconnected and reconnected without having to call out a gas engineer every time.

Example

Sleeping on site
A mine has a security guard who sleeps on the site and has a domestic dwelling for his quarters. If it has gas appliances such as a cookers it is covered by the gas regulations. This is to ensure that he is protected under the regulations.

Part B – Gas fittings – general provisions

Regulation 3: Qualification and supervision

3(1) No person shall carry out any work in relation to a gas fitting unless he is competent to do so.

All persons working on a gas system must be competent to do so, including:

- an employed person
- a self-employed person
- any other person carrying out DIY activities.

3(3) No employer shall allow any of his employees to carry out work in relation to a gas fitting or service pipework and no self-employed person shall carry out any such work, unless the employer or self-employed person, as the case may be, is competent to do so and is a member of a class of persons approved for the time being by the HSE.

The 'class of person' is the installer who has to be registered with any approved, Gas Safe-registered body. The HSE approves the class of person and Gas Safe keep the register and check competence.

Case study

A joiner working in a house is asked by the householder to remove a gas fire as they no longer use it. He doesn't have any gas qualifications but has seen plumbers doing it, so decides to have a go himself. He turns off the gas, disconnects the gas fire and blanks the gas pipe with a plastic push-fit stop-end he has in his toolbox. The customer is happy with his work and the joiner is happy to have earned some extra cash.

Was the joiner competent to do the work? And was it legal for him to have accepted payment for working on a gas fitting? No. He wasn't competent – a qualified gas fitter would know that plastic fittings aren't allowed. The joiner was working illegally as he was not on the Gas Safe Register and was therefore not a 'class of person' approved by the HSE.

3(5) An approval given pursuant to paragraph (3) above (and any withdrawal of such approval) shall be in writing and notice of it shall be given to such persons and in such manner as the Health and Safety Executive considers appropriate.

If you do not keep your qualifications up to date, you will receive a letter informing you that you have been withdrawn from the Gas Safe Register and must no longer carry out gas work.

3(6) The employer of any person carrying out any work in relation to a gas fitting or gas storage vessel in the course of his employment shall ensure that such of the following provisions of these regulations as impose duties upon that person and are for the time being in force are complied with by that person.

According to the regulations, the employer is responsible for any gas work carried out by their employee.

3(7) No person shall falsely pretend to be a member of a class of persons required to be approved under paragraph (3) above.

It is illegal to claim to be a member of class of person approved by HSE if you are not a member of the Gas Safe Register.

3(8) Notwithstanding sub-paragraph (b) of regulation 2(4), when a person is carrying out work in premises referred to in that sub-paragraph in relation to a gas fitting in a vehicle, vessel or caravan:

(a) paragraphs (1), (2) and (6) of this regulation shall be complied with as respects thereto; and

(b) he shall ensure, so far as is reasonably practicable, that the installation of the gas fittings and flues will not contravene the provisions of these regulations when the gas fittings are connected to a gas supply, except that this paragraph shall not apply where the person has reasonable grounds for believing that the vehicle, vessel or caravan will be first used for a purpose which when so used will exclude it from the application of these regulations by virtue of sub-paragraphs (a), (c) or (e) of regulation 2(5).

Figure 2.01 Registered installers must carry out all gas work

Installing a gas fitting into a car, boat or caravan should be done in accordance with the gas safety regulations.

Regulation 4: Duty on employer

Where an employer or self-employed person requires any work in relation to a gas fitting to be carried out at any place of work under his control or where an employer or self-employed person has control to any extent of work in relation to a gas fitting he shall take reasonable steps to ensure that the person undertaking that work is, or is employed by, a member of a class of persons approved by the HSE.

This regulation places a duty on an employer/owner of a business premises to ensure the work is done by a class of person recognised by the HSE (a registered installer on the Gas Safe Register).

Regulation 5: Materials and workmanship

5(1) No person shall install a gas fitting unless every part of it is of good construction and sound material of adequate strength and size to secure safety and of a type appropriate for the gas with which it is to be used.

You must ensure the fitting is fit for purpose. For example, on site, always check that the valves in your toolbox are for use with gas not water. If not, you could accidently break the law and put people's lives in danger. Always check that your fittings are of good structure and sound material.

Figure 2.02 A gas valve will be clearly marked

5(2) (1) No person shall install in a building any pipe or fitting for use in the supply of gas which is:

(a) made of lead or lead alloy

(b) made of a non-metallic substance unless it is

(i) a pipe connected to a readily movable gas appliance designed for use without a flue

(ii) a pipe that is entering a building and is placed inside a metallic sheath

Installation of lead pipes/fittings is prohibited and controls are placed on the use of non-metallic pipes/fittings. Rubber hoses can be used on movable appliances, such as gas cookers.

Inside the building a plastic pipe should be encased in a metal pipe to prevent any gas escaping if the plastic pipe was to fail.

5(3) No person shall carry out any work in relation to a gas fitting or gas storage vessel otherwise than in accordance with appropriate standards and in such a way as to prevent danger to any person.

Figure 2.03 Typical cooker hose connected to bayonet fitting

Figure 2.04 Stop-end for a pipe

The appropriate standards you should refer to are the British Standards, British Gas publications, the Institute of Gas Engineers and Managers technical publications, as well as Building Regulations.

Regulation 6: General safety precautions

6(1) No person shall carry out any work in relation to a gas fitting in such a manner that gas could be released unless steps are taken to prevent the gas so released constituting a danger to any other person.

This includes purging procedures to ensure gas is released safely and avoid explosion.

6(2) No person carrying out work in relation to a gas fitting shall leave the fitting unattended unless every incomplete gas way has been sealed with the appropriate fitting or the gas fitting is otherwise safe.

When you are working on gas pipe always ensure that you close any ends that could allow gas to escape by capping any open ends with the correct cap, such as a brass stop-end or soldered cap-end.

6(4) No person carrying out work in relation to a gas fitting which involves exposing gasways which contain or have contained flammable gas shall smoke or use any source of ignition in such a manner as may lead to the risk of fire or explosion.

As a gas fitter you must do all you can to avoid risk of fire or explosion, for example you should never smoke on the job.

6(5) No person searching for an escape of gas shall use any source of ignition.

The use of naked flame such as a lighter or match to identify the source of escaping gas is illegal and extremely dangerous to your own health and safety, and to those around you.

6(6) Where a person carries out any work in relation to a gas fitting which might affect the gas tightness of the gas installation he shall immediately thereafter test the installation for gas tightness at least as far as the nearest valves upstream and downstream in the installation.

If you disconnect an appliance or break into the pipework you must test the system for tightness in case you have caused a leak.

Case study

A plumber was working under the floor on a heating system when he accidentally fell onto the gas pipe. On first inspection, the pipe seemed sound enough so he carried on with his work. Unfortunately, in falling against the gas pipe, he had dislodged a fitting which was letting a small amount of gas escape. As the plumber lit his blowtorch the gas caught alight and set the house on fire. Everyone managed to escape unhurt but the house burnt down. Any movement of the gas pipe could lead to failure so the plumber should have carried out a tightness test before carrying on with his work.

Regulation 7: Protection against damage

7(1) Any person installing a gas fitting shall ensure that it is properly supported and so placed or protected as to avoid any undue risk of damage to the fitting.

This section deals with the protection of the gas system, stipulating that it is the duty of all installers to fit an installation in such a way that it avoids any damage occurring to the installation.

Case study

A plumber left a pipe unsecured to the wall and when someone leaned on it the pipe moved causing the fitting to be dislodged and gas to leak. This is a reportable offence as he had left an 'at risk' situation in the property. You should ensure all clips are fixed to surface material correctly and are secured within the correct clipping distances.

7(2) No person shall install a gas fitting if he has reason to suspect that foreign matter may block or otherwise interfere with the safe operation of the fitting unless he has fitted to the gas inlet of, and any airway in, the fitting a suitable filter or other suitable protection.

If you know the gas system is likely to have foreign matter in, it is essential that filters are fitted to prevent the foreign matter from clogging up the gas equipment, as this could cause malfunction of the controls and lead to a potentially unsafe situation.

Figure 2.05 A pipe correctly clipped in place

Case study

A fitter was working on a new boiler installation with an incoming meter gas pipe in mild steel. He fitted the new boiler and carried out the tests correctly but, after a day, the boiler stopped working. On investigation he found the filter to the boiler was blocked with steel shale which had been in the steel pipe and become dislodged when work was carried out on it. The fitter cleared out all the steel shale and fitted a filter at the inlet to the meter. This meant the equipment would be protected and the filter could be easily cleaned out.

7(3) No person shall install a gas fitting in a position where it is likely to be exposed to any substance which may corrode gas fittings unless the fitting is constructed of materials which are inherently resistant to being so corroded or it is suitably protected against being so corroded.

Corrosion of pipework and fittings can lead to them failing and gas escaping. Always run pipes and fittings away from such problems or, if this is not possible, protect them from corrosion by using corrosion-resistant material or protecting them with a corrosion-resistant tape, bandage or sleeve.

Regulation 8: Existing gas fittings

8(1) No person shall make any alterations to any premises in which a gas fitting is fitted if that alteration would adversely affect the safety of the fitting in such a manner that, if the fitting had been installed after the alteration, there would have been a contravention of, or failure to comply with these regulations.

When working on/near existing systems, never interfere with the safety of the system, for example avoid making alterations that will make the system contravene the regulations.

Case study

A conservatory was installed at the back of a house where a boiler flue was situated. The builder did not take any action to prevent the flue from discharging its product of combustion (POC) straight into the conservatory. The home owner was subjected to these waste products as he sat in the conservatory one night fell. He fell asleep in his chair and never woke up.

The builder was as fault for not re-siting the flue and was prosecuted. The builder contravened the regulations and was responsible for creating a dangerous situation.

Regulation 9: Emergency controls

9(1) No person shall for the first time enable gas to be supplied for use in any premises unless there is provided an appropriately sited emergency control to which there is adequate access.

The emergency control valve is usually fitted by the gas supplier when first installing the gas supply to a house, but you should always check to see if the system has an emergency control valve that is fit for purpose and accessible.

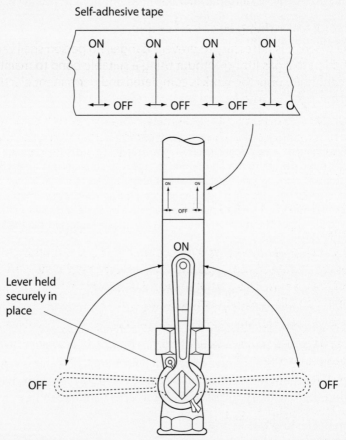

Figure 2.06 Example of an emergency control valve – note the tape which must be provided to indicate the on/off position

9(2) Any person installing an emergency control shall ensure that:

(a) any key, lever or hand-wheel of the control is securely attached to the operating spindle of the control;

(b) any such key or lever is attached so that

(i) the key or lever is parallel to the axis of the pipe in which the control is installed when the control is in the open position; and

(ii) where the key or lever is not attached so as to move only horizontally, gas cannot pass beyond the control when the key or lever has been moved as far as possible downwards.

When installing an emergency control, ensure the handle is in line with the pipe when in the open position.

9(3) Where a person installs an emergency control which is not adjacent to a primary meter, he shall immediately thereafter prominently display on or near the means of operating the control a suitably worded notice in permanent form indicating the procedure to be followed in the event of an escape of gas.

If a gas meter is fitted more than 6 m away from the building, a further emergency control valve should be fitted inside the building. It also should have the correct notices attached, showing the open and closed position and use the term EMC (emergency control for customer use) and gas escape procedure, and refer to emergency contact number written on the label.

Regulation 10: Maintaining electrical continuity

In any case where it is necessary to prevent danger, no person shall carry out work in relation to a gas fitting without using a suitable bond to maintain electrical continuity until the work is completed and permanent electrical continuity has been restored.

Figure 2.07 Temporary bonding being applied between two pipes and meter

You should always use temporary continuity bonds when cutting through a gas pipe or removing a meter in order to comply with this regulation. This is both for your safety and the safety of others, to prevent electrocution and protect the earth path.

Part C – Meters and regulators

Regulations 11–17

Any work on meters and regulators is usually carried out by utilities companies or specialist companies and refers to the meter cupboards and installation. You can use this part of the regulations to check that a meter installation is compliant with the regulations.

Regulation 14 stipulates that primary meters should always be fitted with a regulator and the seal of the regulator cannot be touched by anyone other than the gas transporter. The seal has to be replaced as soon as the work on the regulator has finished.

Part D – Installation pipework

Regulation 18: Safe use of pipes

18(1) No person shall install any installation pipework in any position in which it cannot be used with safety having regard to the position of other pipes, pipe supports, drains, sewers, cables, conduits and electrical apparatus and to any parts of the structure of any premises in which it is installed which might affect its safe use.

Figure 2.08 An external gas meter box

This regulation requires installation pipework to be installed in a safe position having regard to factors that might affect safety.

Case study

When installing a gas pipe through a wall, a gas fitter neglected to sleeve the pipe correctly. The wall was then plastered, covering the pipe in plaster. Gypsum plaster contains substances that can corrode copper pipes. Several years later, a pinhole appeared in the pipe, leading to a gas leak. The gas fitter was taken to court for contravening Regulation 18(1).

It is also not unusual to find cables lying on the gas pipes or water pipes, which is potentially dangerous if the insulation of the wire is breached as electric current could flow down the copper pipes making it live.

18(2) Any person who connects any installation pipework to a primary meter shall, in any case where equipotential bonding may be necessary, inform the responsible person that such bonding should be carried out by a competent person.

Any person connecting installation pipework to a meter is required to inform the person responsible for the premises of the need for equipotential bonding. This requires a form (Advice notice) to be filled in and signed by the customer and installer to prove such notice was given.

Regulation 19: Enclosed pipes

19(1) No person shall install any part of any installation pipework in a wall or a floor or standing of solid construction unless it is so constructed and installed as to be protected against failure caused by the movement of the wall, the floor or the standing as the case may be.

British Standards and codes of practice specify restrictions and protective measures for pipes passing through solid walls and floors, cavity walls and building foundations. A typical example is using a pipe sleeve through a wall to allow for movement in the structure.

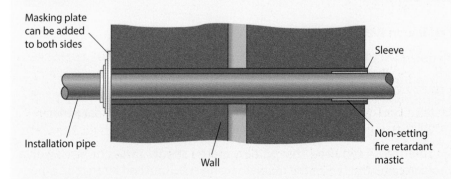

Figure 2.09 A pipe sleeve through a wall

19(4) Paragraph (3) shall not apply to the installation of installation pipework connected to a living flame effect fire provided that the pipework in the cavity is as short as is reasonably practicable, is enclosed in a gas tight sleeve and sealed at the joint at which the pipework enters the fire; and in this paragraph a 'living flame effect gas fire' means a gas fire:

(a) designed to simulate the effect of a solid fuel fire;

(b) designed to operate with a fanned flue system; and

(c) installed within the inner leaf of a cavity wall.

Conditions are stipulated whereby pipework associated with 'living flame effect fires' may be run in a wall cavity. Ducts and voids accommodating installation pipework have to be adequately ventilated. The fire also has to be fanned flued and be installed in the inner leaf of the property.

19(6) Where any installation pipework is not itself contained in a ventilated duct, no person shall install any installation pipework in any shaft, duct or void which is not adequately ventilated.

The possible build-up of gas has to be prevented so venting enables the gas to escape without building up a dangerous atmosphere within the duct.

Regulation 20: Protection of buildings

No person shall install any installation pipework in a way which would impair the structure of a building or impair the fire resistance of any part of its structure.

Pipes that pass from one room to another can break the fire barrier of that room and so require precautions to prevent the spread of fire through the sleeve. The pipes must therefore be sealed at one end with fire retardant mastic (see Figure 2.09).

Regulation 21: Clogging precautions

No person shall install any installation pipework in which deposition of liquid or solid matter is likely to occur unless a suitable vessel for the reception of any deposit which may form is fixed to the pipe in a conspicuous and readily accessible position and safe means are provided for the removal of the deposit.

This regulation refers to a more specialist area of gas installation work that is not usually found in domestic situations. Gas produced from landfill or by anaerobic digestion may, in certain circumstances, require the use of collection chambers.

Regulation 22: Testing and purging of pipes

22(1) Where a person carries out work in relation to any installation pipework which might affect the gas tightness of any part of it, he shall immediately thereafter ensure that:

(a) that part is adequately tested to verify that it is gastight and examined to verify that it has been installed in accordance with these regulations; and

(b) after such testing and examination, any necessary protective coating is applied to the joints of that part.

This identifies the requirements for gas tightness testing after work has been done on installation pipework, and for the purging/sealing of such pipework, where gas is being supplied to the premises where it is installed. Before purging, it is essential you test the pipework that is to be buried before covering it with a protective coating to ensure it cannot be corroded.

22(2) Where gas is being supplied to any premises in which any installation pipework is installed and a person carried out work in relation to the pipework they call also ensure that:

(a) immediately after complying with the provisions of sub-paragraphs (a) and (b) of paragraph (1) above, purging is carried out throughout all installation pipework through which gas can then flow, so as to remove safely all air and gas other than the gas to be supplied

(b) immediately after such purging, if the pipework is not to be put into immediate use, it is sealed off at every outlet with the appropriate fitting

(c) if such purging has been carried out through a loosened connection, the connection is retested for gas tightness after it has been retightened; and

(d) every seal fitted after such fitting is tested for gas tightness.

Yellow ochre	Primrose yellow	Yellow ochre

Figure 2.10 British Standards gas pipe colour code

22(3) Where gas is not being supplied to any premises in which any installation pipework is installed:

(a) no person shall permit gas to pass into the installation pipework unless he has caused such purging, testing and other work as is specified in sub-paragraphs (a) to (d) of paragraph (2) above to be carried out;

(b) a person who provides a gas supply to those premises shall, unless he complies with sub-paragraph (a) above, ensure that the supply is sealed off with an appropriate fitting.

This often happens when new buildings are being constructed: the gas main is put into position before the gas system in the house is ready. The gas supplier will need to cap off the gas pipe to prepare for the system being completed and tightness testing.

Regulation 23: Marking of pipes

23(1) Any person installing, elsewhere than in any premises or part of premises used only as a dwelling or for living accommodation, a part of any installation pipework which is accessible to inspection shall permanently mark that part in such a manner that it is readily recognisable as part of a pipe for conveying gas.

Gas pipework needs to be colour coded in premises *other than* dwellings or living accommodation, such as flats. This is to say pipework in domestic houses/flats does not require markings.

23(2) The responsible person for the premises in which any such part is situated shall ensure that the part continues to be so recognisable so long as it is used for conveying gas.

The responsible person for the building should always maintain the colour coding on the pipework system.

Regulation 24: Large consumers

This regulation applies where a service pipe has an internal diameter of 50 mm or more, or service pipework has an internal diameter of 30 mm or more, and divides to supply more than one floor, or separate areas on one floor. This situation is most often encountered in industrial and commercial premises. This is out of the scope of this qualification so there is no need to look in detail at this regulation.

Part E – Gas appliances

Regulation 25: Interpretation of Part E

In this part:

'Flue pipe' means a pipe forming a flue but does not include a pipe built as a lining into either a chimney or a gas appliance ventilation duct;

This section defines the term 'flue pipe' as being different from a brick-built chimney or a ventilation duct.

'Operating pressure', in relation to a gas appliance, means the pressure of gas at which it is designed to operate.

This section is referring to the pressure at which the appliance is designed to work at and we call that the operating pressure.

Regulation 26: Gas appliances – safety precautions

26(1) No person shall install a gas appliance unless it can be used without constituting a danger to any person.

This section imposes on the installer of an appliance the necessity to ensure it is safe to use when they have completed the work.

26(2) No person shall connect a flued domestic gas appliance to the gas supply system except by a permanently fixed rigid pipe.

If a domestic gas appliance has a flue fitted, then it must be piped up with rigid pipe material, such as copper.

26(3) No person shall install a used gas appliance without verifying that it is in a safe condition for further use.

It is the responsibility of the installer to check that second-hand gas appliances are safe to use.

26(4) No person shall install a gas appliance which does not comply with any enactment imposing a prohibition or restriction on the supply of such an appliance on grounds of safety.

New appliances should conform to the Gas Appliances (Safety) Regulations. These regulations require the relevant appliances and fittings to conform with specified mandatory requirements and to be safe when in normal use. Supply of these products is prohibited unless they bear the CE marking and their safety is underpinned by valid certification/declaration of conformity.

26(5) No person carrying out the installation of a gas appliance shall leave it connected to the gas supply unless:

(a) the appliance can be used safely; or

(b) the appliance is sealed off from the gas supply with an appropriate fitting.

You need to commission the appliance to ensure it is safe to use or cap it off so it can't be used until it is ready to be commissioned.

26(6) No person shall install a gas appliance without there being at the inlet to it means of shutting off the supply of gas to the appliance unless the provision of such means is not reasonably practicable.

All gas appliances should have a control valve fitted on the inlet of the appliance so that the gas supply can be isolated. Only in exceptional circumstances (e.g. if it's not possible to fit one) can you install an appliance without such a valve.

26(7) No person shall carry out any work in relation to a gas appliance which bears an indication that it conforms to a type approved by any person as complying with safety standards in such a manner that the appliance ceases to comply with those standards.

You are not allowed to modify or bypass any controls on a gas appliance that will make it unsafe.

26(8) No person carrying out work in relation to a gas appliance which bears an indication that it so conforms shall remove or deface the indication.

The 'indication' mentioned here is the appliances data badge, which should not be removed or defaced.

26(9) Where a person performs work on a gas appliance he shall immediately thereafter examine:

(a) the effectiveness of any flue;

(b) the supply of combustion air;

Figure 2.11 CE European conformity mark

Figure 2.12 Make sure you are always following the correct test procedures

(c) its operating pressure or heat input or, where necessary, both;

(d) its operation so as to ensure its safe functioning.

And forthwith take all reasonably practicable steps to notify any defect to the responsible person and, where different, the owner of the premises in which the appliance is situated or, where neither is reasonably practicable, in the case of an appliance supplied with LPG, the supplier of gas to the appliance, or, in any other case, the transporter.

This section identifies the testing procedure of an appliance that satisfies the regulation in terms of its safe use.

- You are required to check the flue is working correctly (see below).
- You need to check that the air supply is adequate, including ventilation sizing.
- You need to check it is working at the correct pressure as identified in the manufacturer's instructions, and calculate the heat input into the appliance.
- You will be required to operate the appliance to check it is in good working order.

Any defects must be reported to the owner so that it can be repaired.

26(10) Paragraph (9) shall not apply in respect of:

(a) the direct disconnection of the gas supply of a gas appliance; or

(b) the purging of gas or air from an appliance or its associated pipework or fittings in any case where that purging does not adversely affect the safety of that appliance, pipe or fitting.

When removing a gas appliance, you do not have to check it is safe.

Regulation 27: Flues

27(1) No person shall install a gas appliance to any flue unless the flue is suitable and in a proper condition for the safe operation of the appliance.

The regulation requires any flue to be suitable and in proper condition for safe operation of the appliance it serves. Any power-operated flue system is required to prevent operation of the appliance if the draught fails. Requirements to enable the inspection and prevent spillage of POCs from certain flues are specified. All flues are required to be installed in a safe position.

Flue testing

Checklist		
PPE	**Tools and equipment**	**Source information**
• Overalls	• Blowtorch	• British Standards
• Boots	• Light	• Manufacturer's instructions
• Safety gloves	• Grips	

① Check flue for obstructions using a torch.	② Pre-warm the flue for 5–10 minutes to ensure flue flow is established.	③ Light smoke pellet and hold it in flue openinig and check for spillage/pull.
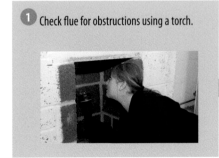		

27(2) No person shall install a flue pipe so that it enters a brick or masonry chimney in such a way that the seal between the flue pipe and the chimney cannot be inspected.

The installation of a flue into a chimney requires the gas fitter to be able to visually inspect it at any time.

27(3) No person shall connect a gas appliance to a flue which is surrounded by an enclosure unless that enclosure is so sealed that any spillage of products of combustion cannot pass from the enclosure to any room or internal space other than the room or internal space in which the appliance is installed.

This is to prevent spillage of POCs from an appliance from spilling into another room, which could go undetected.

27(4) No person shall install a power operated flue system for a gas appliance unless it safely prevents the operation of the appliance if the draught fails.

A fan flue has to be wired to the appliance so that, if the fan fails, it will cut off the gas valve on the appliance.

Regulation 28: Access

No person shall install a gas appliance except in such a manner that it is readily accessible for operation, inspection and maintenance.

You should always allow for access to the gas appliance for maintenance and inspection purposes.

Regulation 29: Manufacturer's instructions

Any person who installs a gas appliance shall leave for the use of the owner or occupier of the premises in which the appliance is installed all instructions provided by the manufacturer accompanying the appliance.

The installer should leave manufacturer's instructions with the owner so that they have the operating instructions and can pass them to a gas engineer when working on the appliance.

Regulation 30: Room-sealed appliance

30(1) No person shall install a gas appliance in a room used or intended to be used as a bathroom or a shower room unless it is a room-sealed appliance.

You should not install appliances that are not room sealed in a bathroom or shower room as tests have proven that moisture in the air can have a detrimental effect on the appliance.

30(2) No person shall install a gas fire, other gas space heater or a gas water heater of more than 14KW heat input in a room used or intended to be used as sleeping accommodation unless the appliance is a room-sealed appliance.

You should not install an appliance in a bedroom/sleeping area if it is larger than 14 kw unless the appliance is room sealed. Any appliance that is smaller than 14 kw must also be fitted with a safety device to turn it off if POC levels in the room become too high.

Figure 2.13 Make sure the owner understands the operating instructions when completing an installation

Case study

Many fatal gas-related incidents involving bedsits and other types of student accommodation are caused by faulty open-flue appliances in rooms where people sleep. Because people are asleep at the time, they do not realise they are being affected by carbon monoxide poisoning and never wake up. Therefore room-sealed appliances are only permitted in bedrooms if the appliance is 14 kw or greater. Appliances less than 14 kw must be fitted with sensing devices that will turn off an appliance if carbon monoxide is detected.

Case study

A gas engineer installs a gas fire to an old brick chimney and finds an old damper plate built into the flue. He decides to wedge it open with a brick and finishes the installation. He tests the flue and all is ok. That winter, gales dislodge the brick and the damper plate closes. The occupier smells fumes but presumes it's the wind blowing the fumes back into the room. They nod off to sleep in front of the gas fire. Fortunately, the fire is fitted with an oxygen depletion device and the occupier is woken as the safety device clicks the gas fire off. A gas fitter is called who finds the damper plate closed and reports it to RIDDOR as it's a potentially dangerous situation that contravenes the gas regulations.

Regulation 31: Suspended appliances

31(1) No person shall install a suspended gas appliance unless the installation pipework to which it is connected is so constructed and installed as to be capable of safely supporting the weight imposed on it and the appliance is designed to be so supported.

This scenario is not often found in a domestic setting, but the pipework for a suspended gas appliance has to be strong enough to withstand any strain put on it by the weight of the appliance or movement that could occur.

Regulation 32: Flue dampers

32(3) No person shall install a domestic gas appliance to a flue which incorporates a manually operated damper unless the damper is permanently fixed in the open position.

A damper is a device often used in brick-built chimney flues designed to partially block off the flue way and reduce the draught, so slowing down the heat from the coal fire. This is not suitable for gas-burning fires, so dampers have to be fixed or locked in the open position to prevent accidental blockage of the flue.

Regulation 33: Testing of appliances

33(1) Where a person installs a gas appliance at a time when gas is being supplied to the premises in which the appliance is installed, he shall immediately thereafter test its connection to the installation pipework to verify that it is gas tight and examine the appliance and the gas fittings and other works for the supply of gas and any flue or means of ventilation to be used in connection with the appliance for the purpose of ascertaining whether:

(a) the appliance has been installed in accordance with these regulations;

(b) the operating pressure is as recommended by the manufacturer;

(c) the appliance has been installed with due regard to any manufacturer's instructions provided to accompany the appliance; and

(d) all gas safety controls are in proper working order.

You must check installation pipework is gas tight, installed in adherence with the regulations, and set to the manufacturer's operating pressure. You must also follow the manufacturer's instructions and check all safety controls are working.

33(2) Where a person carries out such testing and examination in relation to a gas appliance and adjustments are necessary to ensure compliance with the requirements specified in sub-paragraphs (a) to (d) of paragraph (1) above, he shall either carry out those adjustments or disconnect the appliance from the gas supply or seal off the appliance from the gas supply with an appropriate fitting.

33(3) Where gas is not being supplied to any premises in which any gas appliance is installed:

(a) no person shall subsequently permit gas to pass into the appliance unless he has caused such testing, examination and adjustment as is specified in paragraphs (1) and (2) above to be carried out; and

(b) a person who subsequently provides a gas supply to those premises shall, unless he complies with sub-paragraph (a) above, ensure that the appliance is sealed off from the gas supply with an appropriate fitting.

This part explains that unless the equipment can be commissioned, it must be left sealed off from the gas supply.

Regulation 34: Use of appliances

34(1) The responsible person for any premises shall not use a gas appliance or permit a gas appliance to be used if at any time he knows or has reason to suspect that it cannot be used without constituting a danger to any person.

The relevant responsible person must ensure gas appliances are safe to use or have them turned off and an engineer called to check or repair them. This applies even if you are a tenant in the house – you must not use a dangerous appliance.

34(2) For the purposes of paragraph 1 the responsible person means the occupier of the premises, the owner of the premises and any person with authority for the time being to take appropriate action in relation to any gas fitting therein.

This part of the regulation is reinforcing who is responsible for the equipment in premises – those who live in the building and those who own the building.

34(3) Any person engaged in carrying out any work in relation to a gas main, service pipe, service pipework, gas storage vessel or gas fitting who knows or has reason to suspect that any gas appliance cannot be used without constituting a danger to any person shall forthwith take all reasonably practicable steps to inform the responsible person for the premises in which the appliance is situated and, where different, the owner of the appliance or, where neither is reasonably practicable, in the case of an appliance supplied with liquefied petroleum gas, the supplier of gas to the appliance, or, in any other case, the transporter.

Even when you are in a building that you are not working in but see an appliance that is dangerous, you must report it to the responsible person for that appliance.

Part F – Maintenance

Regulation 35: Duties of employers and self-employed persons

35(1) It shall be the duty of every employer or self-employed person to ensure that any gas appliance, installation pipework or flue installed at any place of work under his control is maintained in a safe condition so as to prevent risk of injury to any person.

This stipulates that the employer/self-employed person is responsible for ensuring that any appliance/installation pipework/flue they or their employees have installed is maintained in good working order.

Case study

Part of a shop containing a gas wall heater which hasn't been used for several years is closed down due to poor trading conditions. The employer decides to use the room for an office for their secretary. The room is cold and damp so he turns on the wall heater. He starts to complain of headaches. When he is signed off work with flu-like symptoms, a fellow worker takes over the secretary's work but eventually also complains of headaches. The employer begins to realise there is something wrong with the room and calls in a qualified plumber. The plumber quickly diagnoses a faulty heater and caps it off. The secretary returns to work and takes the employer to court, suing for compensation for endangering his life.

Figure 2.14 Make sure that landlords are fully briefed on installations

Regulation 36: Duties of landlords

36(1) In this regulation – 'landlord' means:

(a) in England and Wales:

(i) where the relevant premises are occupied under a lease, the person for the time being entitled to the reversion expectant on that lease or who, apart from any statutory tenancy, would be entitled to possession of the premises; and

(ii) where the relevant premises are occupied under a licence, the licensor, save that where the licensor is himself a tenant in respect of those premises, it means the person referred to in paragraph (i) above;

This section deals with the definitions of 'landlord' and 'tenant', going into great detail before moving onto the roles below.

36(2) Every landlord shall ensure that there is maintained in safe condition:

(a) any relevant gas fitting; and

(b) any flue which serves any relevant gas fitting,

so as to prevent the risk of injury to any person in lawful occupation of relevant premises.

36(3) Without prejudice to the generality of paragraph 2 above, a landlord shall:

(a) ensure that each appliance and flue to which that duty extends is checked for safety within 12 months of being installed and at intervals of not more than 12 months since it was last checked for safety (whether such check was made pursuant to these regulations or not);

(b) in the case of a lease commencing after the coming into force of these regulations, ensure that each appliance and flue to which the duty extends has been checked for safety within a period of 12 months before the lease commences or has been or is so checked within 12 months after the appliance or flue has been installed, whichever is later.

Regulation 36 was introduced to ensure rented accommodation would be kept in good order by landlords. Prior to this regulation being in force, many incidents of death and injury had been caused in rented properties (especially flats and student accommodation) where landlords were not maintaining the gas appliances within their ownership. A further problem could happen with tenants who had caused damage to gas equipment. Now the landlords not only have to have yearly checks carried out on their gas appliances but also check at each change of tenant.

The GSIUR 1998 place duties on gas users, installers, suppliers and landlords. Where an employer or self-employed person has a gas appliance installed in the workplace or maintenance work carried out on existing gas appliances or fittings, they must make sure that the engineer is qualified to work on that particular type of equipment. At present, the engineer can prove this by being registered on the Gas Safe Register.

When a gas appliance is installed, it must be located in a position that is easily accessible for use, inspection and maintenance. Employers, the self-employed or anyone responsible for business premises must not allow a gas appliance to be used if they suspect it may be dangerous. The GSIUR 1998 also place specific duties on landlords to:

- ensure gas appliances and flues are maintained in a safe condition
- have annual safety checks carried out by an appropriately qualified Gas Safe engineer
- retain records of these checks for at least two years and issue them to tenants within 28 days of the checks being carried out but does not need to have the signature of who checked it if it after more than 28 days.

For further information on gas appliances and the responsibilities of landlords and tenants, visit:

- www.GasSafeRegister.co.uk/tenant
- www.GasSafeRegister.co.uk/landlord

Part G – Miscellaneous

This is for the parts of the regulations that don't fit into any other category.

Regulation 37: Escape of gas

37(1) Where any gas escapes from any pipe of a gas supplier or from any pipe, other gas fitting or gas storage vessel used by a person supplied with gas by a gas supplier, the supplier of the gas shall, within 12 hours of being so informed of the escape, prevent the gas escaping (whether by cutting off the supply of gas to any premises or otherwise).

In case of a gas leak, the gas suppler must turn the gas supply off within 12 hours of notification.

37(2) If the responsible person for any premises knows or has reason to suspect that gas is escaping into those premises, he shall immediately take all reasonable steps to cause the supply of gas to be shut off at such place as may be necessary to prevent further escape of gas.

This places a duty on the property owner or tenant to turn off the gas at the emergency control valve in the case of a gas escape.

37(3) If gas continues to escape into those premises after the supply of gas has been shut off or when a smell of gas persists, the responsible person for the premises discovering such escape or smell shall immediately give notice of the escape or smell to the supplier of the gas.

If after turning off the gas supply they can still smell gas, then the responsible person must inform the gas supplier.

37(4) Where an escape of gas has been stopped by shutting off the supply, no person shall cause or permit the supply to be re-opened (other than in the course of repair) until all necessary steps have been taken to prevent a recurrence of such escape.

It would be very dangerous to turn back on the gas without first having the gas escape repaired.

37(5) In any proceedings for an offence under paragraph (1) above it shall be a defence for the supplier of the gas to prove that it was not reasonably practicable for him effectually to prevent the gas from escaping within the period of 12 hours referred to in that paragraph, and that he did effectually prevent the escape of gas as soon as it was reasonably practicable for him to do so.

This is a get-out clause for gas suppliers if it took more than 12 hours to turn off the gas supply, provided they can prove it was outside their control.

37(6) Nothing in paragraphs (1) and (5) above shall prevent the supplier of the gas appointing another person to act on his behalf to prevent an escape of gas supplied by that supplier.

This section allows gas suppliers to subcontract the work out.

37(7) Nothing in paragraphs (1) to (6) above shall apply to an escape of gas from a network (within the meaning of regulation 2 of the Gas Safety (Management) Regulations 1996[11]) or from a gas fitting supplied with gas from a network.

This section stipulates that a network gas escape has to be dealt with by the gas supplier.

37(8) In this regulation any reference to an escape of gas from a gas fitting includes a reference to an escape or emission of carbon monoxide gas resulting from incomplete combustion of gas in a gas fitting, but, to the extent that this regulation relates to such an escape or emission of carbon monoxide gas, the requirements imposed upon a supplier by paragraph (1) above shall, where the escape or emission is notified to the supplier by the person to whom the gas has been supplied, be limited to advising that person of the immediate action to be taken to prevent such escape or emission and the need for the examination and, where necessary, repair of the fitting by a competent person.

The gas supplier should advise the customer to have the appliance or fitting checked by a Gas Safe engineer.

Regulation 38: Use of antifluctuators and valves

38(1) Where a consumer uses gas for the purpose of working or supplying plant which is liable to produce pressure fluctuation in the gas supply such as to cause any danger to other consumers, he shall comply with such directions as may be given to him by the transporter of the gas to prevent such danger.

38(2) Where a consumer intends to use for or in connection with the consumption of gas any gaseous substance they shall:

(a) give to the transporter of the gas at least 14 days notice in writing of that intention; and

(b) during such use comply with such directions as the transporter may have given to him to prevent the admission of such substance into the gas supply; and in this paragraph 'gaseous substance' includes compressed air but does not include any gaseous substance supplied by the transporter.

38(3) Where a direction under paragraphs (1) or (2) above requires the provision of any device, the consumer shall ensure that the device is adequately maintained.

38(4) Any direction given pursuant to this regulation shall be in writing.

This fluctuation will only really occur on an industrial scale, so is not within the domestic scope.

Regulation 39: Exception as to liability

No person shall be guilty of an offence by reason of contravention of regulation 3(2) or (6), 5(1), 7(3), 15, 16(2) or (3), 17(1), 27(5), 30 (insofar as it relates to the installation of a gas fire, other gas space heater or a gas water heater of more than 14 kilowatt gross heat input), 33(1), 35 or 36 of these regulations in any case in which he can show that he took all reasonable steps to prevent that contravention.

If you can prove in court that, in defence of you contravening certain regulations, you took reasonable precautions to prevent breaking these regulations, you may be found not guilty by the court of law.

Regulation 40: Exemption certificates

This regulation relates to an exemption certificate for training and assessment centres to allow them to put faults on equipment and systems within their centres. So for the majority of fitters this piece of legislation does not apply.

Regulation 41: Revocation and amendments

This regulation is used to revoke and amend all previous versions of the gas regulations up to 1998.

For revoking and amending regulations

This regulation details previous gas regulations that have been replaced or amended by the GSIUR 1998.

OTHER RELATED REGULATIONS

The gas regulations (GSIUR, 1998) are not the only regulations related to gas installations. Other regulations include those listed in Table 2.01.

Regulation	Year
The Health and Safety at Work Act (HASAWA)	1974
The Management of Health and Safety at Work Regulations (MHSWR)	1999
Pipelines Safety Regulations (PSR)	1996
Gas Safety (Management) Regulations (GSMR)	1996 (as amended)
Workplace (Health, Safety and Welfare) Regulations (WHSR)	1992
Provision and Use of Work Equipment Regulations (PUWER)	1998
Construction (Design and Management) Regulations (CDM)	2007
Pressure Systems Safety Regulations (PSSR)	2000
Health and Safety (Safety Signs and Signals) Regulations (SSR)	1996
Dangerous Substances Reporting of Injuries, Diseases and Dangerous Occurrences Regulations (RIDDOR)	1995
Explosive Atmospheres Regulations (DSEAR)	2002
Gas Acts (GA)	1986 and 1995
Gas Appliances (Safety) Regulations (GASR)	1995
Building Regulations	2000 and 2010

Table 2.01 Other relevant regulations

Commissioning records

A gas commissioning record form is where gas operatives record checks carried out while commissioning/servicing a gas system or component. The form must contain the following information:

- installer's and customer's details
- Gas Safe registration details
- appliance and control details
- whether or not the system has been flushed and if so with what

- whether or not an inhibitor has been used and if so which
- heat input
- burner operating pressure
- central heating flow and return temperature
- hot and cold water flow and temperature (combi or water heater)
- whether or not a scale reducer has been fitted
- whether or not a condensation drain has been fitted to the manufacturer's instructions
- whether or not it is building regulations compliant
- whether or not you have demonstrated the operation of controls to the customer
- whether or not you have left the manufacturer's literature with the customer.

Knowledge check

1 The Gas Safety (Installation and Use) Regulations define an 'appropriate fitting' as:

a Tees and elbows
b Copper tube
c A fitting that is used to affect a gas-tight seal
d Mild steel pipe

2 Which of the following installations are covered by the Gas Safety (Installation and Use) Regulations?

a Natural gas powered cars
b Natural gas powered forklift trucks
c Natural gas powered buses
d A house connected to metered natural gas supply

3 The Gas Safety (Installation and Use) Regulations specifically state 'No Person shall carry out work in relation to gas fitting unless they are':

a A plumber
b A gas fitter
c Competent to do so
d A pipe fitter

4 The primary aim of the Gas Safety (Installation and Use) Regulations, while work is being carried out, is to:

a Ensure customer satisfaction
b Satisfy Gas Safe
c Prevent a danger to any person
d Prevent a danger to property

5 When completing work on a gas supply that may affect the gas tightness of the installation, the installation must be immediately:

a Vented
b Purged
c Serviced
d Tested for tightness

6 When an installer working on the existing gas supply leaves the property for more materials, they must ensure:

a A temporary continuity bond is left in place
b An 'At Risk' notice is displayed
c A 'Work in Progress' notice is displayed
d All open ends of the pipework are sealed and tightness tested

7 In relation to an installation, no person shall make any alterations to any premises which could:

a Decrease the operational efficiency of an installation
b Adversely affect the safety of an installation
c Reduce the amount of natural light
d Increase the flow of air within a room

8 Landlords are responsible for ensuring all the gas equipment in rented dwellings are safety inspected at intervals of not more than:

a 6 months
b 12 months
c 18 months
d 24 months

9 Following the installation of an appliance, which of the following statements best describes the checks an installer must carry out:

a Check the installation is sound; check the effectiveness of the flue; check there is an adequate supply of combustion air; and check the operating pressure and heat input are as required by the manufacturer and that the appliance can function safely
b Check that there is an adequate supply of combustion air; check the extract fan is operating satisfactorily; and check the operating pressure and heat input
c Check the effectiveness of the flue; check there is an adequate supply of combustion air; and check all extract fans operate satisfactorily
d Check the operating pressure and heat input; check there is an adequate supply of combustion air; and check it has been serviced every year

10 A second-hand appliance can only be installed if it:

a Has been suitably modified
b Is verified to be in safe condition
c Comes with guarantee of purchase
d Has a modern flame supervision device fitted

Chapter 3

Communication, documentation and responsibilities

INTRODUCTION

Communication is the passing and receiving of information between two or more people. Knowing how to communicate effectively is an essential skill for everyone involved in the building industries. Making sure you pass on the right information to the right people at the right time using the most suitable form of communication will help guarantee success in the job you're undertaking.

CUSTOMER CARE

As a gas operative, you are at the front line when it comes to customer care issues, as it will usually be you on site, dealing with the customer. There are certain things that are important for good customer care, such as effective communication.

To give effective customer care as a gas operative, you need to know that there will be things on the job that you will not be able to influence, such as if the customer were to ask for work or alterations in addition to the original works that were quoted for; this would be the responsibility of your line manager or employer.

You needn't be the line manager to have a customer-caring attitude. You can show you care if you:

- Understand what the customer's real needs are. Taking a little time to explain what you are going to do on a job before it starts could avoid wasting time later. Misunderstandings happen – try to deal with any mix-ups before you start.
- Respect customer concerns. Do not put customers down. Try to solve any issues raised. If you're not able to deal with it, make sure the issue is passed back to your line manager or supervisor as quickly as possible.
- Show you are committed to your job. Remember to be punctual, thorough, careful and tidy. Always be honest and polite.

Providing good customer care is not only good for your job satisfaction but is great for business. The easiest and most cost-effective way of getting work is through repeat business or recommendations from happy customers. The success and reputation of your business will soon be affected if you develop a reputation for poor customer care. With a poor reputation, the business will have to rely on getting work from new customers (through marketing and advertising) and if the number of customers drops because of a negative image, a business could ultimately be forced to cease trading.

Improving customer care

There will always be ways in which a company can improve its customer care. Being in the front line means you may be able to see problems with the care of your customers.

Your company may invite you to suggest ways of improving their customer care, such as:

- asking the customer to complete a feedback form
- following up the visit with a courtesy call to check that all is working okay and there are no problems with the work done
- acting on queries immediately.

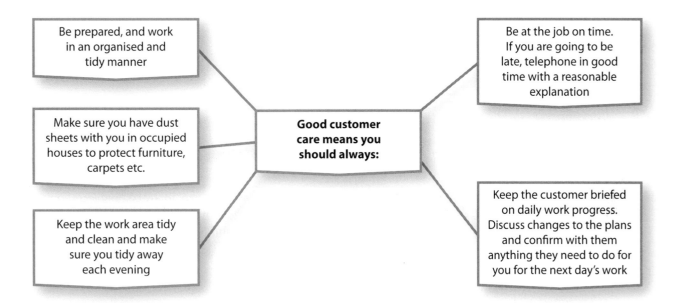

Figure 3.01 Good customer care – do

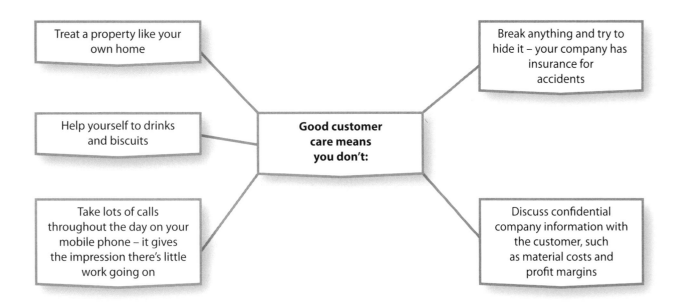

Figure 3.02 Bad customer care – don't

COMMUNICATING WITH THE CUSTOMER

Communication can take place in various forms. Each mode of communication, whether it be email, phone, letter or face to face, has benefits and drawbacks, and can be more suitable for conveying certain types of information than others. Poor communication can lead to dispute and disagreement on a contract so your ability to effectively communicate with customers and colleagues is important.

Mode of communication	Benefits	Drawbacks
Electronic		
Email	• Quick • Formal or informal writing style • Can send attachments (text, photos etc.) • Message recorded (date, time, content, receipt)	• Have to be online (but can send from some mobile phones and other hand-held devices) • Not guaranteed an immediate response
Fax	• Quick • Can send drawings, schematics etc.	• Outdated – not many companies/people have them anymore • Size limit per page – up to A4
Text message (SMS)	• Quick • Good for brief messages to inform of delivery and expected arrival times, availability etc.	• Limited size of message/data • Not guaranteed an immediate response
Verbal		
Face to face	• Can go into greater detail if there are any concerns with the contract	• No record of what's been said • Can lead to spontaneous disagreement with the other party
Word of mouth (second party passes on information)	• Good for passing on recommendations	• Can be unreliable if messages get mixed up if not remembered/passed on correctly • Time delay
Phone	• Instantaneous • Can be used anywhere (indoors/outdoors, providing there's a signal)	• No record of what's been said
Other		
Letter	• Formal • Content recorded (date, signatures, details of agreements etc.)	• At least a one-day delay between sending and receipt
Visual	• Allows you to interpret body language • Immediate • Instantaneous feedback from customer	• Some customers may not understand what is being passed on
Graphic	• Can use drawing and images to aid customer understanding	• May take time for customer to decide

Table 3.01 Modes of communication – benefits and drawbacks

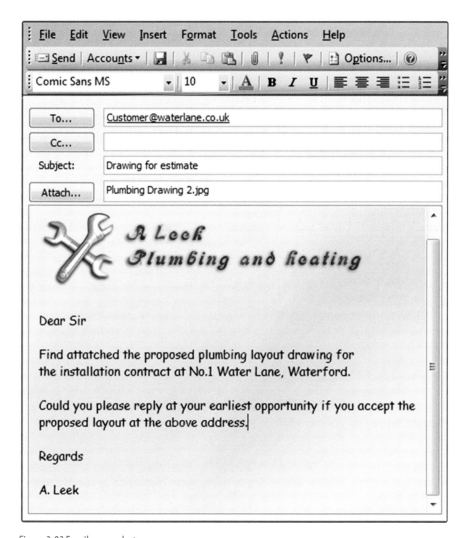

Figure 3.03 Email screenshot

The email shown contains:

To... Customer@waterlane.co.uk

Cc...

Subject: Drawing for estimate

Attach... Plumbing Drawing 2.jpg

A Leek Plumbing and heating

Dear Sir

Find attatched the proposed plumbing layout drawing for the installation contract at No.1 Water Lane, Waterford.

Could you please reply at your earliest opportunity if you accept the proposed layout at the above address.

Regards

A. Leek

Figure 3.04 Verbal communication for fact finding and simple issues

Effective communication

Effective communication comes about by listening to any points raised and trying to understand your customer's requests/requirements. You should then ask effective questions to get a full picture of the issues.

When communicating with your customer:

- be polite and keep to the point – this applies to both speaking and writing
- use positive **body language** – don't look bored and uninterested; be confident and look the person in the eye
- don't assume anything – base your communication on facts
- don't interrupt a person when they are talking to you
- talk at the right level – avoid technical jargon and give fuller explanations if you need to
- look for a customer's reaction – you can often tell by their body language what they are thinking.

It is important to use the right type of communication for the right circumstance. Verbal communication should be used for fact-finding and simple issues. Written communication should be for more formal activities.

Key term

Body language
A form of non-verbal communication through conscious or subconscious gestures and movements.

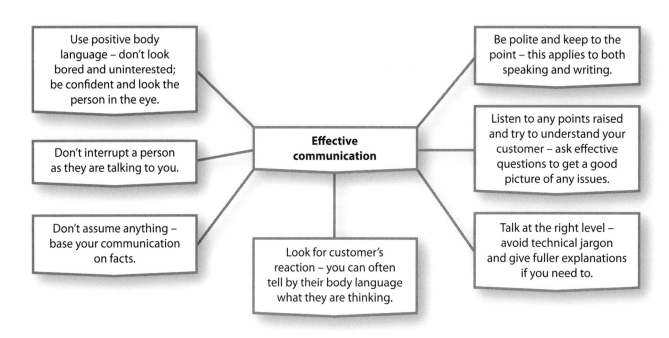

Figure 3.05 The key to effective communication

Use verbal communication for:	Use written communication for:
confirming the location of components before they're installed	giving a job quotation
establishing the initial job requirement	confirming the work (this is done by the customer)
things that need to be carried out before you commence the work, e.g. emptying the airing cupboard	major alterations to the original quotation or specification
solving simple problems and straightforward complaints	confirming quotation/specification alterations (this is done by the customer)
	commissioning reports or job records (usually kept by the customer for future use)
	dealing with more complicated problems and complaints

Table 3.02 The appropriate uses of verbal and written communication

Remember that written communication leaves a permanent record. If you use verbal communication, you may not be able to prove your discussions ever took place.

Talking with customers

You are employed to give a service that the customer either can't or doesn't want to do themselves. As a specialist, you are expected to know your job thoroughly. To show your competence you should:

- know your subject matter, but be prepared to seek advice and guidance if you need support
- communicate well: show you have a clear understanding of your trade and explain things to the customer in terms they will understand

- be positive about your trade: it shows you are in control and enjoy what you do
- plan the job properly: mistakes may be made if you make things up as you go along and the customer may lose confidence in your abilities.

Showing confidence in yourself is all about having a positive attitude to work.

- Learn to deal with problems that don't go your way. Never let a discussion with a customer turn into an argument – avoid shouting and confrontation. Refer serious problems to your employer to deal with.
- Develop work routines. Apply yourself to your work consistently – if a job works well one way, do it that way again in future. But be open to trying different and better ways of doing the job.
- Do what you say you're going to do. Always follow things up – nothing annoys customers more than empty promises.
- Never set unrealistic targets.

Surveying the site and presenting designs

Often customers will want to be presented with a plan of what you are going to undertake for them. During your initial survey you will have asked the client their requirements; from this you will have designed a system that meets their needs.

After these initial discussions you will then need to check the services: is there a gas meter in the dwelling and a safe, appropriate place to install any new appliances? Check that water and electrical services are available, and that there is an equipotential bonding connection at the meter to use.

Make a note of proposed pipe runs and measure up the distances required from the meter, you will need these for the sizing of your installation. See Chapter 5 for further information about the sizing of gas installations.

Once all of the site measurements are taken from site the information can be taken back to the office for drawings to be developed for the next stage of the contract.

The schematic drawings and designs for your job can be presented in several ways depending on the nature of the work. For new build contracts, the design will be presented on a **scale drawing**. For modifications to an existing dwelling, the design will be presented on a **single line drawing** (not to scale).

For a small maintenance job you would normally present the design plan verbally, and this can be followed up with a written quote for the job on completion.

When you have **tendered** for a larger job the presentation of your quote to undertake the work can be done using word-processing software or, where calculations have been made, a spreadsheet. Spreadsheets can also be used for invoices (see Figure 3.18 on page 75) as you can itemise the cost of components and give a total cost for the job.

Key terms

Scale drawing
A drawing that shows a real object with accurate proportions, reduced or enlarged (scaled) from actual size. Scale is shown a ratio, e.g. scale of 1:10 means that for every measurement of 1 on the drawing, the life-size object is 10 times that measurement.

Single line drawing
A drawing composed of narrow lines varying in width and density to produce tone and shading.

Tender
A supplier's offer to carry out work or a service for a fixed price. Tendering a contract is a common legal process for bigger projects.

STARTING AND FINISHING A JOB

Starting a job

It is important for you to understand the extent of your responsibilities as a Level 3 gas operative. You are being prepared to take responsibility for running a domestic job under your own initiative. Before you start a job, you and your line manager need to work out what your responsibilities are when dealing with customers and the types of issues that should be handed over to your superiors. A bigger company may have a customer-service policy that you should adhere to.

Due to the differing circumstances under which you may be employed as a gas operative, the level of responsibility you have will vary from business to business. For example, in a company specialising in service and maintenance you may be expected to give a price for the job and collect payment on completion. With service and maintenance work it is normal for a customer to require a reasonable **estimate** of the cost before work begins. In order for you to provide an estimate, your company should have procedures in place for you to obtain the price of any parts you will need. They will also have a standard cost of labour calculation for you to use and include in the estimate.

If your company specialises in installation, then it is likely you will install full systems. When working on these jobs it is important to have clear communication right from the start. You may not have a full system layout plan. It is not standard with private customers to give a highly detailed job specification. However, details of pipe runs etc. will have been discussed at the initial survey visit. The survey write-up should end with a quotation that gives brief details of the job (such as the specification and placement of major components).

Before commencing any work, check the dwelling for any previous damage to furniture, carpets, wall coverings and any fixed units or appliances. Point these out to the customer, make a note of them, and ask the customer to sign the note in acknowledgement that the damage was done prior to your work commencing. This avoids any potentially unpleasant disputes with the customer as to what damage was done, when and by whom.

Before starting major jobs you need a good briefing from your line manager, or the person who did the survey/pricing, on what was originally agreed with the customer. Your supervisor/employer will go through the details of the system layout, explaining where pipe runs are to go, how you are to tackle the job and what rooms you'll need to enter. It's vital that this planning process is completed before the work begins so that any misunderstandings about the details of the job can be sorted out in good time. Identifying and dealing with problems and misunderstandings before installation or maintenance has taken place avoids wasting time, resources and money, and helps guarantee a job well done.

You may also be responsible for any work colleagues who are on a job with you. You should make sure their work is of an acceptable standard.

While you probably do not have the responsibility for hiring or firing staff, you do have the responsibility of running the job and you are the front line

Key term

Estimate
The approximate cost of job (also known as a 'quote' or 'quotation').

Figure 3.06 The correct paperwork is important on the job

in customer care. Make sure you have procedures in place to deal with any problems that may arise. You should be prepared for minor problems (e.g. the customer wasn't happy because your colleague hadn't tidied up) and more serious problems (e.g. your colleague had an argument with the customer and swore at them). Serious problems should be referred to your supervisor/employer for them to resolve.

Finishing a job

Try to leave the customer's property as you found it – in a tidy state with all rubbish and materials cleared away. As part of the job completion stage it is very important you commission the installation properly. If you do not take your time in doing so, you may be called back to complete the work, which may taint the customer's opinion of you and your company.

Before you leave, the customer will need a thorough briefing in how to operate any new installations. You need to complete any commissioning records and leave the manufacturer's instructions for everything you've installed on site.

Identify any future service and maintenance needs (you can use these to get future work for the company). Some companies may ask you to get a satisfaction statement from the customer as part of the job completion phase. Others may follow up in a few days' time with a phone call to ask for feedback. It is common for your employer to check up on your work performance.

Paperwork to be left with the customer

On completion of the job, it is a requirement of the regulations that you must leave with the customer any manufacturer's instructions which it was supplied with. Be sure to tell the customer to keep them in a safe place for future use.

On completion of a gas installation or appliance servicing work you must complete some paperwork, relevant to the task you've just undertaken. If the appliance is new, then the engineer who completed the paperwork should register the appliance with Gas Safe. There is a cost involved with this, which would be included in the quote for the job, and the customer will receive a certificate from Gas Safe to confirm that the appliance has been registered.

Figure 3.07 You will need to make sure the customer fully understands all paperwork

With new appliances, complete the 'Benchmark' certificate which comes with the appliance and leave this with the customer. Where the job is in multiple dwellings or a new build, this information should be passed on to the responsible person in charge of the contract. This information should then be passed on to the householder/landlord by this responsible person.

As a trainee gas operative you would not be able to sign off an installation or appliance as you would not be registered with Gas Safe. The Gas Safe registered operative you are working with will take full responsibility for any works that they sign off.

See Chapter 11 for further information about situations which are unsafe and may cause harm to the customer, and the procedures to follow.

NVQ Diploma Level 3 Gas

DEALING WITH CUSTOMER COMPLAINTS

By following good customer care principles, you shouldn't receive too many complaints. However, complaints do occur. Your full job description should tell you how much responsibility you have for dealing with customer complaints.

When responding to an employer's check, the customer may not remember your technical abilities but will remember the way you treated them.

Reasons why customers complain	
1	The customer didn't get what they had been promised
2	A company employee was rude
3	The service was regarded as poor – the customer felt no one was going out of their way to be helpful
4	Nobody listened to the concerns or issues the customer raised
5	The gas operative had a 'can't do' or 'couldn't care' approach

Table 3.03 The top five reasons why customers complain

Dealing with a rogue customer

On rare occasions you may have to deal with a rogue customer. A rogue customer may be:

- A customer whose expectations of a job are far higher than the price they will pay (usually due to the quality of materials used). For example, a customer may not accept their new fire because it is not the same as the one in the showroom. They then refuse any replacement fires.
- A customer who is determined to get your company to do work without paying for it. For example, a customer may deliberately damage components after installation and then refuse to pay, claiming it was your fault.

Rogue customers are few and far between but you will no doubt come across them. As a gas operative it is helpful to notice signs that may indicate a rogue customer and report these back to your supervisor/employer at the earliest possible opportunity. For such purposes, employers tend to prefer formal written communication rather than verbal. Your employer should then resolve the matter with the aid of effective communication.

RISK ASSESSMENTS

As a gas operative you are surrounded by potential **risks** to your health and safety. It is essential that you are aware of the **hazards** in your workplace and know how to deal with them.

What is risk management?

Risk management is a process that involves assessing the risks that arise in your workplace, putting *sensible* health and safety measures in place to control them and then making sure they work in practice. These measures need to be **reasonably practicable.**

Key terms

Risk
The likelihood that a hazard will harm someone or something, e.g. if there are no guard rails on the scaffolding it is likely that a construction worker will fall and break a bone.

Hazard
Anything (equipment, environment) with the potential to cause harm, e.g. working at height on scaffolding.

Reasonably practicable
A health and safety measure is deemed reasonably practicable if the level of risk involved is equal to the extent of the measures required to avoid that risk (whether it be in terms of cost, time or effort). In other words, if you think certain measures are worth the time, cost and effort to avoid a certain risk, it is reasonably practicable (i.e. reasonable and feasible) to put those measures in place. You can work this out for yourself, or you can simply apply accepted good practice.

What is a risk assessment?

A risk assessment is a careful examination of your work and workplace to identify anything that could cause harm to you and those around you, so that you can weigh up whether you have taken enough precautions or need do more to prevent harm.

Identifying the levels of work-related risk – ALARP and SFAIRP

In compiling a risk assessment you may find these terms useful: ALARP stands for 'as low as reasonably practicable' and SFAIRP stands for 'so far as is reasonably practicable'. In essence, these are the same; however, SFAIRP is the term most often used in the Health and Safety at Work Act and in regulations, and ALARP is the term used by risk practitioners.

Identifying work-related hazards

When compiling your risk assessment you will need to be able to define the hazards that you may come across on a particular job. Examples of such hazards are:

- working at height
- soldering
- using solvents
- uneven surfaces.

Writing a risk assessment

When you did your Level 2 plumbing or gas qualification you needed to understand about risk assessments and why they are needed, but your supervisor wrote the assessments themselves. Now you are a Level 3 plumber or gas operative, you will be the one writing the risk assessments.

To write a risk assessment, follow the Health and Safety Executive's five-step process.

Step 1: Identify the hazards

Walk around the site and look to see what hazards there are.

Step 2: Decide who might be harmed and how

Consider all those who may enter the workplace while you are working there – colleagues with different jobs to do, members of the public, residents, other tradespeople. Taking into account what you and others will be doing on site, who might be at risk of harm and how might they be harmed?

Step 3: Evaluate the risks and decide on precautions

Now that you have spotted the risks you need to decide what to do about them. The law requires you to do everything that is reasonably practicable to protect people from harm.

Can you get rid of the hazard altogether? Removing the hazard is the simplest way of avoiding any harm it may cause; but not all hazards are removable and those that are need to be moved to somewhere safe where they will no longer pose a risk.

If the hazard cannot be removed, how can you control the risks so that harm is unlikely?

When you are controlling the risks you need to apply the following principles:

- try a less risky option
- prevent access to the hazard (e.g. by guarding it)
- organise the work to reduce exposure to the hazard
- issue personal protective equipment (e.g. clothing, footwear, eye protection)
- provide welfare facilities (e.g. first aid, washing facilities).

Step 4: Record your findings and implement them

When writing down your results, keep it simple so that everyone who needs to can understand it. For example 'Fume from lead welding: local exhaust ventilation used and regularly checked'.

Your risk assessment must be suitable and sufficient. You will need to show that:

- a proper check was made; you asked all those who might be affected
- you dealt with all the significant hazards, taking into account the number of people who could be involved
- the precautions are reasonable, and the remaining risk is low
- you involved your staff or their representatives in the process.

Step 5: Review your risk assessment and update if necessary

No workplace remains the same from day to day, week to week, so be sure to review your risk assessment regularly to ensure any new hazards are accounted for.

Risk calculation formula

Risk factors are calculated using a simple formula:

Likelihood × consequence = risk

The outcome of the likelihood of an accident occurring and the maximum consequences should it happen, will reveal a risk factor of between 1 and 25, with 25 being the highest risk level.

Using Tables 3.04 and 3.05 you can assess the risk for any tasks which are undertaken.

Likelihood	Scale value
No likelihood	0
Very unlikely	1
Unlikely	2
Likely	3
Very likely	4
Certainty	5

Table 3.04 Likelihood of an accident occurring

Injury or loss	Scale value
No injury or loss	0
Treated by first aid	1
Up to 3 days off work	2
Over 3 days off work	3
Specified major injury	4
Fatality	5

Table 3.05 Maximum consequences of an accident

A figure between 1 and 6 = minor risk, can be disregarded but closely monitored.
A figure between 8 and 15 = significant risk, requires immediate control measures.
A figure between 16 and 25 = critical risk, activity must cease until risk is reduced.

Other elements that make up a risk assessment form include:

- **risk exposure** which describes the individuals or groups of people that may be affected by the work activity or process – control measures must take account of all those people

- **safeguards hardware** which describes the in-built safety features of work equipment, for example, on powered machines this would include machine guards or trip switches

- **control measures** which describe the additional safeguards that underpin your arrangements; where these are identified they must be followed through, and a record kept of any outcomes etc.

The list below shows some of the most common risks you will face during your career:

- working at height – the most common cause of construction accidents

- soldering, including high-temperature brazing using oxy-acetylene bottles

- using solvents – ventilation must be available where the installation is being put in

- using specialist tools, such as a power threading machine, press-fit tool and hydraulic bender.

Here is an example of a risk assessment for working with specialist tools. It is based on the model developed by the British Plumbing Employers Council (BPEC).

Task
Manual handling of loads
Specialist tools

Application of equipment	**Application of substances**
Pipe-bending machines, stilsons, ropes, lead dressers, bending springs, block and tackle, spanners etc.	N/A

Associated hazard
Risk of muscle strains
Risk of sprains
Risk of musculo-skeletal injury

Likelihood	**Consequence**	**Risk factor**
3	3	9

Risk exposure	**Safeguards**
Gas operative	Nil

Control measures
1 Specific training and instruction to gas operatives – kinetic lifting
2 Individual assessment to be performed for all tasks
3 Workplace inspections conducted at 3 monthly intervals
4 Random safety inspections
5 Suitable and sufficient personal protective equipment
6 Medical screening for staff at risk

Figure 3.08 A generic risk assessment form

METHOD STATEMENTS

Companies nowadays would be expected to have available method statements for different tasks which they may be involved with. These can be tasks which are performed on a regular basis, for example core drilling for a flue from a boiler. Method statements link to the risk assessment for the task being undertaken.

On larger contracts the company would be requested to supply the principal contractor with their method statements for the contract in question.

What is a method statement?

A method statement is a document detailing how a particular process will be carried out and is sometimes called a 'safe system of work'. The first task is to carry out a risk assessment, as outlined above. The method statement should outline the hazards involved and include a step-by-step guide for methods of controlling that particular task. It should also detail which control measures have been introduced to ensure the safety of anyone who is affected by the task or process.

Information a method statement should contain

Such a statement is commonly used to describe how construction/installation works can be carried out safely and should include:

- background details of your company
- site address
- overview of the project
- the type of activity
- details of the hazards associated with that activity
- details of how the work will be managed safely, including the possible dangers/risks associated with your particular part of the project and the methods of control to be established.

A method statement for the removal of a boiler flue is shown in Figure 3.09 opposite.

Background information	
Company Details	A Leek Ltd 58 Gas Street, Tel: 555678 Fax: 555670 Email: aleek@aservice.co.uk
Site Address	Contact name: Mrs A User Address: 60 Pump Lane Contact No: 555789
Activity/Risk	To access and work safely during installation of a flue for a boiler
Implementation and control of risk	
Hazardous task risk	**Method of control**
Access roof space	Where access to the roof space is required then use a suitably secured stepladder or loft ladder of the correct height for the task.
Access working area	Access to work area to be kept clear and free from obstructions. Electric lead lights shall illuminate all access and work areas.
Replacement of an existing flue	Replacement of flue is to be made with products which will not be hazardous to health, e.g. stainless steel or high-temperature plastics.
Use of blow lamp – hot soldering	All pipe lagging within 500 mm is to be removed. A suitable fire extinguisher will be adjacent to the working area.
Removal of existing flue	The existing flue will be carefully removed from dwelling. Where existing flue is made from asbestos cement materials, the safe removal of asbestos regulations should be adhered to. Refer to the risk assessment for removal of asbestos cement flue products. For more information about asbestos flues refer to HSE publication a35 Replacing an asbestos cement (AC) flue or duct.
Safe disposal of existing flue	Where an existing flue is made of asbestos cement then the material is to be double bagged and sealed with label attached and taken to a registered tip for disposal. Metal flues are to be disposed of safely ensuring that they cannot cause harm to others.
Site control	
Personal protective equipment (PPE)	Appropriate PPE is to be supplied by the employer.
Inspection of equipment	All equipment such as stepladder, ladders and blow lamp shall be regularly inspected before commencement of work. Ensure electrical equipment has a current PAT test certificate.
Customer awareness	The customer will be notified of all potential dangers throughout the contract. The customer will be notified of any delays to the contract.
Protection of customer floor coverings	Ensure that all carpets are protected where access is required with dust sheets.

Figure 3.09 Method statement

PLANNING WORK PROGRAMMES

In the building services industry different types of job will require different types of work programme. More complex jobs always need more careful planning and documentation. Simpler tasks may be easy to organise, but there may still be benefits from drawing up a short work programme, for you and your customer.

What is a work programme?

	Week commencing		1/01/2013					08/01/2013					15/01/2013				
Plot Number	Activity	1	2	3	4	5	1	2	3	4	5	1	2	3	4	5
Plot 1	Guttering and rainwater pipes	▓														
	First fix gas and plumbing		▓	▓	▓											
	Second fix								▓	▓	▓					
	Commissioning											▓				
	Snagging											◪				
Plot 2	Guttering and rainwater pipes			▓												
	First fix gas and plumbing				▓	▓	▓									

Figure 3.10 Work programme

To begin with you need to produce a grid similar to that shown in Figure 3.09, which is for plumbing and gas services on a multi-dwelling new-build site. The site manager would hold a master work programme on which activities for all trades would be identified.

From the master copy you can glean key dates (otherwise known as milestones) such as expected handover dates or payment dates. You can see from this small extract that there is an overlap of work between plot 1 and plot 2 so from the work programme you should be able to identify your labour requirements for the contract.

Types of work programme

The type of work programme required depends on the type of work undertaken.

Private installation work

Your line manager/supervisor is likely to make a mental note of the work programme for a private installation, based on an agreed start and finish date with the customer.

Private service/maintenance work

Work programmes for private service or maintenance work would be on a contract sheet provided by the employer detailing the contract and date/time to start. Where preventative maintenance, servicing and landlord safety checks are being undertaken, to ensure these are completed as planned, a schedule similar to the one in Figure 3.10 would be implemented.

New-build installation contract work

On larger contracts involving more dwellings, such as new-build installation work, the approach is more scientific and a contract programme will be provided. This might consist of an overall programme for all site trades, as well as a separate programme for each trade, including plumbing. These can be produced using project planning software such as Microsoft Project. Work programmes are often shown as a Gantt chart (Figure 3.09), which is similar to a bar chart.

Figure 3.11 is an example of a programme for a plumbing installation.

Service/maintenance contract work

On large service/maintenance contract work you would normally follow a work programme (similar to that in Figure 3.11) especially if this is preventative maintenance, which follows a standard company procedure set out in a maintenance/service document.

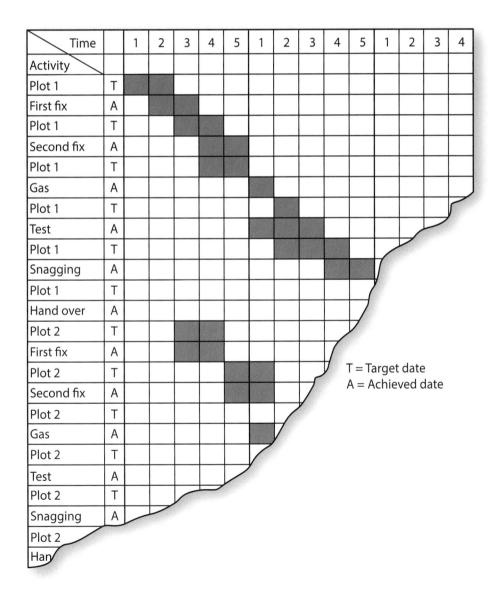

Figure 3.11 A work programme for a plumbing installation for a new build

A bill of quantities

Part of the process for planning work activities against contract specifications on larger installation contracts requires a bill of quantities (B of Q). This is produced to specify how the work will be carried out, as well as the quality and quantity of the materials. The B of Q is compiled by measuring all the quantities, based on a drawing, referred to as 'taking off'. Figure 3.12 is a typical example of an extract from a B of Q.

Item No.	Description	Quantity	Unit	Rate (£)	Cost (£)
A	All sanitary ware supplied by a manufacturer 'Hiline' pedestal-mounted wash basin in white vitreous china to BS 3402, 67 cm x 53 cm	10	Item	55.00	550.00
B	'Starly' bath in cast white acrylic sheet 170 cm x 70 cm, and slip-resistant base	10	Item	250.00	2500.00
C	'Space' close-coupled WC suite with washdown bowl and box flushing rim in vitreous china to BS 3402, 39 cm wide x 79 cm high	10	Item	150.00	1500.00
D	Table 15 mm copper tube grade X half-hard to EN 1057, supplied in 6 m lengths	120	M	1.10	132.00
E	Table copper integral solder ring fittings to EN 1254, 15 mm				
	Tees	20	Item	1.20	24.00
	Elbows	40	Item	0.80	32.00
	Tap connectors	30	Item	1.80	54.00
	Straight connectors	40	Item	0.40	16.00

Figure 3.12 Bill of quantities

The B of Q is produced by a quantity surveyor, usually on behalf of the architect, and its main purpose is to cost a contract in great detail. It will also be used throughout the contract to control costs and provide milestones for contractors' payments. The architect, on behalf of the client, will use the B of Q before a contract starts as part of the tender documentation. It is sent out to contractors with the rate and cost columns left blank for them to fill in when tendering for the work.

A material schedule contains information similar to the above, but would be used by a gas operative on site as a working document to provide details of what materials would be specified for a particular dwelling. It is also unlikely to have any costings in it.

Component and appliance details

Generally speaking, most gas operatives use manufacturers' instructions for specific details of components or appliances. This type of information is supplied with the component or appliance in the delivery packaging. If working on an existing appliance on a maintenance contract (particularly boilers), you must have access to manufacturer's instructions. You should be able to obtain copies of instructions from most manufacturers.

The scope, purpose and requirements of the work

The scope, purpose and requirements of the work to be undertaken are usually described using drawings. Drawings for contracts are available from the architect and copies would be sent to contractors and sub-contractors during the tendering process of the contract. You need to ensure that the installation meets industry standards and the requirements as set out in the contract specification.

Materials used in plumbing installations should be to the relevant European Norm (EN) or British Standards (BS) number. BS also makes recommendations on design and installation practice, for example BS 6891:2005 +A2:2008 Installation of low pressure gas pipework of up to 35 mm (R1¼) in domestic premises (2nd family gas) – Specification.

Figure 3.13 You need to refer to plans throughout the building process

In addition to British Standards, the following legislation places statutory responsibilities on gas operatives. Many of the requirements laid down are absolute, so if they are not complied with, severe penalties can follow.

- Water Supply (Water Fittings) Regulations 1999
- Gas Safety (Installation and Use) Regulations 1998
- Electricity Supply Regulations 1998 and Electricity at Work Regulations 1989
- Building Regulations 2000
- Health and Safety at Work Act 1974.

Factors that affect timeframe

Delays to the progress of works being undertaken can lead to losses for the contracting company. These delays can cause a dispute with the main contractor or client and, with some contracts, penalties can be incurred.

Delays can be caused in one of three ways:

- by the main contractor
- by the employer (taking personnel away from the contract)
- by other events that neither the main contractor nor contractor can control, such as delayed delivery of materials or bad weather.

Delivery requirements and non-availability of materials

For security reasons and storage problems on site you cannot always take delivery of all materials needed to complete a job. Take as an example working on a new housing estate where each new property requires full gas, water and waste installation; you can at this point use the work programme to organise delivery of materials and do so in stages to avoid any security and storage problems.

You could arrange with your supplier to take delivery of a first fix kit for each plot number which would contain all materials required for the first fix of hot and cold water pipework, central heating and, depending upon the construction of the building, the radiators, soil and waste pipes and materials for the guttering and rainwater pipes (which need to be installed before dismantling the scaffolding).

For the second fix kit the suppliers would deliver the appliances, boiler, radiators (if not included in the first fix) and all the necessary fittings and pipework to complete the second fix.

These supplies would be ordered by a buyer if you are employed by a large company or by your immediate line manager if a small company. If there is a delay from the supplier this will affect progress of the contract and so the site manager (or customer, if a small contract) would need to be informed.

If a long time has elapsed from when a contract was first designed and assembled to it actually being tendered for, accepted and started, some materials may no longer be available. When this occurs, other materials, to the same standard and specification, need to be sourced and approved by the architect and customer/client. This may incur costs and cause delays.

Job specifications

Larger contracts will always have a job specification which has been prepared by the quantity surveyor. These help to select the required resources against contract specification materials. To ensure that you are sourcing the correct materials you will need to consult the job specification and this details the type and often the manufacturer of a product.

Job specification		
Contract: Single dwelling, 60 Well Lane, Waterford		
Item	Type	Standard
Copper tube	All copper tube to be grade X half-hard manufactured to BS EN 1057.	Installed to current water regulations and to BS 8000: part 15 Pipework to be tested to BS 6700:2006+A1:2009 for metallic pipework
Copper fittings	Soft soldered solder ring fitting to BS EN 1254. Manufacturer: Yorkshire Imperial Fittings	All solder to be lead free to BS EN 29453:1994/ISO 9453:1990 Soft solder alloys – Chemical compositions and forms
Fluxes	Water soluble Self-cleaning Manufacturer: Everflux	Flux is not to be used excessively; excess must be cleaned and removed on completion of joint as per the manufacturer's instruction.

Figure 3.14 Job specification example

To find out how many of a specified appliance you need you would consult the B of Q. The total amount, which your estimator would have priced against, is stated in this document. Some B of Qs are structured in a room-by-room format, for example Kitchen: 1 Left-hand drainer S/S sink.

Plant

Plant includes such things as scissor lifts, cherry pickers and mobile scaffolds. These all have to be taken into consideration when planning for the contract. Details of plant required for a job would have to be included at the tendering stage of the contract. Plant should be hired for the shortest possible period in order to keep costs down.

Vehicles

Transport for a small company is essential as you may only be on a contract for a short period of time and have several jobs in a day, especially when undertaking servicing and maintenance work. You may need to visit the merchants for materials and components which you do not carry in your van stock; without a vehicle you would need to contact your line manager to visit the merchants for you, adding time and labour to the final cost of the job. This may cause the customer to complain, as they may be expected to pay the bill.

For larger contracting companies it would be uneconomical to have many vehicles standing around all day. They would have at least one vehicle which could be of the mini-bus type. The vehicle can be used for visits to merchants for small items that may be needed to complete work, this could be in the case where a variation to the works has been ordered and the work may need to be done urgently. Without this transport delays can occur and, in some cases, delays incur penalties.

Equipment

Specialist equipment such as threading machines and oxy-acetylene brazing equipment can be hired on an 'as and when needed' basis to keep costs down.

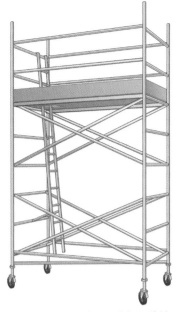

Figure 3.15 Plant, such as mobile scaffold towers, must be planned for at an early stage

Figure 3.16 A threading machine is one of many pieces of equipment you will need to use

IDENTIFYING WORK RESPONSIBILITIES AND ROLES

Gas operative

It will be the Level 3 qualified gas operative who will take overall control of the contract. They would organise the work on a daily basis while following the work programme. The Level 2 qualified gas operative would be working with minimum supervision.

Customer

As an employee of a business, the first step is to recognise who your customers are. You will encounter **private**, **contracting** and **internal** customers.

If your company has been invited to do a contract directly by the customer, they are a **private customer**. Usually they will not have any technical knowledge of the work to be carried out and will put their trust in you and

Key terms

Private customer
Arranges and pays for work themselves.

Contracting customer
Employs a specialist business to carry out the work.

Internal customer
Is from within your company.

your company. However, they will have certain expectations of the work. Trying to meet their expectations before they need to state them is a sign of good customer care.

Your company may do contract work for organisations such as property developers, housing associations or local authorities. These are **contracting customers**. Do not assume that the customer is different for this type of work – your customer is the organisation. That organisation will have various staff representatives who take the lead in running the contract, such as a site agent or clerk of works. These people can be thought of as your front-line customer. They too will have expectations of what they need from you and your company, which may not be very different from those of a private householder. However, the customer's representative may have an in-depth technical knowledge of the service you are providing.

You may work for a plumbing company that is part of a larger building services or construction company. If carrying out work for them, they would be an **internal customer**. In this situation it may be easy to forget customer care issues because the customer feels very distant from the work. Your customer is the parent company representative, who, in these days of competitive contracting, can go outside your company for services if customer care is lacking.

Figure 3.17 The architect will be involved throughout the building process

The site management team

When working on a construction site, whether you are the owner of a small company or you are running a contract for a larger company, at some point you will have to communicate with different members of the site management team. The way you communicate with the site management and your personnel is essential to the smooth running of the contract.

Architect

The architect's role is to plan and design buildings. The range of their work varies widely and can include the design and procurement (buying) of new buildings, alteration and refurbishment of existing buildings, and conservation work. An architect's work includes:

- meeting and negotiating with clients
- creating design solutions
- preparing detailed drawings and specifications
- obtaining planning permission and preparing legal documents
- choosing building materials
- planning and sometimes managing the building process
- liaising with the construction team
- inspecting work on site
- advising the client on their choice of contractor.

On larger contracts it would be rare for you to communicate directly with the architect. The architect has a representative on large sites and this is usually the clerk of works or project manager. It is essential that you get written confirmation when the clerk of works requests any work to be done which has been instructed by the architect. Ensure that you save all written communication whether it is by email, fax or letter.

On a small contract such as an extension to a dwelling, the architect could also be the project manager. Direct communication with the architect is essential and you should ensure any verbal agreements or requests for extra works are followed up with written confirmation for your records. *Do not* begin any of the extra works until you receive a written confirmation.

Quantity surveyor

A quantity surveyor:

- advises on and monitors the costs of a project
- allocates work to specialist subcontractors
- manages costs
- negotiates with the client's quantity surveyor on payments and final account
- arranges payments to subcontractors.

To be paid by the quantity surveyor for **interim payments** you would have to submit an **invoice** for works completed since the last interim payment. For payment of any works other than the works that are in the B of Q they would expect a written **variation order** from the architect confirming this.

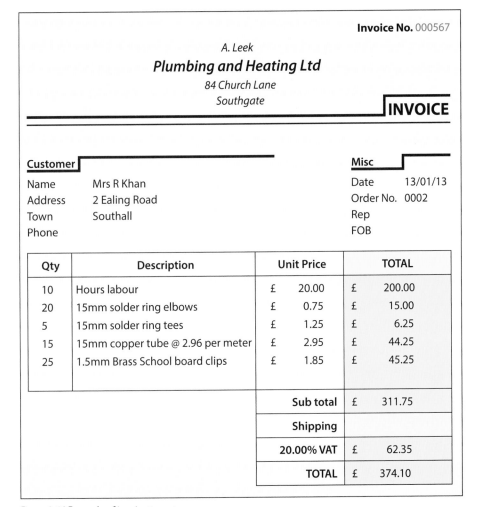

		Invoice No. 000567
	A. Leek	
	Plumbing and Heating Ltd	
	84 Church Lane	
	Southgate	**INVOICE**

Customer		**Misc**	
Name	Mrs R Khan	Date	13/01/13
Address	2 Ealing Road	Order No.	0002
Town	Southall	Rep	
Phone		FOB	

Qty	Description	Unit Price	TOTAL
10	Hours labour	£ 20.00	£ 200.00
20	15mm solder ring elbows	£ 0.75	£ 15.00
5	15mm solder ring tees	£ 1.25	£ 6.25
15	15mm copper tube @ 2.96 per meter	£ 2.95	£ 44.25
25	1.5mm Brass School board clips	£ 1.85	£ 45.25
		Sub total	£ 311.75
		Shipping	
		20.00% VAT	£ 62.35
		TOTAL	£ 374.10

Figure 3.18 Example of invoice to customer

Buyer

The buyer's or procurement officer's role is to:

- identify suppliers of materials
- obtain quotations from suppliers
- purchase all the construction materials needed for a job
- negotiate on prices and delivery dates
- resolve quality or delivery problems
- liaise with other members of the construction team.

Estimator

An estimator:

- calculates how much a project will cost including plant, materials and labour
- identifies the most cost-effective construction methods
- calculates cash flows and margins
- seeks clarification on contract issues affecting costs.

If you are to plan your workforce's activities successfully, good communication with the buyer and estimator is essential. The buyer is purchasing the materials that you are going to use, so will require a clear, written order for the items you will need. The buyer is also responsible for delivery issues, so will need to keep you informed about delivery dates and any delays as they arise.

So as the buyer for the contract is buying all materials for the contract they would need a written purchase order. These orders are raised by the procurement officer or, in a small company, by the line manager/supervisor. These orders are essential so that all purchases can be accounted to individual contracts, costs can be tracked and jobs monitored to check that they are not, in the worst case scenario, losing money. The purchase order is given to the supplier. This is so that invoices can be attached to a purchase order, which is essential for the calculation of costs against the price of the contract and the calculation of your profit and losses.

Case study

Lucas has discovered an unforeseen problem when he went for a site visit to survey the dwelling. He discovered that he would be unable to install the pipework as drawn on the installation plan. Lucas needs to take into consideration the following questions:

- Who should he discuss the alterations with?
- Which member of the site team would give the authority for the alterations?
- What paperwork would he need so that the company will get paid for the alterations?

Discuss with your tutor and classmates the actions that should be taken.

Building surveyor

A building surveyor is a professional, trained in understanding and interpreting building law. They are authorised to assess building plans with a view to ensuring they are compliant with the building regulations. In addition to having recognised qualifications, a building surveyor must be registered.

Building surveyors are involved in the maintenance, alteration, repair, refurbishment and restoration of existing buildings. A building surveyor's work includes:

- organising and carrying out structural surveys
- legal work, such as negotiating with local authorities
- preparing plans and specifications
- advising people on building matters, such as conservation and insulation.

They will resolve any queries on the building regulations and allied legislation presented by staff or other persons on the contract.

Your communication with the surveyor could be regarding the building regulations for the installation of heating and hot water systems. You may wish to clarify some aspects of the building regulations or associated legislation before the contract starts.

Project manager/clerk of works

On larger jobs, the project manager/clerk of works is the site representative of the architect so your dealings with them will involve them checking that you have installed your appliances/equipment as specified by the architect.

The project manager/clerk of works will keep records of:

- daily weather conditions
- plant and materials that have been delivered or removed from site
- any stoppages, including industrial disputes (strikes)
- official visitors to the site, for example the building control officer
- the quantity of personnel on site, on a daily basis
- drawings received from the architect
- any variation orders passed on to contractors
- the progress of works undertaken for the week.

The project manager prepares **snagging lists** like that in Figure 3.20 for remedial works prior to sign-off at the end of the contract.

Figure 3.19 The surveyor will need to take a nunber of onsight observations

Key term

Snagging list
A form used to identify small defects with the work that has been installed.

Plot/Room	Type of snag	Completed
Kitchen	Sink plug missing	✓
	No handle on cold washing machine valve	✓
Cloak room	Insufficient clips on hot and cold pipework	✓
	Cold water tap drips, staining wash basin	
	Flux and solder not cleaned off pipework	
Bedroom 1	Thermostatic valve loose	
Bedroom 2	Radiator loose	

Figure 3.20 Plumbing snagging list

You may be called upon to come to some agreement as to the scope of works to be undertaken when resolving the snagging list issues. Some of the snags that have been itemised may only be snags in the clerk of works' own personal opinion. You will have to prove that you have installed equipment and pipework to the contract specification and to British Standards. For example, for the snag 'Insufficient clips on hot and cold pipework' in Figure 3.20, you would have to prove that you have clipped to correct distances as specified in standard BS 8000: part 15 for hot and cold supplies.

Structural engineer

Structural engineers are involved in the structural design of buildings and structures, such as bridges and viaducts. The primary role of the structural engineer is to ensure that these structures function safely. They can also be involved in the assessment of existing structures, perhaps for insurance claims, to advise on repair work or to analyse the viability of alterations and adaptations.

You will most likely work with a structural engineer when working on the safe installation of pipework, equipment and systems within the fabric of the building, such as when:

- Drilling through floor slabs for service pipework where holes have not been left in the concrete when cast.
- Checking if the structure is capable of taking the weight of equipment. If the structure is not strong enough to take the weight then it would be the structural engineer's job to come up with a solution to the problem. This could also involve the architect, building services engineer, contracts manager and construction manager.

Building services engineer

The building services engineer is responsible for designing, installing and maintaining water, heating, lighting, electrical, gas, communications and other mechanical services such as lifts and escalators in domestic, public, commercial and industrial buildings. A building services engineer's work includes:

- designing the services, mostly using computer-aided design packages
- planning, installing, maintaining and repairing services
- making detailed calculations and drawings.

Most building services engineers work for manufacturers, large construction companies, engineering consultants, architects' practices or local authorities. Their role often involves working with other professionals as part of a team on the design of buildings, for example with architects, structural engineers and contractors. When working for a consultant, this job is mainly office based at the design stage. Once construction starts, there will be site visits to liaise with the contractors installing the services.

When working for a contractor, the building services engineer may oversee the job and even manage the workforce and is therefore likely to be site based. When working for a services supplier, the role will require being involved in design, manufacture and installation, and may involve spending a lot of time travelling between the office and various sites.

The building services engineer is the person who designs the mechanical services systems, especially on a large contract. You would communicate with them:

- where it is not possible to carry out the installation as it has been drawn on the installation plan
- when an appliance is not able to be sourced that has been specified in the contract specification – the building services engineer would have to agree to an alternative, ensuring that the specification is the same as for the original appliance that was specified.

Contracts manager

The contracts manager's role is to:

- be responsible for running several contracts
- work closely with the construction management team
- be the link between the other sections of the business and the upper management staff
- make sure the job is running to cost and programme
- record any alterations to a contract.

As the contracts manager would not be permanently on site during the contract, communication would be done through a third party such as the construction manager. You may meet face to face if the contracts manager attends site meetings.

Ensure that any alterations to the contract are recorded and saved.

Construction manager

The construction manager's role is to:

- be responsible for running the construction site or a section of a large project
- develop a strategy for the project
- be able to plan ahead to solve problems before they happen
- make sure site and construction processes are carried out safely
- communicate with clients to report progress and seek further information
- be a workforce motivator.

The construction manager is also known as site manager, site agent or building manager, and they have the overall responsibility for the running of the contract. When communicating with the construction manager (regarding e.g. alterations, progress, delays to the contract) it is always preferable to do so in writing, either by fax, email, letter or any form of communication that can be saved and recorded.

In summary, the best form of communicating with the site management team is written. This is essential as written communication can be tracked throughout the contract and you *must* keep all evidence of this including emails, letters, variation orders, and a log of your telephone conversations (when, who with and what was discussed/agreed).

Workplace craft operatives

A great deal of plumbing work involves communication and at every stage in your career you will need to understand your level of responsibility for this. Most companies have different ways of dealing with this, but a general overview is given below.

Apprentice gas operatives/plumbers

- Work directly with a qualified member of the staff.
- Should be given the necessary level of work instruction and supervision to undertake the work that is set for them.
- Are not usually be directly involved in communication with customers and co-contractors.
- Should pass all problems and work issues to their supervisor.

Level 2 qualified operatives

- Do not usually take full responsibility for the contract.
- Would normally work under minimal supervision.
- Respond to queries from customers and co-workers.
- Would usually forward requests for additional work to their supervisor to action.

Level 3 qualified operatives

- Usually take full responsibility for the contract.
- Deal with queries from customers and co-contractors.
- Are responsible for dealing with requests for additional work.
- Are responsible for confirmation of the work, pricing and complaints from customers.

Supervision

Apprentices are under the supervision of a plumber qualified to Level 2. The plumber will have the responsibility of guiding the apprentice in all aspects of plumbing work. They will have to ensure that the apprentice is working safely. If working on a construction site, they must abide by the site rules.

The job supervisor, which would be the Level 3 qualified plumber, would monitor the on-site personnel ensuring that progress is as per the work programme. If the work progress falls behind this could result in more labour being required to complete the contract, which impacts on profits.

It would be their responsibility to report back to the senior management any problems that have arisen on site, such as:

- safety issues
- trade disputes
- discipline
- disputes between client and contractor.

Where the dispute is regarding quality, it would be best practice to get a third party involved who is independent of both client and contractor to establish the quality of the installation. This can, in many cases, settle the dispute, especially with a private customer.

Figure 3.21 The job supervisor should provide opportunities for sharing knowledge and experience

Knowledge check

1 How should the plumbing supervisor handle the supplier if two deliveries of sanitaryware have failed to come up to specification?

a Speak calmly to the supplier and ask them not to do it again

b Make demands of the supplier by using a firm and assertive tone to explain the situation

c Use a low, calm voice to explain the problem and seek a solution

d Demand talks with the managing director and refuse to take no for an answer

2 Good customer relations for a business can start with

a Acting promptly to requests for information

b Polite dealings with potential customers

c Deposits before work commences

d Both A and B

3 On a new large building project, to whom is the plumbing company ultimately accountable for the quality of customer service?

a Client

b Public

c Architect

d Building control officer

4 What is the responsibility of the clerk of works?

a To check that the work complies with the specification laid down by the architect

b To check if the safety regulations have been complied with

c To inspect the work on behalf of the local authority

d To co-ordinate suppliers and subcontractors

5 A plumbing subcontractor's representative may sometimes visit the site to measure the amounts of materials that have been installed. What is the purpose of this exercise?

a To check the plumbing materials to complete the job

b To assess the level of theft

c To create materials orders to suppliers

d To prepare interim valuations (invoices) for stage payments

6 When constructing a large office building, with all mechanical services this involves, any changes or alterations to the plumbing specification will be made by which party?

a Client

b Clerk of works

c Building control officer

d Architect

7 Which of the following is the principal role(s) of the quantity surveyor?

a To measure all materials used on site

b To carry out valuations

c To organise the subcontractors required for a job

d All of the above

8 Another name for a procurement officer in a plumbing or contractor company is

a Planner

b Site engineer

c Buyer

d Estimator

9 Which is the preferred method for checking and chasing deliveries of materials to a plumbing job?

a Semaphore

b Email

c Phone

d Fax

10 Which of the following best describes a work programme?

a It identifies when plumbing jobs should start and be completed

b It is used to help the contractor plan the number of plumbers needed for each task

c It sets out dates when payments are due for each bit of work completed

d It gives the plumber an indication of material quantities

The combustion process and testing

INTRODUCTION

The safe and correct combustion of gas in appliances such as boilers and ovens is essential to ensure the safety of the public both in their homes and in the workplace. As an engineer working in these settings you must therefore thoroughly understand the principles of combustion and be able to diagnose faults to ensure the safe installation and running of gas appliances.

WHAT IS GAS?

The word 'gas' comes from the Greek word for 'chaos' and refers to the way atoms in a gas move in a chaotic manner.

A gas has a lot more space between its molecules than a liquid or solid, which allows them to move about much more freely and quickly. This space also means they are lighter than liquids or solids, and the energy with which the molecules bounce around causes pressure to build up if the gas is contained in a pipe for example.

Molecules move in all directions so they create equal pressure on all of the walls of their container. Gases must be kept in sealed containers otherwise the particles escape and diffuse (mix) into the air.

Natural gas is a made up of a mixture of hydrocarbons. The main hydrocarbon in natural gas methane, but there are also small amounts of ethane, propane and butane.

Natural gas comes from organic matter like trees and small sea creatures that died many millions of years ago, which decayed and was covered with layers of silt and clay, and eventually turned into rock. Over millions of years, the heat of the earth and the pressure caused by the weight of rocks turned this organic matter into the fossil fuels gas, coal and oil.

Types of gases used to supply domestic appliances

There are three types of gases used in the gas industry: natural gas mostly made from methane gas; and propane and butane gases, which are classed as liquid petroleum gases (LPG).

Constituent	Chemical symbol	Percentage forming natural gas
Methane	CH_4	94.40
Ethane	C_2H_6	3.14
Propane	C_3H_8	0.60
Butane	C_4H_{10}	0.19
Pentane	C_5H_{12}	0.22
Hydrocarbon	C_9H_2O	1.40
Carbon dioxide traces	CO_2	0.00
Nitrogen	N_2	0.05
Sulphur traces	S	0.00
TOTAL		**100**

Table 4.01 Typical constituents of natural gas and their chemical symbols

Table 4.02 shows the individual characteristics of natural gas, propane and butane.

Characteristic	Natural gas*	Propane	Butane	Notes
Specific gravity (SG of air = 1.0)	0.6	1.5	2.0	Methane will rise but propane and butane will fall to low level
Calorific value (mega joules per cubic metre)	39 MJ/m^3	93 MJ/m^3	122 MJ/m^3	Appliances are designed to burn a particular gas
Stoichiometric air requirements	10:1	24:1	30:1	Methane requires 10 volumes of air to 1 volume of gas – LPG requires more
Supply pressure	21 mbar	37 mbar	28 mbar	Appliances must be matched to the gas used
Flammability limits	5 to 15 % in air	2 to 10 % in air	2 to 9 % in air	Ranges within which gas/air mixtures will burn
Flame speed	0.36 m/sec	0.46 m/sec	0.45 m/sec	This is the speed at which a flame will burn along a gas mixture
Ignition temperature	704 °C	530 °C	408 °C	Approximate temperatures
Flame temperatures	1930 °C	1980 °C	1996 °C	Approximate temperatures

* = Methane

Table 4.02 Key properties of gases

Odour

Odorants are added to odourless gases such as natural gas to aid detection. Odorants now in use contain diethyl sulphide and butyl and ethyl mercaptan (C_2H_5SH). Mercaptans are a group of sulphur-containing organic chemical substances that have a strong smell like rotting cabbage. This means that if mercaptans are in the air, even at low concentrations, they are very noticeable and will help alert you to and detect a gas leak.

Viscosity

The **viscosity** of natural gas depends on its temperature and the pressure it is under. In contrast to liquids, the warmer the gas, the more viscose it is. An increase in a gas's viscosity leads to a resistance in the flow of the gas, so when the temperature of the gas increases, the quantity of gas flowing through a pipe will decrease. This means more gas will flow in a cold gas pipe.

Families of gases

There are three main families of gases:

- manufactured gases
- natural gas
- LPG.

Manufactured or 'artificial' gases are generally made from coal, oil feed-stocks and also include LPG/air mixtures.

Natural gas, as the name suggests, is naturally occurring and has been supplied to the UK from the North Sea and Irish Sea reserves, with excess gas being supplied to Continental Europe. However, more recently the UK has become a net importer of gas.

Key terms

Stoichiometric
This is the exact mix of oxygen to natural gas required to achieve complete combustion/reaction, which ensures only carbon dioxide and water vapour are produced. This is nearly impossible to achieve outside a laboratory and so most manufacturers allow for this by designing the burner to take in more oxygen to ensure the complete reaction takes place.

Viscosity
The viscosity of a substance is its resistance to flow. The more viscous a substance, the slower the flow. Viscosity is increased in pipes with the friction of the pipe walls. Gases, unlike liquids, become less viscous the colder they are and the less pressure they are under.

Specific gravity (relative density)

When we compare the weight of natural gas to air, which has a specific gravity (SG) of 1.0, we find that natural gas is 0.6, just over half the weight of air. Therefore natural gas will rise. Propane has an SG of 1.5 and butane has SG of 2.0, so both are heavier than air and will fall to a low level. This will have an effect on where you look for leaks: natural gas will be concentrated at the ceiling and LPG (propane and butane) around your feet.

Explosive mixtures

Gas can explode if mixed with the right amount air/oxygen. Everything must be done to prevent this from happening by accident.

Igniting a flammable gas/air mixture in a closed container (e.g. a gas meter) will cause the gas to burn faster than a similar mixture in an open container. This is because the heat generated by the burning mixture will cause the gas volume to increase, leading to an increase in pressure and an explosion.

The larger the container, the faster the flame will spread. The more the flammable gas/air combination is mixed, the larger the explosion will be. The way a gas is ignited, whether by flame or spark, does not affect the power of the explosion.

When removing a gas meter from a gas installation, you will need to make sure that the gas system is correctly isolated and capped of. The meter also has to be capped off to prevent an air/gas mix from occurring within the meter chambers as this will cause an explosion if it came into contact with a source of ignition.

THE COMBUSTION PROCESS FOR GASES USED IN DWELLINGS

Combustion is a chemical reaction that needs three elements: fuel + oxygen + ignition. The reaction of this process causes heat and products of combustion (POC). The most common type of combustion that occurs in a domestic property is that of natural gas.

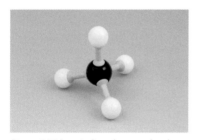

Figure 4.01 A typical methane molecule

The combustion equation

Combustion equations help you to work out whether a certain combination of gases, in particular quantities, will combust completely or incompletely. The quantity of heat produced is determined by how active/energised the atoms of a substance are.

A methane molecule (CH_4) consists of one carbon and four hydrogen atoms. This readily reacts with oxygen in the air making it a good fuel to use with a source of ignition.

A simple equation of combustion with methane is shown in Figure 4.02. To ensure complete combustion, one volume of methane must react with two volumes of oxygen. The reaction is complete because the correct quantity of oxygen is present to complete the reaction.

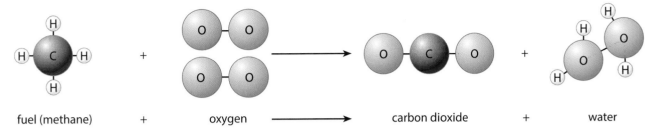

Figure 4.02 Combustion reaction of methane

An incomplete reaction will lead to the production of carbon monoxide (CO), as illustrated in the equation below, because oxygen reacts with hydrogen before carbon so you are left with CO not CO_2.

methane + oxygen = carbon monoxide + water
$2CH_4 + 3O_2 = 2CO + 4H_2O$

Here, poor combustion is caused by a lack of oxygen to react with the methane. As well as producing CO, soot deposits will form from the flame burning incorrectly.

Air requirements for combustion

When you light burners, it is important to know about the principles of combustion and the reasons why the correct mixture of gas and air is needed. This correct mixture is called 'complete combustion', and this is essential for gas safety. Gases such as natural gas (methane) and LPG (propane and butane) are carbon-based gases, and if the combustion process is not correct, CO can be produced. CO is a highly toxic gas and can kill. It is therefore absolutely essential that the correct amount of oxygen is mixed with the gas to ensure complete combustion.

Figure 4.03 demonstrates the correct combustion process and includes the products of combustion.

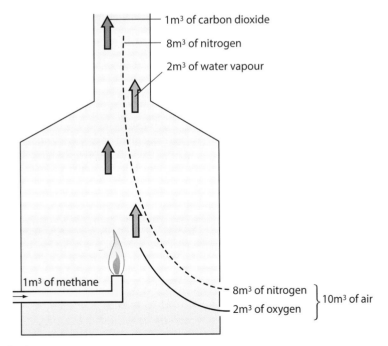

Figure 4.03 Combustion process

Air is made up of only about 20 per cent oxygen; the other 80 per cent is mainly nitrogen, which doesn't burn so plays no important part in the process and just adds to the bulk of products going out of the flue. This ideal mixture of gas and air is sometimes referred to as the 'stoichiometric mixture'.

Flammability limits

Flammability limits are the limits at which gas and air will burn. If there is too much or too little gas or air in the mix it will not burn. Natural gas will only burn if there is between 5 and 15 per cent natural gas in air.

If the gas is not burned it will also become an explosive mix. LPG has a lower limit of flammability: 2 to 10 per cent of propane gas in air and 2 to 9 per cent of butane gas in air.

Safe working

Always avoid build-up of natural gas in any space as it will soon get to the explosive limit. To do so, ensure the space is well ventilated and no ignition sources are present when working with gas.

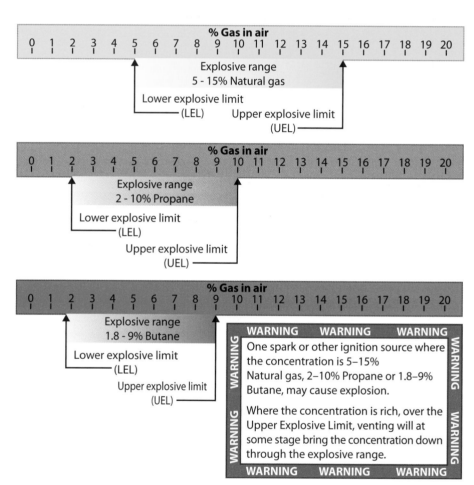

Figure 4.04 Combustion flammability range

Calorific value

The calorific value (CV) of a fuel is the amount of heat given when a unit quantity of fuel is burnt. It is the amount of energy released and is expressed as mega joules per cubic metre (MJ/m^3). Calorific value can be expressed as gross CV or net CV.

As gas is burnt a percentage of the heat is given up into the water vapour in the products of combustion; this is known as 'latent heat' and is used in condensing boilers. In non-condensing appliances the latent heat is lost to the air outside. The term for all the heat generated by burning a set volume of gas in a gas appliance is 'gross heat input' (gross CV) and the term for the **heat energy** that passes into the heat exchanger is known as 'net heat input' (net CV) or 'sensible heat'. All manufacturers of gas appliances are required to quote net heat input to reflect the heat that actually passes into the appliance heat exchanger.

Typical calorific value of natural gas is 38.79 MJ/m^3 gross or 34.9 MJ/m^3 net and is variable dependent on the source.

Heat rate (power)

The most often used units to measure power are:

- watts (W)
- kilowatt (kW) – the kilowatt is the usual unit for an appliance power
- mega joule/hour (MJ/h).

Sometimes imperial measurements are used due to the age of the appliance, although this is rare these days. See Table 4.03 below for conversions.

Key term
Heat energy
A unit of energy is called a joule (J) or mega joule (MJ). The amount of heat energy is determined by how active/energised the atoms of a substance are. This energy is passed on by atoms colliding and transferring that energy to the next atom, which is how we sense heat energy.

1 Btu	=	1055 J	1000 Btu/h	=	1.055 Mj/h
1,000 Btu	=	1.055 MJ	1 Horsepower	=	745.7 W
1,000 Btu	=	0.2931 kWh	1 Horsepower	=	2.685 MJ/h
1 Therm	=	100 000 Btu	1 Horsepower	=	33,000 ft lb/min
1 Therm	=	105.5 MJ	1 Horsepower	=	42.41 Btu/min
1 Therm	=	29.307 kWh	1 W	=	1 J/sec
1 Btu	=	0.252 kcal	1 W	=	3.412 Btu/h
1 kcal	=	3.968 Btu	1 kW	=	3.6 MJ/h
1 kWh	=	3.6 MJ	1 kW	=	3412 Btu/h
1 kWh	=	3412 Btu	1 MJ/h	=	0.2778 kW
1 cal	=	4.1869 J	1 MJ/h	=	947.8 Btu/h
1 kWh	=	0.03412 Therm	1 kW	=	1.341 Horsepower
1 J*	=	0.000947 Btu	1 J	=	1 W/sec
1 J	=	0.2388 cal	1 J	=	0.2388 cal
1 MJ**	=	0.2778 kWh	1 J	=	0.0009478 Btu
100 MJ	=	0.9478 Therm	1 cal***	=	Quantity of heat required to raise the temperature of 1 gram of water by 1°C
1000 Btu/h	=	0.2931 kW	1 Btu	=	Quantity of heat required to raise the temperature of 1 lb of water by 1°C

* British thermal units (Btu)

** joule (J) / mega joule (MJ)

*** calorie (cal) / kilocalorie (kcal)

Table 4.03 Metric conversion table

Wobbe number of gases

The Wobbe number is an indication of the amount of heat produced from a burner for a particular gas. It is found by dividing the CV by the square root of the SG:

$$\frac{CV}{\sqrt{SG}} = \text{Wobbe number}$$

The Wobbe number for natural gas is between 48.2 and 53.2 (metric).

The amount of heat that a burner will produce depends on:

- the amount of heat in the gas CV
- the rate at which the gas is burned – the gas rate (see the section on 'Gas rate/pressure setting' on page 118).

The factors that affect the heat output are:

- the jet or injector size
- the gas pressure, forcing it out of the injector
- the SG of the gas, which affects how easily the pressure can push out the gas due to the energy in the gas molecules.

Stoichiometric air requirements

This is the amount of air (in cubic metres) required for complete combustion of one cubic metre of gas. The perfect mixture of a fuel gas and air is termed the "stoichiometric mixture" and is virtually impossible to achieve in the customer environment.

Theoretical stoichiometric calculations

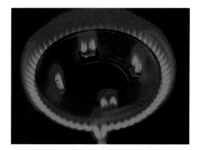

Figure 4.05 Flame on a burner

Theoretically, CO will stop being produced when 9.58 volumes of air per volume of gas is present. Due to the design of burners and the path that combustion air takes, it is impossible to supply the exact stoichiometric conditions. It may work under laboratory conditions but when air is supplied to a burner, air for combustion will not always reach every flame – some air will rise due to convection. More air needs to be supplied to ensure complete combustion.

For stoichiometric combustion the volume of air required to burn 1 unit volume of natural gas is 9.58 volumes (assumes air is 21 per cent O_2 and 79 per cent N_2).

If there is 8.5 per cent CO_2 in the flue gases, then the air required to burn 1 volume of gas will be 13.2 volumes.

For G20 gas (100 per cent methane) the air required per kW of heat input is calculated to be $3.89 \times 10{-}4$ m^3/sec.

Combustion efficiency

As a gas operative, your aim is to achieve, as near as possible, the stoichiometric mixture that will produce the best combustion efficiency. To do this you need to understand the relationship between the amount of air (oxygen) supplied and the effects this has on the flue products.

$$CH_4 + 2O_2 = CO_2 + 2H_2O$$

This formula burns 2 volumes of methane with enough oxygen, resulting in a complete reaction, producing CO_2 and water.

$$2CH_4 + 3O_2 = 2CO + 4H_2O$$

This formula has insufficient oxygen so results in incomplete combustion, which produces CO not CO_2. CO is poisonous if inhaled.

Levels of CO2 and the potential effects on the gas combustion process

Poorly installed open-flued appliances can lead to CO_2 entering a room through blockages or down-draught conditions. Products of combustion spilling into the room can be at first unpleasant but, if left, can create dangerous conditions. A CO_2 concentration of 1.5 per cent to 2 per cent of the room's volume can start to interfere with an open-flued appliance's combustion.

The result will cause **vitiation** of the open-flued appliance, turning the CO_2 entering the room into CO, which will have a greater risk to the occupants. Carbon dioxide on its own is only dangerous as an **asphyxiant** as it will replace the oxygen in the room so as to cause the death of the occupant. It is important that open flued and flueless appliances are regularly serviced and checked to ensure that no spillage can take place and that purpose provided ventilation is adequate.

Incomplete combustion

If the combustion process and conditions are not correct then it will result in incomplete combustion and harmful products can be released from the flame.

The most dangerous product of incomplete combustion is the toxic gas CO.

Causes of incomplete combustion

There are several causes of incomplete combustion. **Overgassing** is when an incorrect burner pressure and/or wrong injector size supplies more gas than the appliance was designed for. **Chilling** occurs when a flame touches a cold surface or is exposed to a cold draught; the flame pattern is disturbed and sooting may occur, causing even more problems.

Flame impingement is when one flame from a burner port is deflected (possibly by foreign matter on the burner) into another flame. The point of contact produces a cold spot in the flame (since there would be no **secondary air** at that spot) resulting in incomplete combustion. Impingement can also occur when the flame touches any part of the combustion chamber, producing a chilling effect on the flame, and allowing carbon deposits to form through incomplete combustion. These deposits can fall on other parts of the burner, causing further impingement. This is why, once a boiler starts to soot up, the process becomes accelerated.

Reduced oxygen levels in a room will cause the air to become **vitiated** (made impure) and will affect combustion. **Under-aeration** can occur even in a room where there is adequate ventilation because maladjustment of the primary ports or blockage to the lint filter will prevent sufficient draw of **primary air**. **Poor flueing** is caused by a partially blocked or blocked flue preventing the products of combustion from leaving the combustion chamber.

Figure 4.06 A yellow flame

Figure 4.07 Sooting on an appliance heat exchanger

Figure 4.08 Staining around a flue

Visual signs of incomplete combustion

There are several signs of incomplete combustion, including yellow flames, sooting and staining. Burners with insufficient air will burn with a **yellow flame**. Note, however, that some gas fires are designed to give yellow flames as a live fuel effect. A match is a typical example of a flame that has insufficient air for complete combustion.

Sooting is a sign of incomplete combustion and is indicated by the presence of unburnt carbon (soot) on the appliance heat exchanger or radiant.

Staining may be seen around the flue or draught diverter, and may also be due to spillage of flue products due to a poor flue.

If you encounter any of these problems, you must investigate them and advise the customer not to use the appliance as it could prove fatal. You must ensure that any appliance showing symptoms of incomplete combustion is not used and follow the correct 'unsafe situations' procedure. See Chapter 11 for more information on responding to unsafe situations.

CARBON MONOXIDE POISONING

Carbon monoxide is produced when gas appliances such as boilers, built-in ovens or freestanding cookers aren't fully burning their fuel. This usually happens if they have been incorrectly or badly fitted, not properly maintained, or if vents, chimneys or flues become blocked.

A concentration of only 0.04 per cent CO in air can be fatal if inhaled for just a few minutes. Carbon monoxide combines much more easily with haemoglobin in the blood stream than oxygen. Haemoglobin collects and carries oxygen around the body in the red blood cells. If the haemoglobin becomes saturated with CO then it cannot take in oxygen, preventing the

% CO saturation of the haemoglobin	Symptoms
0–10	No obvious symptoms
10–20	Tightness across the forehead, yawning
20–30	Flushed skin, headache, breathlessness and palpitation on exertion, slight dizziness
30–40	Severe headaches, dizziness, nausea, weakness of the knees, irritability, impaired judgement*, possible collapse
40–50	Symptoms as above with increased respiration and pulse rates, collapse on exertion**
50–60	Loss of consciousness, coma
60–70	Coma, weakened heart and respiration
70 or more	Respiratory failure and death

* Mental ability is affected so that a person may be confused and on the verge of collapse without realising that anything is wrong.
** Any sudden exertion would cause immediate collapse, and therefore an inability to escape from the situation.

Table 4.04 Effects of CO intake on adults

body from receiving oxygen. This quickly leads to poisoning causing serious brain damage and sometimes death.

Table 4.04 opposite shows the effects of CO on adults, with the saturation of haemoglobin shown as a percentage.

Carbon monoxide is colourless, tasteless and has no smell, making it difficult to recognise – but there are ways you can spot a potential risk.

Safety measures to avoid CO exposure

Along with fitting a CO detector/alarm (see the section on 'Fitting CO detectors' page 94), there are some tell-tale signs of CO emissions:

- yellow or orange cooker flames – gas flames should be blue
- signs of sooting or staining around appliances
- pilot lights which will not stay lit
- more than normal amounts of condensation on windows.

Primary measures to prevent CO exposure

To prevent CO from endangering lives, all appliances must be correctly installed. Gas appliances should be maintained regularly and serviced every 12 months.

Secondary measures to prevent CO exposure

An audible CO alarm should be fitted to alert householders to any CO in their home.

CO detectors

The CO detector should have:

- an audible alarm (not just an indicator tool)
- a British Standard EN 50291 mark (BS EN 50291) or shown with the CE mark
- a British or European Kitemark
- Loss Prevention Certification Board (LPCB) or equivalent approval mark.

CO concentration (parts per million)	Without alarm before	With alarm before
30 ppm	120 mins	–
50 ppm	60 mins	90 mins
100 ppm	10 mins	40 mins
300 ppm	–	3 mins

Table 4.05 CO concentration to alarm activation time (EN 50291 Table 4)

As soon as the CO alarm is activated it should sound at CO concentrations above 50 ppm.

There are two basic types of alarm – battery operated and mains power supplied. You should always follow the manufacturer's instructions.

Figure 4.09 Typical CO alarm

Fitting CO detectors

CO alarms can be activated by aerosols so this should be considered in the positioning of the alarm to be away from areas where this could cause accidental activation. It is also important to know that CO is a gas and can be carried into the building some distance from the source. A normal flue gas analyser will not be good enough to be used continually for detecting CO in a room and is a short-term measurement only.

Setting up and installing the CO detector is a straightforward task. The alarm needs to be placed in a central location, such as a landing or hallway – do not place in a cupboard or near to an outside door. Fit the alarm to a wall at head height or onto shelves at a similar height. The alarm should be at least 1 m from any gas appliances but in the same room as the appliance. The alarm should be regularly tested using the test button and the responsible person should renew the batteries every year or whenever the low battery alarm sounds.

Maintenance requirements of CO detectors

Always check with manufacturer's instructions on the maintenance requirements of the alarm. The average life span is around five years, but this is not industry standard.

Keep the detectors clean using a vacuum cleaner – never use detergents or solvents as chemicals can contaminate the detector. Keep the detectors out of the reach of young children.

Causes of activation of CO detectors

Gas-powered cars and equipment should never be used in a home or garage, even if the house/garage doors are open. This is because the air pressure in most houses is typically at a lower pressure than that outside, so the gas can be drawn into the home. Charcoal grills are not designed to be used indoors and can easily lead to CO being produced in the room.

It is a requirement to have an EN 50291-compliant domestic CO alarm in the room where a solid fuel appliance is installed. Part J of the Building Regulations states that CO alarms reduce the risk of poisoning from other types of combustion appliance, i.e. non-solid fuel appliances (gas- and oil-fired).

DEALING WITH DOMESTIC CO GAS ESCAPES

The following section contains information from BS 7967 part 1 relating to the investigation of potentially lethal CO emissions in a building. When you're called to a report of fumes in a building you should carry out the full procedure outlined in this standard and as required by the Gas Safety Regulations below.

The Gas Safety (Installation and Use) Regulations

Part G37 (8) of the regulations says:

> In this regulation any reference to an escape of gas from a gas fitting includes a reference to an escape or emission of carbon monoxide gas resulting from incomplete combustion of gas in a gas fitting, but, to the

extent that this regulation relates to such an escape or emission of carbon monoxide gas, the requirements imposed upon a supplier by paragraph (1) above shall, where the escape or emission is notified to the supplier by the person to whom gas has been supplied, be limited to advising that person of the immediate action to be taken to prevent such escape or emission and the need for the examination and, where necessary, repair of the fitting by a competent person.

In the event of a gas escape, if a fatality or injury has occurred the incident will be investigated by the appropriate authority within the emergency services and reported to the Health and Safety Executive (HSE). Only after such an investigation has been completed should you become involved.

When investigating any escape of fumes or gas it is important to take into consideration the safety of everyone concerned, including yourself. BS 7967 part 1 sets out the steps to take:

1 protect life
2 protect property
3 locate all fuel-burning appliances
4 locate any escape of gas, fumes, smells or spillage/leakage of combustion products
5 confirm the safe installation and operation of all suspect gas appliances
6 advise the customer of any remedial action that is required
7 complete all necessary reports, documentation and actions as advised in the Gas Industry Unsafe Situations Procedure.

Protection of life and property from CO gas escapes

If CO levels above 30 ppm (parts per million) are noted at any time during the working visit, you must turn off all fuel-burning appliances wherever possible and advise the occupants to leave the dwelling immediately. Open windows and doors to ventilate the dwelling before you leave. Note that a cooker, when first turned on, can produce a reading of above 30 ppm (usually from the grill) in the first few minutes of operation.

On re-entering the building, check the atmosphere continually with an analyser from the point of access inwards, throughout the dwelling. The dwelling should not be considered safe to fully re-enter to carry out any further investigation until CO concentrations have fallen below 10 ppm, preferably down to normal outdoor background levels. Lower levels of CO exposure may be used as evacuation criteria for the gas operative, based on a risk assessment of the specific circumstances by the competent gas engineer. Risk assessments should take into account the length of exposure to the CO in the building.

One of the main combustion products from gas appliances is CO_2, which is regarded as an asphyxiant but is also a toxic substance which could be present in the air in sufficient quantity to prove harmful to humans and animals. For guidance on the occupational health considerations of CO_2, refer to the HSE guidance note on Workplace Exposure Limits (EH40).

Investigation of CO gas escapes

As a gas operative asked to carry out tests for a gas escape by the house owner or through Gas Safe, you should keep the occupant fully informed of procedures and updated on how the investigation is progressing. On arrival, you could encounter:

- people who have experienced or are experiencing nausea, dizziness, chest pains, headaches and/or palpitations when appliances are or have been in use. Some common illnesses have similar symptoms to those resulting from exposure to CO such as flu
- the presence or report of unusual smells, oppressive atmospheres, strange tastes and/or condensation particularly on cold surfaces when appliances are in use
- sooty marks/stains on or around appliances and flues
- the presence of CO and/or CO_2
- CO detector having been activated
- a phone referral from an emergency service provider or an agent acting on behalf of the customer.

CO investigation procedure in a dwelling

When conducting a CO investigation, consider the questions and guidance below.

1.	Does the problem only happen when appliances, including mobile/portable appliances, are or have been in use? Are there any safety warnings attached to the installation/appliances? Do the occupants feel unwell in the property but better when outside or away from the property? If so, what are their symptoms? Is there a pattern to the occurrences, for example have they been observed once, more than once or many times. In what circumstances has the pattern emerged? Is it during particular weather conditions or domestic activities? While using a particular household chemical? In a particular room?
2.	If fuel-burning appliances other than gas appliances are working in the room, where possible, check them for CO production and spillage. If these appliances are thought to be the cause of the problem, you should recommend the customer seek further advice.
3.	Review information continuously and determine which appliances are suspect. Concentrate on suspect appliances first.
4.	If no faults are found on suspect appliances, progressively check other gas appliances. If you still detect no faults, consider other possible CO sources (e.g. barbeques or car exhaust from the garage).

continued

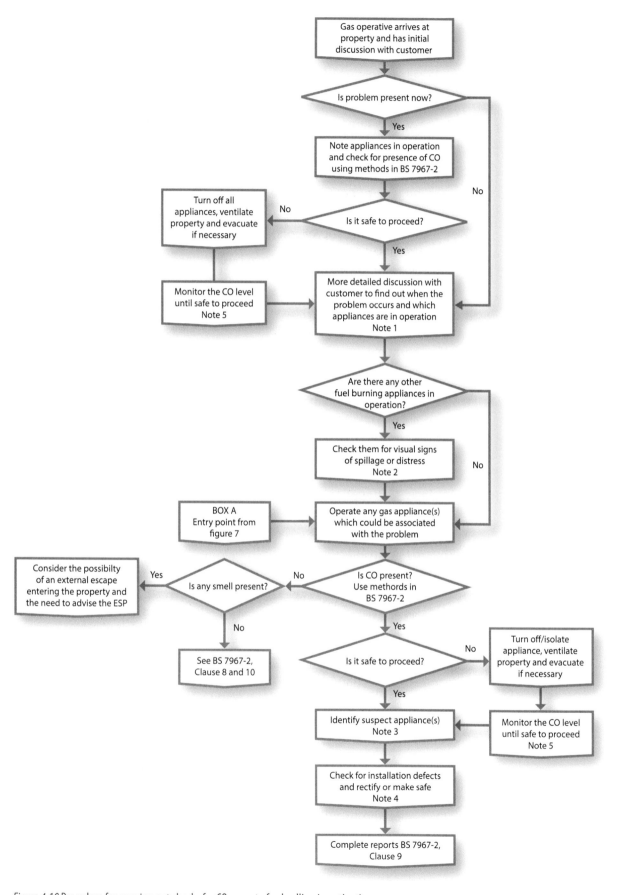

Figure 4.10 Procedure for carrying out checks for CO as part of a dwelling investigation

> 5. If for any reason CO levels do not fall sufficiently to allow the investigation to proceed, it is recommended that the property be evacuated/remain evacuated. If sources of CO other than fuel-burning appliances are suspected you should ensure that the customer and, where possible, any other persons potentially affected (e.g. neighbours if it is suspected that the source of CO originates from a neighbouring property) are aware of such concerns and advise them to ensure appropriate follow-up action is taken. All such advice should be documented.

Checking domestic CO detectors

When the customer becomes alerted by an activated **CO indicator card**, check with them how quickly and in what circumstances, the colour change took place. If the indicator card has changed colour, use an electronic portable combustion gas analyser in accordance with BS 7967-2 to confirm the CO levels and rectify any faulty appliance installations.

If the checks indicate that there is no CO present, investigate the possibility of other causes for the card changing colour, such as aerosol use nearby. This is especially relevant where the customer has suffered no ill effects and where subsequent checks confirm that appliances are operating satisfactorily with no installation faults.

Where indicator cards are used they should be positioned in accordance with the manufacturer's instructions. You should be aware of the fact that:

- Indicator cards are not considered suitable as detector in sleeping accommodation since they do not provide an audible alarm that could waken a person who is asleep.
- Most cards have a useful time limit, after which they should be replaced to remain effective.
- Other products such as aerosol sprays can cause them to change colour without CO being present.

Where an **electrical CO alarm** is installed, confirm whether it has been activated and for how long, how frequently and in what circumstances. Check that the alarm is working in accordance with the manufacturer's instructions to ensure it has been set up correctly. Be aware that electrical alarms will emit a range of different sounds depending on the alarm manufacturer and the particular circumstances of the alarm going off.

If the alarm activation is genuine, inspect all appliances in the building and use a portable electronic combustion gas analyser in accordance with BS 7967-2 to confirm the CO levels. Rectify any faulty appliance installations.

If these checks indicate that there is no CO is present, suspect that the alarm is faulty or has been activated by other products, such as aerosol sprays. This is especially relevant where the customer has suffered no ill effects and where subsequent checks confirm that appliances are operating satisfactorily with no installation faults.

If subsequent investigations suggest a CO source at a level that could have activated the alarm, but there is no evidence that it has been activated, consider that it may be faulty or incorrectly positioned.

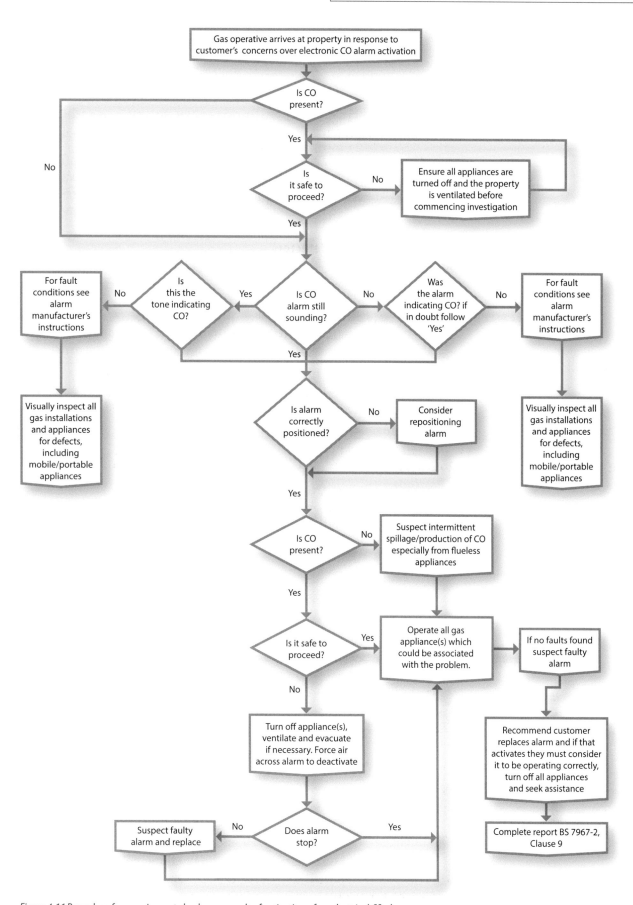

Figure 4.11 Procedure for carrying out checks as a result of activation of an electrical CO alarm

For flueless appliances (e.g. cookers), if subsequent investigations suggest a CO source at a level that should not have activated the alarm, but there is evidence that it has been activated, consider that the alarm may be faulty or incorrectly positioned.

In the absence of manufacturer's instructions for the location of the alarm, refer to guidance given in the BRE Good Building Guide 30 on CO detectors.

In all cases where a faulty alarm is suspected, recommend to the customer that it is tested for correct installation and operation by the supplier/manufacturer and replaced if necessary. The procedure for carrying out checks as a result of the activation of an electrical CO alarm is shown in Figure 4.11 on page 99.

When following the procedure outlined in Figure 4.11:

- review information continuously to assess whether this is a true CO alarm activation
- ensure that CO levels do not build up from an unknown source while you work
- where appropriate refer to the flow chart in Figure 4.10 on page 97 that shows the procedure for carrying out checks as part of a dwelling investigation.

BS 7967 Dwelling investigation report		
Job address	**Client details (if different)**	**Registered business details**
Name: **Address:** **Postcode:** **Tel no:** **Report left with:**	**Name:** **Address:** **Postcode:** **Tel no:** **Is the property rented?**	**Company:** **Address:** **Postcode:** **Tel no:** **Gas Safe reg no:** **Gas operative's name:** **Gas Safe ID card serial no:**
Problem as reported by occupier		
What is the problem? When does it happen/is there a pattern to the occurrences? Is it happening now? Which gas appliances were on at the time? Were there any other fuel burning appliances in operation? Who was affected? What were the occupiers' symptoms? Were there any extreme weather conditions at the time?		
Observations	**Conclusions**	
Initial CO readings on entering property (ppm): Which appliances were in operation? How many appliances were in operation? No. of appliances examined: Which appliances were checked? Gas tightness test on whole installation: Pass/Fail	Problems were indentified and the installation/appliances have been made safe: Yes/No Rectification work completed/required is detailed in the attached sheets: Yes/No The installation is safe to use and no defects were identified: Yes/No	
	Gas operative's signature	**Date**

Figure 4.12 Example format of a dwelling investigation report

Dwelling investigation reports

A dwelling investigation report is the form provided in BS 7967-1 that you should complete when carrying out CO checks. The general content of this form is shown in Figure 4.12.

CO investigation procedure for an appliance

Visually check all gas appliances for signs of spillage, combustion problems, incorrect flueing and ventilation, and general condition.

Visually check all other fuel-burning appliances, such as oil and coal appliances, for signs of spillage and/or combustion problems.

Consider the findings from these visual examinations together with all other information gained since arriving on site, for example to identify suspect appliances that require further investigation or rectification. More information on investigating CO_2 gas escapes can be found on page 94.

If a fuel-burning appliance other than a gas appliance is identified as being suspect, advise the customer to seek the relevant expert assistance.

It is essential to investigate all gas appliances initially identified as suspect as there could be more than one faulty appliance/installation. The most likely appliance should be investigated first before progressively checking other suspicious appliances.

You should carry out the investigation following the steps below:

1.	Check the room atmosphere for the presence of CO in accordance with BS 7967-2:2005, Clause 5 before operating any suspect appliance.
2.	Put the suspect appliance into operation. If appropriate for the appliance design, carry out a spillage test. Check for a build-up of CO in the room in accordance with BS 7967-2. If the CO level in the room exceeds 30 ppm (except when investigating a cooker – see Protection of life and property from gas escapes on page 95), you should immediately turn off the appliance, open all doors and windows, and vacate the room or space where this level has been measured. Do not re-enter the room or space, other than for the reason of sampling the atmosphere. Advise anyone else not to enter the building until the level has dropped below 10 ppm and preferably is down to outdoor background level.
	If spillage or CO build-up is occurring, only operate the appliance for short periods to avoid exposure to potentially harmful combustion products.
	Unless the investigation requires the maximum CO level to be determined, once spillage has been confirmed it is not necessary to continue testing until the maximum level that can be generated in the room or space is confirmed.
	Where spillage is found, investigate the installation/appliance further and rectify any faults as necessary. Part of the investigation should include the checks in point 3 below.

continued

3. Check the burner operating pressure and/or gas rate and check flame picture following the appliance manufacturer's instructions. Where possible, check the combustion performance following the appropriate method specified in BS 7967-2.

4. If the combustion performance ratio (CO/CO_2) is greater than that allowed in BS 7967-3 or where it is not possible to check the combustion performance of an appliance, perform a detailed examination of the appliance.

5. When no spillage is found and the combustion performance ratio is within the limits allowed by BS 7967–3, carry out the following checks:

   a Visually examine appropriate areas of the appliance for signs of spillage, heat stress, corrosion or damage. It might be necessary to remove the outer casing to do this.

   b Visually examine all heat exchangers for obstruction or surface deposits, and clean as necessary.

   c Replace any parts necessary to restore operation.

6. If no fault can be identified, consider other potential sources of CO or odours.

Detailed examination of a suspect appliance

The following steps should be taken in full whenever it has not been possible for you to check the combustion performance ratio (CO/CO_2) of a suspect appliance in accordance with the appropriate methods in BS 7967-2 or when the combustion performance ratio (CO/CO_2) is greater than allowed in BS 7967-3.

1. Ensure gas and electrical supplies are isolated.

2. Ensure that all components that could affect combustion are undamaged and replace as necessary.

   A high combustion performance ratio (CO/CO_2) reading might be due to damage or the ageing of components of the appliance. If this is the case then the relevant components should be replaced and the combustion test repeated. However, it should also be noted that certain components will contain volatile compounds for which it will be necessary to 'burn off the newness' before reliable measurements can be obtained. Examples include a gas-fire ceramic fibre fuel effect, a cooker grill fret, insulation and adhesives. If such a component is fitted as a replacement, it will also be necessary to burn off the newness before a reliable measurement can be obtained.

3. Inspect and, if necessary, clean the injectors, venturis, burners, lint guards, air path to the combustion chamber and any other items recommended by the appliance manufacturer's instructions.

4. Inspect and clean any pilots/injectors.

5. Inspect and, if necessary, clean the heat exchanger, fluehood and flueways, ensuring any baffles etc. are correctly positioned.

6. Ensure that any seals and fastenings are present, in good condition and secured in accordance with the manufacturer's instructions.

7. Reinstate gas and electricity supplies.

8. Ensure the flame picture is satisfactory.

9. Decide if it is possible to carry out a combustion performance test in accordance with BS 7967-2.

Where it is possible to carry out a combustion performance test, if the test result conforms to the combustion performance ratios allowed in BS 7967-3, perform the final checks listed under 'Final checks on a suspect gas appliance' below.

If the CO/CO_2 ratio result is above the limit allowed then recheck steps 1 to 8 above. If the combustion ratio is still too high then seek help from the appropriate technical support, for example from the appliance manufacturer or Gas Safe.

When a new part has been fitted, turn the appliance to full rate and take a reading after 10 mins. If the reading is not correct or still rising, take readings at 20-min intervals until it stabilises at an unacceptable level. If the level does not fall within 20 mins, check the appliance again to determine the cause of the high reading.

Where it is not possible to carry out a combustion performance test, perform the final checks listed under 'Final checks on a suspect gas appliance' as follows.

Final checks on a suspect gas appliance

The following final checks should be performed before confirming that the gas appliance is no longer suspect or dangerous to use.

1. Check all disturbed gas connections. Test for gas tightness where a gas escape was previously identified/repaired.

2. Confirm that the burner operating pressure and/or gas rate is correct.

3. Where possible, carry out a final combustion performance test in accordance with the methods in BS 7967-2 to confirm that the combustion performance is within the limits allowed in BS 7967-3:2005, Clause 5.

4. Check the operation of any flame supervision device using the manufacturer's approved method.

5. Ensure all seals or fastenings are present, in good condition and secured in accordance with the manufacturer's instructions. Be especially vigilant with room-sealed positive fan pressure appliances.

    Refer to the HSE's published test method for this type of appliance, Industry guidance for the checking of case seals and the general integrity of room-sealed fan assisted positive pressure gas appliances.

6. Where appropriate for the appliance design, carry out a spillage test.

7. Ensure any warning labels necessary to ensure safe use of the appliance are present and correct.

Further information on the testing of individual appliance types can be found in BS 7967-1 Clause 7.

Be sure to fill out an investigation report (see Figure 4.13), give the customer a copy and explain the information on it.

Appliance investigation reports

BS 7967 Investigation report	Job ref no:	
Field	**Response**	
Appliance type		
Gas family type		
Location		
Make		
Model		
Serial number		
Flue type	O/F / R/S / Flueless	
Visual condition of the flue and termination satisfactory	Yes / No / Not applicable	
Is flue to current standard?	Yes / No / Not applicable	
Flue flow test	Pass / Fail / Not applicable	
Spillage/leakage test	Pass / Fail / Not applicable	
Weather conditions during test		
Condition of appliance on visual inspection	Good / Poor / Sooty / Other If other state which:	
Combustion test readings (BS 7967-2) CO (ppm) CO_2 (%) CO/CO_2 ratio		
CO measured in atmosphere (BS 7967-2) Initially (%) After appropriate BS 7967-2 test (ppm)		
Ventilation satisfactory	Yes / No / Not applicable	
Burner operating pressure (mbar) and gas rate		
Appliance satisfactory after detailed investigation	Yes / No	
List the faults identified		
Note of rectification work required		
Is the appliance safe to use?	Yes / No	
If appliance not checked, state why		

Figure 4.13 Example format of a gas appliance investigation report

Additional investigation considerations

There are some additional investigation considerations you should take into account, including movement of CO and the generation of CO, smells and fumes.

Movement of CO

When conducting a CO investigation, always bear in mind the possibility of CO being carried around the dwelling via rising warm air or combustion products. CO could be present in rooms containing no fuel-burning appliances or flue systems, especially in upper storeys, as a result of tracking through pipe ducts, suspended wooden floors or false chimney breasts.

Be alert to the possibility of CO entering the property from adjacent properties or re-entering from flue outlets, even where the terminal positions conform to appropriate standards. Consider the potential for different weather conditions to adversely affect the flue and/or appliance performance. Extended flues and/or special flue terminals on adjacent properties can be an indication of a 'local' problem.

If in doubt about the spillage or leakage of CO or CO_2 from flued appliances, seek additional technical support, for example from the appliance manufacturer or Gas Safe.

Where it is not practicable to correct all situations that are not to current standards (NCS), it is essential to take great care in exercising a judgement to reinstate the installation. In these cases the advice given in the Gas Industry Unsafe Situations Procedure should be followed.

Generation of CO, smells and fumes

Where no faults have been found with the gas installation, consider the following possibilities.

CO could also be generated by:

- smoking
- other fuel-burning appliances in the dwelling
- vehicles or generators in attached buildings/dwellings
- engines on boats.

Other potential causes of smells and fumes could be from:

- gas escapes on appliances
- poorly cured fibreglass log effects
- paint smells, particularly on new appliances
- dust on appliance surfaces
- recent cavity wall insulation
- recent painting and decorating activity
- use of solvents and adhesives
- damp proofing or timber treatment
- from outside the dwelling, such as barbecues or bonfires
- drains.

Investigation reports

Whenever you have completed a CO investigation, you should leave a report that provides sufficient detail to enable the customer to understand the nature of the tests carried out, the results, the status of all relevant appliances and any remedial action required.

It is recommended that a standard reporting format (as shown in the example forms on pages 100 and 104) is used to ensure that all necessary information is conveyed to the appropriate persons. If the customer will not authorise the investigation of a suspect appliance, this should be documented.

Where other sources of CO are suspected, you should ensure that the customer and, where possible, any other persons potentially affected (e.g. neighbours if it is suspected that the source of CO originates from a neighbouring property) are aware of such concerns and advise them to ensure that appropriate follow-up action is taken. All such advice should be documented.

Outstanding work or recommendations for additional items such as CO alarms, servicing arrangements and advice on the correct use of appliances should also be documented and left with the customer. It is important to leave a written copy of this information and advice with the customer as not all of them will be in a position to adequately advise the appropriate people, such as other members of the household or a landlord, with regard to important safety information if this has only been conveyed orally.

Completion and leaving the property

Where an unsafe situation is found, you should make every attempt to leave the appliance working safely.

Be sure to correct the defect or instigate the Gas Industry Unsafe Situations Procedure when any defect has been noted that could affect:

- the effectiveness of the flue
- the supply of combustion air
- the operating pressure or heat input of the appliance
- the safe functioning of the appliance.

If hazardous levels of CO remain in the property that are not attributable to gas appliances, advise that the property remains evacuated until an appropriate authority, such as the local Environmental Health Department, advises otherwise.

Where this guide has been followed in full and no faults are found which could have resulted in the occurrence as reported by the customer, the installation can be reinstated. However, where no fault has been found on this and a previous occasion, reinstate the appliance and recommend to the customer further investigation and long-term monitoring. In both circumstances advise the customer that while no sources of CO have been identified, if they continue to feel unwell it is recommended that they consult their doctor and ask if the doctor thinks that it would be appropriate for them to have a blood test in order to establish whether they have been exposed to CO.

TYPES OF GAS BURNER

There is a wide variety of burners used for domestic gas appliances, such as pre-mix, forced-draught, radiant, and simplex and duplex burners.

Burners can be further classed as:

- atmospheric – natural draught domestic burners
- forced-draught – a growing number of domestic burners operate on this principle, using a air/gas ratio valve.

Progress check

1 What are the visual checks to find problems with a gas appliance/installation?

2 Once activated, the CO alarm shall remain in operation at CO concentrations above what level?

3 How can CO fumes be carried around a house?

4 When CO has been detected, advise people not to enter the building until the CO has dropped below what level?

5 If an unsafe appliance is found, what steps should you take?

Most burners work on the principle of the 'pre-aerated' flame, as shown in Figure 4.14. This is identical to a Bunsen burner, as it has the primary air port open.

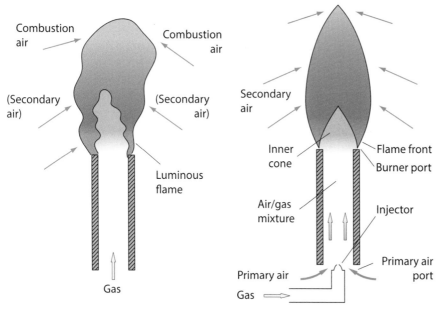

Figure 4.14 Post-aerated flame and pre-aerated flame

Case study

A CO alarm was regularly going off at a house during the winter. It would always sound in the morning. On investigation it was discovered that, first thing in the morning, the owner would run his car engine to warm it up and to defrost the windows. He would leave the garage door open and, as this was attached to the house, the fumes would migrate into the house triggering the alarm.

What other factors do you think could trigger an alarm from outside?

A pre-aerated flame has four zones. The first zone is made up of pre-mixed air and gas – this area is called the **inner cone**. The second zone is created by the speed of the gas mix being equal to the flame speed and is called the **flame front**. The third zone is known as the **reaction zone** where most of the gas is reacted with to release its heat energy. The fourth zone is called the **outer mantle** where complete combustion occurs after air surrounding the flame is used up to complete the reaction.

Figure 4.14 also shows a post-aerated flame, which has a yellow ragged flame. It is searching for oxygen and in doing so becomes elongated and is therefore unsuitable for burners.

The process is shown in Figures 4.15 and 4.16 and is outlined below:

- A to B is the burner and burner port. Inside this opening the temperature of the mixture continues to rise as the burner heats up.
- B to C is the flame front. This is where the mixture temperature rises most quickly. Because air is drawn in the average mixture strength falls. There is some unburnt gas inside the cone.

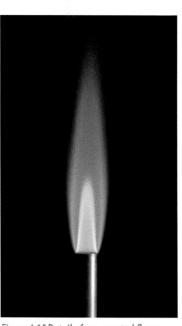

Figure 4.15 Detail of pre-aerated flame

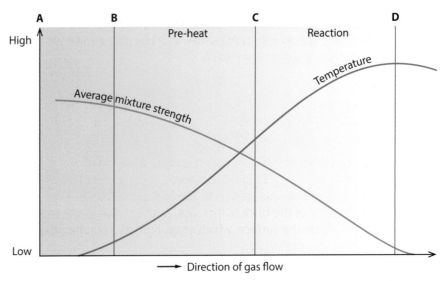

Figure 4.16 Air/gas mixture and temperature through the four flame zones of a pre-aerated flame

■ C to D is the reaction zone. Here the temperature continues to rise and the drawn-in air continues to cause the mixture strength to fall.

■ D is the outer mantle. This is where the combustion is completed and is the hottest part of the flame.

Atmospheric burners

Pre-mix burners

A pre-mix burner is a burner that mixes the gases inside the burner before ignition. The gas needed to provide the heat input is mixed with more air than is required (i.e. 10–15 per cent) to burn it completely.

The air and gas are mixed together before passing into the burner. This can be before or after the centrifugal fan, depending on the boiler design. If it is before the fan, the air/gas mix will pass through the fan unit itself and is therefore in a combustible condition.

The head of the burner (as shown in Figure 4.17) is made with perforated metal and it is designed to fit into the appropriate shape inside the boiler.

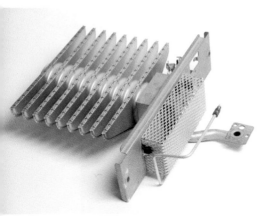

Figure 4.17 A typical burner assembly

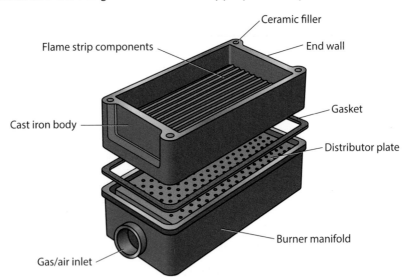

Figure 4.18 Fully pre-mix ribbon burner

Flame chilling

A flame that is vitiated becomes longer and this can cause the flame to touch the combustion chamber surface resulting in flame chilling. Incomplete combustion will result from the flame chilling as it is cooled below its complete reaction temperature. When flames touch each other or a cold surface, they starve each other of oxygen leading to CO production.

Light back

Light back is caused when the speed of the gas/air mixture is reduced by partial blockage or incorrect settings. Light back is very dangerous. You must adjust the gas pressure and remove any blockages to reduce light back and the risk of explosion at the meter. For more information see the section on 'lighting back' (page 112).

Safe working

Vitiated air is also dangerous to humans due to the lack of suitable levels of oxygen. Always keep rooms well ventilated if you are working in vitiated air.

Knowledge check

1 The specific gravity of natural gas is

a 1.0
b 3.0
c 0.6
d 0.8

2 What is the term for the heat energy that passes into the heat exchanger?

a Net heat input or sensible heat
b Gross heat output
c Latent heat
d Condensing heat input

3 What is vitiation?

a Reduced oxygen levels in a room
b Excess oxygen in a room
c Ignition problems
d Vent too large

4 What is the approximate amount of oxygen available in air?

a 40 per cent
b 20 per cent
c 50 per cent
d 10 per cent

5 What does the retention flame do?

a Relight the main flame as it tries to lift off
b Turn off the flame
c Relight the pilot flame
d Relight the gas fire

6 Energy convertors changing potential energy to kinetic energy are known as?

a Injectors
b Air/gas ratio valves
c Transformers
d Venturi

7 Aldehydes are formed by partial oxidation of

a Primary alcohols
b Secondary carbons
c Zealites
d Carbon monoxide

8 Chilling occurs when a flame

a Touches a hot surface
b Touches a cold surface
c Is turned off
d Is turned on

9 In pre-aerated burners, flames and inner cones must be

a Bigger than the outer mantle
b Smaller than the port
c Of the correct length to ensure complete combustion
d Yellow in colour

10 Burners should be cleaned regularly to prevent

a Maintenance being needed
b The burner functioning correctly
c Combustion occurring
d Lint build-up

11 A methane molecule consists of

a 1 carbon and 4 hydrogen atoms
b 3 carbon and 2 hydrogen atoms
c 1 carbon and 2 methane atoms
d 5 silicon and 3 nitrogen atoms

12 The Wobbe number for natural gas is between

a 42.2 and 43.2 (metric)
b 48.2 and 53.2 (metric)
c 28.2 and 59.2 (metric)
d 1.2 and 3.2 (metric)

13 The percentage of nitrogen in the air is taken to be about

a 20 per cent
b 41 per cent
c 89 per cent
d 79 per cent

14 The visual checks to find problems on appliances are

a Signs of spillage, combustion problems, incorrect ventilation
b Incorrect ventilation, signs of spillage, incorrect burner pressure
c Signs of spillage, gas rate, flame speed
d CO/CO_2 ratio, gas rate, input pressure

Pipework, fittings, purging and testing

INTRODUCTION

This chapter covers the installation requirements of pipework and fittings, and their appropriate materials. It also details how to correctly size your installation so that there is sufficient gas feeding each type of appliance, ensuring they work efficiently and safely.

Instructions and guidance on purging the installation of air so that gas reaches all parts of the system and testing the system for tightness are also featured.

This chapter covers some essential skills and knowledge that all gas operatives should learn and understand to ensure all installation work is carried out correctly and in accordance with the current standards.

REGULATIONS AND STANDARDS ON PIPEWORK AND FITTINGS

The Gas Safety (Installation and Use) Regulations (GSIUR) require you to ensure that gas installation pipework and fittings are installed so that they are:

- properly supported or placed so as to avoid any undue risk of damage to the fitting
- installed safely with regard to other services, drains, sewers, cables, conduits and electrical appliances (switches)
- installed so that foreign matter does not block or otherwise interfere with the safe operation of the fitting unless suitable protection is fitted
- installed so that the structure of the premises in which they are installed does not affect use.

Where your work requires you to install pipework to a primary meter without electrical cross-bonding, you need to inform the responsible person that cross-bonding may be required and the installation should be carried out by a competent person.

This chapter covers the requirements of British Standards (BS) 6891 (Installation of low pressure gas pipework of up to 35 mm (R1¼) in domestic premises (2nd family gas) – Specification) and the Institute of Gas Engineers (IGE) Utilisation Procedures (UP)1B (Tightness testing and direct purging of Natural Gas installation).

SUITABILITY OF PIPEWORK MATERIALS

When you consider the selection of the type of material you are using for a gas installation you have to take into consideration its strength, appearance, the location in which it is going to be installed, and the cost of the material. Remember that it will also need protection against corrosion depending on its installation location.

Most new domestic installation use copper tube and fittings but you may still come across steel pipe installations in older properties.

Type of material	British or European Standard
Copper pipes	BS EN 1057
Copper capillary fittings	BS EN 1254-1
Copper compression fittings	BS EN 1254-2
Steel pipes	BS 1387
Malleable iron fittings	BS 143 and BS 1256
Rigid stainless steel	BS EN 10216-5, BS 3605, BS EN 10312
Corrugated stainless steel	BS 7838
Polyethylene pipe	BS EN 1555
Polyethylene fittings	BS 5114 or BS EN 1555-3
Flexible hoses	BS 669-1 & 2
Ball valves	BS EN 331
Solder	BS EN 29453
Jointing compound	BS EN 751-2
PTFE tape	BS EN 751-3

Table 5.01 Suitability of pipework materials and fittings

Copper tube

Copper tube to BS EN 1057 is used for both gas and water supplies, and comes in a range of sizes from 6 mm and (for the installations covered in this book) 35 mm. The fittings used are manufactured to BS EN 1254 for both capillary and compression fittings. You should keep the use of fittings to a minimum on copper tube and, where aesthetically and practicably acceptable, bends should be used in preference to elbows.

Compression joints

Compression joints should only be used where they will be readily accessible, allowing the nut to be tightened to make a gas-tight joint. You should square cut and deburr the ends of any pipe to be joined by a compression fitting. Pipes under floor or in shafts, channels, ducts or **voids** without removable covers are not considered to be readily accessible. The use of jointing compounds is allowable for compression fittings. It should comply with BS EN 751 and be a non-hardening compound (gas jointing paste). Where PTFE jointing is used it should comply with BS EN 751-3 (gas approved thread tape).

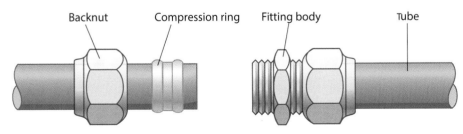

Backnut   Compression ring   Fitting body   Tube

Figure 5.01 Jointing of copper using compression fitting

Soldered joints

Soldered joints are the most common and most reliable type of joints used. The joints rely on **capillary action** between the outer wall of the pipe and

Key terms

PTFE
PTFE stands for polytetrafluoroethylene and is a synthetic fluoropolymer of tetrafluoroethylene that is used for numerous applications, such as on threads for connections to gas fires.

Void
An enclosed space through which an installation pipe may be run.

Safe working

BS 6891:2005 7.6 states that PTFE and jointing compound should not be used in conjunction with each other.

Key term

Capillary action
Capillary action or capillarity, is the ability of a liquid to flow in narrow spaces (e.g. a thin tube) without the assistance of, and in opposition to, gravity.

the inner wall of the capillary fitting. It occurs by two forces coming together, cohesion and adhesion, and will only occur when sufficient heat has been applied to both surfaces.

There are two types of capillary joints:

- end feed, to which you need to add solder
- solder ring fittings, where the solder is already inside the fitting.

To make a joint:

- square and deburr both ends
- clean both surfaces using cleaning mats – do not use steel wool
- apply flux to both surfaces – flux prevents the joint from oxidising and breaks down surface tension on the surfaces to assist the flow of the solder
- if using end feed fittings you will also need to introduce solder to the fitting – do not overload with the solder
- after soldering the joint, clean off excess solder
- remove any excess flux from the joint – failure to remove the flux could cause corrosion to the pipework.

Not all self-cleaning fluxes are ok to use with gas installation pipework, check with manufacturer's data sheet before you apply it.

When you've completed the joint, turn off the blow lamp and place it in a safe position to avoid damaging the surrounding area by scorching or burning. Check the joint visually. Do not apply any protective coverings until the pipework has been tightness tested.

Press-fit fittings

Press-fit fittings can be used on gas installation pipework. The joint for these are made by compressing a special fitting onto the pipe. You should only use the press-fit fittings designed purposefully for the installation of gas pipework. These are normally identified by the colour of the sealing ring inside the fitting. The example below shows a Pegler ®Xpress fitting which has been fitted by a hydraulic press-fit machine (see Figure 5.03). The jaws are designed to fit specific manufacturer's fittings.

Safe working

Apply flux to pipework and fittings using a brush as it is corrosive.

Figure 5.02 Press-fit fitting

Figure 5.03 Press-fit tool

How to assemble a press-fit joint

- Cut the pipe to length using a pipe cutter. Do not use a hacksaw as these produce a rough end on the pipe which can damage the 'O' ring.

- Ream out the pipe to ensure that there is no restriction to the flow of gas.

- Insert to full depth of fitting and mark pipe, do not press fit until absolutely sure that measurements and positioning are correct.

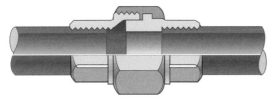

Figure 5.04 Union connector

Steel pipework

Malleable iron fittings should conform to BS 143 and BS 1256. Steel pipework is no longer used as much on domestic properties nowadays as it was during the 1970s and 1980s, but you will still come across it in older properties. It is still an acceptable material for the supply of gas in domestic dwellings, as the material is robust and can stand up to many years of wear and tear. However, where it has been installed in older properties within screeds and without protection, the pipework will corrode and leak.

Jointing

Types of joints are screwed, ground face unions or longscrews. The joints are threaded using either a hand threading stocks and dies or a pipe threading machine. The threading machine in Figure 5.05 is a portable pipe threading machine.

Figure 5.05 Electric pipe threading machine

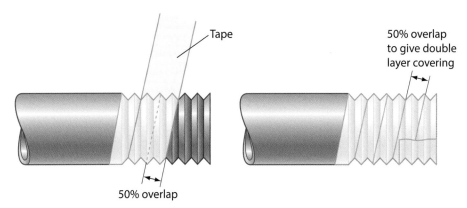

Tape

50% overlap to give double layer covering

50% overlap

Figure 5.06 Thread wrapping method

After threading, remove all cutting oil/compound from the thread and inside the pipe before you apply jointing compound. This is normally done using either jointing compound or PTFE tape. Apply the compound or tape to the male thread of the pipe and wind with a 50 per cent overlap in a direction counter to the thread form. Then tighten the pipe into the fittings using pipe wrenches. The PTFE used should be suitable for gas installations.

Do not use hemp on any threaded joint except in conjunction with thread sealing compounds for long screw back-nut seals.

Corrugated stainless steel

Corrugated stainless steel to BS 7838

This type of pipe is now being used for gas installations the following extracts from the 'Tracpipe' brochure should help you to understand correct methods of jointing.

CUT-TO-LENGTH: Determine proper length. Cut through plastic cover and stainless steel pipe using a tube cutter with a sharp wheel. Cut must be centred between two corrugations. Use full circular strokes in one direction and tighten roller pressure slightly (a quarter turn) after each revolution. DO NOT OVERTIGHTEN ROLLER, which may flatten pipe.

Figure 5.07 Tracpipe joint

Figure 5.08 Gas PTFE tape (onewrap)

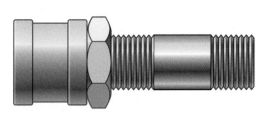

Figure 5.09 A long screw

STRIP COVER: Using a utility knife, strip back the plastic cover about 25 mm from the cut end to allow assembly of fittings.

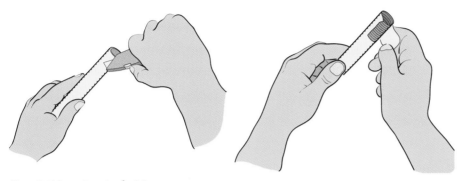

Figure 5.10 Preparing pipe for joint

INSTALL BACK NUT: Slide the back nut over the cut end; place the two split rings into the first corrugation next to the pipe cut. Slide the back nut forward to trap the rings.

FIT AutoFlare® FITTING: Place the AutoFlare® fitting into the back nut and engage threads. Note that the AutoFlare® fitting is designed to form a leak-tight seal on the stainless piping as you tighten the fitting. (The piloting feature of the insert will not always enter the bore of the piping before the tightening operation, but will centre the fitting when tightened.)

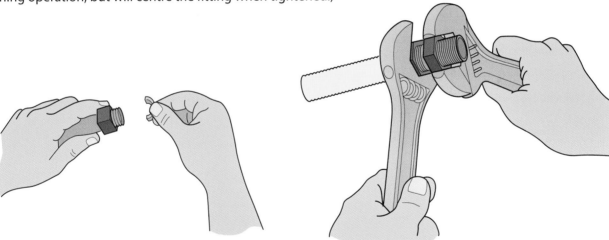

Figure 5.11 Joint preparation

Figure 5.12 Tightening Tracpipe joint

On completion of the joint test the installation for tightness then inspect the joint to check that no stainless steel is visible. Where the stainless steel is visible behind the nut of the fitting, wrap it with self-bonding silicone tape.

Fittings

Threaded steel fittings and copper alloy fittings may be used in conjunction with corrugated stainless steel pipe. Jointing paste and PTFE may be used on the male thread of the joints in accordance with BS 6891. Under no circumstances must jointing compound be used on the metal-to-metal seal between the pipe end and the fitting, as this could impair the seal of the fitting.

Plastics

Polyethylene (PE) pipes and fittings are used for exterior work only. This type of installation is a specialist area and should not be attempted by anyone without the expertise. Specialised courses and qualifications are available for this area of work.

Polyethylene pipe and fittings should only be used to extend underground polyethylene installation pipe above ground level for entry into a building. The polyethylene pipework above ground level should rise vertically to the point of entry into the building, which should be as close as practicable to the external ground level. The length of pipework above ground level should be protected against daylight and mechanical damage.

Do not use solvent weld joints and only use fusion and electrofusion welding if you are trained and competent to do so.

It is preferable to use metal gas pipe at entries into domestic buildings. The GSIUR require that where polyethylene pipe is used to enter the building, the part within the building is placed within a metal sheath, which is constructed and installed so as to prevent, as far as practicable, gas escaping into the building if the pipe should fail.

Lead (lead composition) pipework

When you encounter lead composition pipes that need connection joints, only use a soldered cup-joint onto copper pipe or a suitable brass union fitting. Compression fittings designed for jointing water-weight lead pipework should not be used, for example 'lead loc'.

You must take extra care when soldering a joint onto lead composition pipework as blow lamps might provide too much heat at the joint. Only use lead solder for this purpose and ensure joints are mechanically strong and gas tight.

Safe working

Lead is poisonous in larger quantities and can be absorbed through the skin. Apply barrier cream before working with lead and lead composite pipe to protect yourself.

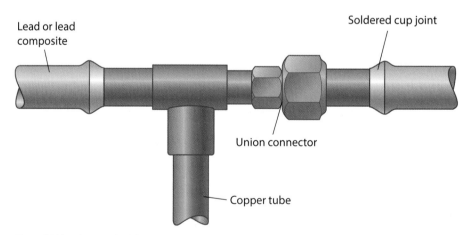

Figure 5.13 Lead composite joint

JOINTING AND SUPPORTS FOR PIPEWORK

Pipes must be adequately supported and correctly jointed for gas installations to function properly and safely.

Jointing

There are special requirements you should take into account when working on pipework jointing for gas:

- any flux must remain active during the heating process only as it cannot be flushed out of the system
- no flux should be allowed to come into contact with stainless steel
- compression fittings should only be used where they are readily accessible, not under floors, in ducts etc.
- push-fit and quick-release fittings are not normally used for gas installations
- union joints for steel pipe should be sited in accessible locations
- hemp should not be used on threaded joints
- jointing pastes should not be used with PTFE tape.

Pipework supports

Pipework should be supported at correct intervals according to its material and size – see Table 5.02. Pipe clips or supports must be of a type not likely to cause corrosion.

Material	Nominal size	Interval for vertical support	Interval for horizontal support
Mild steel	Up to DN 15 DN 20 DN 25	2.5 m 3.0 m 3.0 m	2.0 m 2.5 m 2.5 m
Copper tube	Up to 15 mm 22 mm 28 mm	2.0 m 2.5 m 2.5 m	1.5 m 2.0 m 2.0 m
Corrugated stainless steel	DN 10 DN 12 DN 15 DN 22 DN 28	0.6 m	0.5 m

Table 5.02 Maximum intervals between pipe supports

Copper tube may be used for external applications providing it is fully supported along its route and identified as conveying gas with marking tape.

Where pipework is installed in a premises not used for domestic purposes then the pipework must be colour-coded yellow ochre or marked with identification tape. The tape in Figure 5.14 would satisfy this requirement.

PIPEWORK INSTALLATION REQUIREMENTS

There are special considerations for pipework laid below floors, in solid floors and in walls.

Figure 5.14 Identification tape

Progress check

1 Which British Standard covers the 'Installation of low pressure gas pipework of up to 35 mm (R1¼) in domestic premises'?

2 What is the British/European Standard of copper pipe that gas installations are installed to?

3 When can compression joints be used?

4 When jointing steel pipework and fittings, what other jointing material should not be used with PTFE?

5 When can polyethylene (PE) pipe be used above ground?

Pipes laid in joisted floors

Where pipes are installed between solid timber joists in floors, intermediate floors or roof spaces, you must ensure they are correctly supported (see Table 5.02).

Where pipes are installed between timber engineered (I) joists, the pipes should be installed through the web of joists in accordance with Table 5.02 and the joist manufacturer's guidance.

When installing pipes between metal web joists, be sure to pass the pipes between the metal webs with pipe supports fixed to the top or bottom of the timber flanges. The flanges of timber engineered joists and metal web joists should not be notched.

Where pipes are laid across solid timber joists fitted with flooring, they should be located in purpose-made notches or circular holes.

Do not notch joists of less than 100 mm and make sure all notches are made in accordance with Figure 5.15 and Figure 5.16.

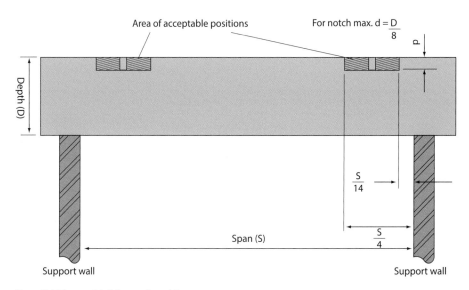

Figure 5.15 Acceptable joist notch positions

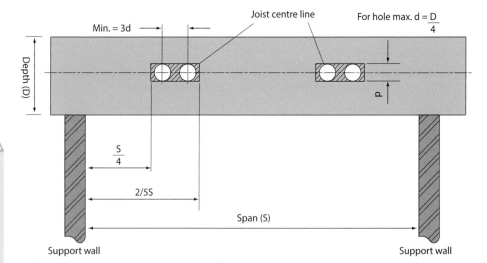

Figure 5.16 Acceptable positions for holes through joists

Safe working

Before running pipework below suspended floors, carry out a visual inspection to note the position of any electrical cables, junction boxes and ancillary equipment, in order to avoid accidental damage or injury when inserting pipework.

Care should be taken when re-fixing flooring to prevent damage to the pipes by nails or screws. Where possible, the flooring should be appropriately marked with the position of the pipework to warn others. Where possible the pipework layout should remove the need for notching solid timber joists.

Pipes laid in solid floors

Pipes must be protected against damage and corrosion to meet the requirements of BS 6891. Figures 5.17, 5.18, 5.19 and 5.20 show acceptable methods of locating pipework in various types of solid floor.

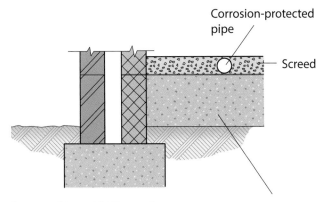

Figure 5.17 Pipework laid in screed

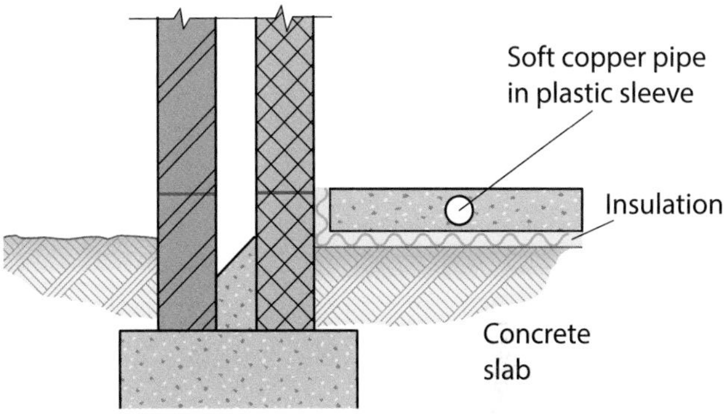

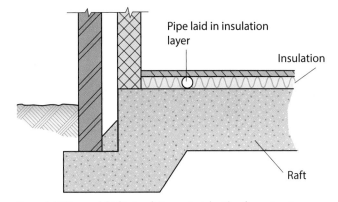

Figure 5.18 Pipework laid in screed with insulating material

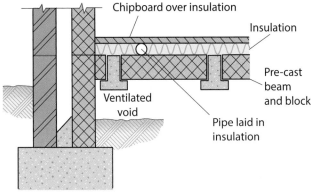

Figure 5.19 Pipework laid in chipboard with pre-cast block and beam

Figure 5.20 Pipework laid in insulation material with raft construction

When laying pipes in solid floors the following must be taken into account:

- keep joints to an absolute minimum
- do not use compression fittings
- use only acceptable pipework protection methods:
  — factory-sheathing and larger-diameter plastic sleeving previously set into the concrete (no joints are to be located in the plastic sleeving) for soft copper pipes
  — factory-sheathing (plastic coated) or appropriate on-site corrosion-resistant wrapping material for pipes laid on top of the base concrete and in the screed
  — pre-formed **ducts** with protective covers
  — additional soft covering material at least 5 mm thick and resistant to the ingress of corrosive materials such as concrete.

> **Key term**
>
> **Duct**
> A purpose-designed enclosure to contain gas pipes.

Protection against corrosion

It is essential that you ensure all pipework is protected from corrosion. Factory-finished protection is preferred, for example the use of plastic coating where pipework is to be routed through corrosive environments. It acceptable to wrap the pipes in protective tape but the pipes must be tested before doing so. You should use stand-off clips to avoid contact with wall surfaces. Remember that soot is very corrosive so pipes in fireplace openings must be suitably protected.

Pipes laid in walls

You should take into account the following considerations when siting pipework within wall surfaces:

- keep pipes vertical that are to be covered in plaster (see Figures 5.21, 5.22)
- provide ducts/access wherever possible

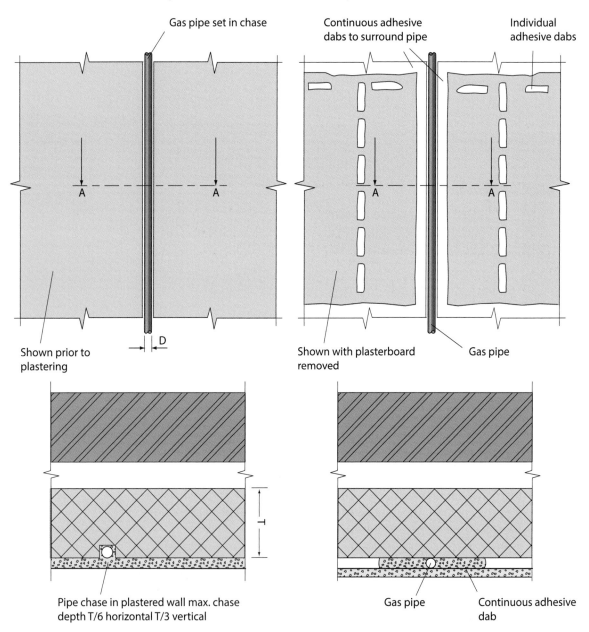

Figure 5.21 Pipes in walls

Figures 5.30 and 5.31 show examples of possible pipework layouts and the necessary protection measures.

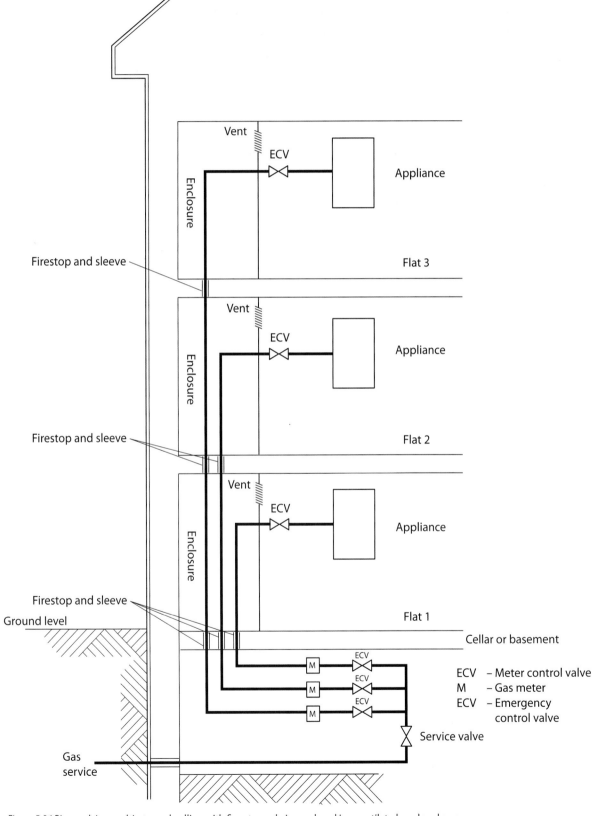

Figure 5.31 Pipework in a multi-storey dwelling with fire-stopped pipework and in a ventilated, enclosed area

Micro point/leisure points

Special fittings such as micro points (also known as 'leisure points' – see example in Figure 5.25) may be used when working on gas to provide a neat finish to the ends of a gas carcass installation. Some even have their own shut-off valve. Small-diameter rigid pipes are then used to connect to appliances, although flexible hoses can be used for cookers or refrigerators.

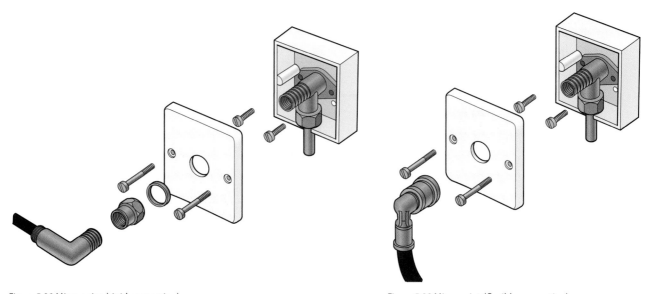

Figure 5.32 Micro point (rigid connection)

Figure 5.33 Micro point (flexible connection)

Uncompleted work

Part B. Regulation 6 (2) of the GSIUR states:

No person carrying out work in relation to a gas fitting shall leave the fitting unattended unless every incomplete gasway has been sealed.

When you need to leave an uncompleted work installation unattended, irrespective of how long it is going to be left for, you need to ensure the installation is made safe before you leave. You must therefore ensure every gas pipe is sealed by capping it off using the appropriate fitting. The type of fitting you decide to use is important; you can use either compression fittings or capillary fittings. Remember that where pipework is not visible (e.g. under floors) then compression fittings must not be used.

Connection and disconnection of pipes and fittings

Where any installation pipe is no longer required, you should disconnect it as close to the point of supply as practicable. All pipe ends should be sealed with an appropriate fitting, e.g. a plug or cap. Where metal installation pipework needs connecting or disconnecting, fix a temporary continuity bond where production of a spark or shock could cause a hazard, whether or not permanent equipotential bonding has been established.

- do not raise to high pressures as this may cause the meter regulator to lock up
- allow one minute for temperature stabilisation
- if necessary, readjust the pressure to 20 mbar – if the supply control valve has been turned on to readjust the pressure then turn off the valve
- do not proceed with the test procedure until a stable reading is obtained
- take a reading and then test for a further 2 mins
- record any pressure loss after the 2 min period has elapsed.

If there is **no perceptible pressure movement/loss** and no smell of gas, then the installation has passed the test. If pressure is lost, the test has failed and the leak must be found or the installation sealed and made safe.

No perceptible pressure movement/loss:

For a fluid gauge – where the movement is 0.25 mbar or less, it is considered 'not perceptible'. Therefore, if the gauge is seen to move, it can be inferred that the pressure within the installation has altered by more than 0.25 mbar.

For an electronic gauge – where an electronic gauge can register less than 0.25 mbar the pass criteria of 'no perceptible movement' has to be considered to be a maximum of 0.25 mbar. If a gauge can read to one decimal place 'no perceptible movement' is considered a maximum of 0.2 mbar.

Testing the complete installation

Once it has been established that the pipework is not leaking then open the isolating valves to all appliances, where a cooker is fitted reconnect the bayonet fitting, and test again to ensure that the complete system is safe:

- if a cooker with a lid is installed then raise the lid to ensure that all valves are being tested
- if a pipework only test is satisfactory and all appliance isolating valves are now open, slowly raise the pressure to 20 mbar and turn off the supply
- allow 1 min for temperature stabilisation
- record any pressure loss in the next 2 mins.

Final checks

Once you have established that the pipework is not leaking:

- turn on the isolating valves to all appliances and test again to ensure that the complete system is safe
- if the pipework-only test was satisfactory and all appliance-isolating valves are now open, slowly raise the pressure to 20 mbar and turn off the supply
- again, do not raise the pressure to more than 20 mbar as this may cause the meter regulator to lock up
- allow one minute for temperature stabilisation
- record any pressure loss in the next two minutes.

Remember, if you are testing pipework only, then no pressure drop is allowed. With new installations there is no allowable pressure drop for new appliances.

Figure 5.49 Manometer at 20 mbar

Figure 5.50 Medium pressure regulator/ release mechanism

For new installations incorporating existing appliances then the pressure drop detailed in Table 5.09 is allowed providing that it can be proven that the loss is *not* on any part of the new installation.

For existing installations with appliances fitted:

* with a U6/G4 meter – pressure loss of up to 4 mbar is acceptable, provided there is no smell of gas
* with an E6 meter – pressure loss of up to 8 mbar is acceptable, provided there is no smell of gas
* where no meter is fitted in the dwelling, such as a flat supplied by a communal meter, pressure loss of up to 8 mbar is allowed.

If the test has failed then either:

* trace and repair the escape(s) and re-test the installation; or
* make safe by disconnecting appliances(s) or the relevant section of the installation, as appropriate, and seal all open ends with an appropriate fitting.

Where the test is successful:

* remove the pressure gauge and re-seal the test point
* slowly turn on the gas supply
* test the pressure test point with LDF
* test the ECV/AECV outlet connection with LDF
* test the regulator connections and, where appropriate, the meter inlet valve connections with LDF.

When you've completed the test, if the installation is new, or if any gasways have been exposed to air, or the work may have allowed air into the installation by any other means, purge the installation following the procedure on page 156. If the installation is existing and there is no possibility that air may have been allowed into the installation, record the test results and, where appropriate, inform the responsible person.

Medium-pressure fed installations

There may be occasions when operatives will come across a medium pressure-fed installation. The configuration for gas control and metering connections is shown in Figure 5.51. If the installation has a meter test valve fitted, then the test procedure is exactly the same as that already covered. However, if there is no test valve included then a different procedure needs to be followed as you need to check both the ECV and the medium- to low-pressure regulator.

This is best done in three stages: test the ECV, test the regulator, test for tightness.

Upstream   Downstream

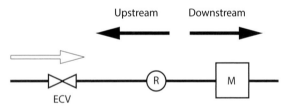

ECV

Low pressure-fed gas meter installation

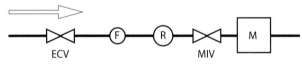

ECV   MIV

Medium pressure-fed meter installation with a meter inlet valve fitted

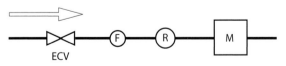

ECV

Medium pressure-fed meter installation without a test valve fitted

ECV – Emergency control valve
R – Regulator
M – Meter
F – Filter
MIV – Meter inlet valve (Test valve)

Figure 5.51 Low- and medium-pressure fed installations

Note: where an appliance has *not* been commissioned then it must be either be:

- disconnected from the gas supply or sealed with an appropriate fitting, e.g. a cap end or iron plug; an appropriate label should be attached which indicates that the appliance has not been commissioned
- commissioned.

For larger installations with a volume up to 1 m³ and with a diameter up to 150 mm then more information should be obtained from the Institution of Gas Engineers publication IGE/UP/1A.

RE-ESTABLISHING GAS SUPPLIES

Having installed, tested and purged each part of the system, you must now put appliances back into use, and once again it is essential that the correct procedure is followed. Whether you are working on an existing system installed several years ago or one that you have just modified yourself, you must still be sure to make several critical checks *before* considering the installation as safe for use.

The general procedure for putting an installation into use is to:

- make a visual check of the appliances
- check there is ventilation available where required
- light each appliance in turn, checking that all user controls are working
- check all open-flued appliances for spillage.

The checks are carried out to ensure that there is no danger to the user. If an appliance proves not to be operating satisfactorily and there is a risk or danger to the user, action should be taken to shut off or disconnect it and a suitable label attached. All appliances must either be commissioned fully or be disconnected from the installation.

Knowledge check

1 Which one of the following pipework materials is unsuitable for gas pipework in a new installation?

a Lead
b Copper
c Mild steel
d Malleable iron

2 Flexible gas hoses need to comply with which British Standard for use on gas installation?

a BS 6700
b BS 3604
c BS 669
d BS 143

3 When using a water-filled manometer, where should the reading for the gas pressure be taken in relation to each leg of the manometer tube?

a One from the bottom and one from the top of the meniscus
b Both from the top of the meniscus
c Both from the bottom of the meniscus
d Both from half way down the meniscus

4 What can fluxes left on the external surface of copper pipes lead to?

a Corrosion
b Malleability
c Porosity
d Elasticity

5 When gas is supplied for use in any premises for the first time, it must not be supplied without the provision of what?

a A means of isolation within 600 mm of the meter inlet
b An appropriately sited emergency control to which there is adequate access
c An appropriately sited balancing valve to which there is adequate access
d An appropriately sited gas meter to which there is adequate access

6 An installation consists of 16 m of 28 mm diameter copper tube, 10 m of 25 mm diameter steel pipe, 30 m of 22 mm diameter copper tube and an E6 meter. What is the installation volume of this system?

a 0.012 m3
b 0.02 m3
c 0.029 m3
d 0.04 m3

7 During a gas system installation a gas fitting is suspected of being blocked – what precautions should be taken according to the regulations?

a A filter to the gas inlet must be fitted
b Under no circumstances should the gas fitting be installed
c The fitting may be installed if deemed safe by the gas fitter
d There are no special precautions

8 What is the principal reason for sleeving a gas installation pipe when passing through a cavity wall?

a To encourage staining of the wall
b To prevent gas escaping to the air
c To facilitate removal of the pipe at a later date without damage to surrounding brickwork
d To prevent gas escaping into the cavity

9 A 'let by' test should be carried out at what pressure?

a 6 mbar
b 8 mbar
c 10 mbar
d 20 mbar

10 Steel pipes and copper tubes used for gas installations in domestic premises must conform to which of the following regulations?

a BS 12
b BS 1387 and BS EN 1057
c BS 5750 and ISO 9001
d All of the above

Ventilation

INTRODUCTION

It is essential that anyone carrying out gas work is competent to do so, and this includes knowing how to check and calculate the ventilation factors for appliances. It is important for the correct combustion of the gas to maintain a good supply of fresh air to an appliance. This chapter deals with the supply of combustion air through air vents in different situations and with different types of appliances. It is essential that you get a good grasp of the calculations required in order to ensure the safe operation of appliances you work on.

REGULATIONS ON VENTILATION

The main gas safety regulation that applies to ventilation is Regulation 26 of the Gas Safety (Installation and Use) Regulations:

26(9) Where a person performs work on a gas appliance he shall immediately thereafter examine –

(a) the effectiveness of any flue;

(b) the supply of combustion air;

(c) its operating pressure or heat input or, where necessary, both;

(d) its operation so as to ensure its safe functioning.

It is your responsibility to check that the supply of air to appliances is adequate to ensure the correct performance of the appliances.

GENERAL PRINCIPLES OF VENTILATION CALCULATIONS

The main considerations for ventilation calculations are the type of appliance, the gas input to the appliance (calorific value – CV), the room size and use, and the free area of any vent required for the given room or space.

Net calorific value

Calorific value is the amount of heat that is released when a gas is burnt. It is measured in Megajoules per cubic metre of gas. Always check the data given with an appliance to establish the basis on which the heat input is given – this is normally on the data badge. The data in this chapter refers to heat input expressed in terms of **net calorific value** (net CV) with conversion given for natural gas gross CV quoted in brackets, where appropriate.

All heat input calculations are now based on net kW ratings, so you may need to convert the figure if the kW rating is given in gross. You can calculate the net kW rating by dividing the gross rating by 1.1.

$$\frac{\textbf{kW gross}}{\textbf{1.1}} = \textbf{kW net}$$

The ratio of gross to net heat input is approximately 1.11:1 for natural gas, 1.09:1 for propane and 1.08:1 for butane.

Adventitious ventilation

Natural ventilation through cracks in floorboards, windows and doors etc. is often called 'adventitious ventilation'. In older properties, even those with weather stripping and double glazing, there will always be the equivalent free area of 3500 mm^2 in adventitious ventilation. It is therefore assumed that a room can provide adequate ventilation via adventitious ventilation for an open-flued appliance up to 7 kW.

The free area of unmarked air vents

The free area of a vent is the space between the actual metal or plastic body of the vent. For flat **air vents** the free area is measured correctly by checking the actual width and length of the slots accurately. For raised or brick air vents, the free area is measured by checking the actual width and length of the slots accurately. Ensure all holes are measured by the smallest cross-sectional dimensions through the thickness of the vent. To measure the free area, take the width of the slots, multiply by the depth of one slot and multiply again by the number of slots.

Key term

Air vent
An air vent is a non-adjustable purpose-provided ventilation arrangement designed to allow for permanent ventilation of a room or compartment.

Example

If the width of the slots on an air vent is 75 mm and the depth is 10 mm, the free area is:

75 mm × 10 mm × 10 slots = 7500 mm^2 or 75 cm^2

Typical terracotta wall ventilator
The unobstructed fraction is about ⅓

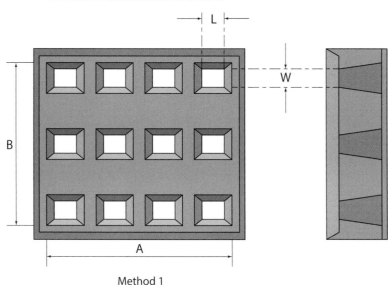

Method 1
Minimum effective area
= W x L x number of openings

Figure 6.01 Sizing of free area of a terracotta vent

Example

It is proposed that a 20 kW (gross boiler) will be ventilated by an air brick with a free area of holes measuring 8 mm × 8 mm. There are 48 holes. Is the air brick suitable?

First, convert the boiler heat input from gross to net: $\dfrac{\text{20 kW gross}}{\text{1.1}}$ **= 18 kW net.**

Free area required: **(18 kW net – 7 kW adventitious air) × 5 cm² = 55 cm².**

Actual free area of air brick: **8 mm × 8 mm × 48 holes = 3072 mm² or 30.7 cm².**

The air brick is therefore unsuitable for the boiler heat input requirement.

Typical sheet metal ventilation grille
Unobstructed fraction is about ⅔

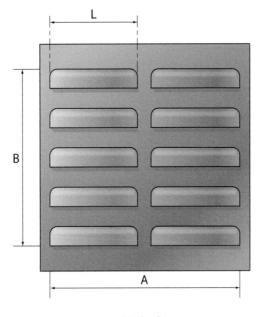
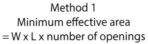
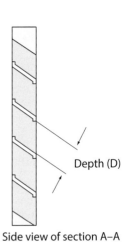

Method 1
Minimum effective area
= W x L x number of openings

Figure 6.02 Calculating free area of a metal grille

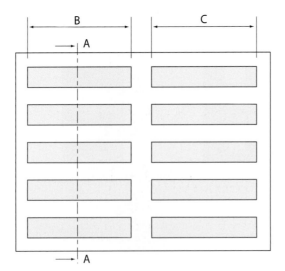

Depth (D)

Side view of section A–A

Figure 6.03 Calculating free area of a plastic type air vent

VENTILATION REQUIREMENTS FOR DIFFERENT TYPES OF GAS APPLIANCE

Air for combustion is not the only reason for ventilation in a dwelling. Ventilation is also needed to:

- remove condensation from cooking, showering, drying cloths and moisture from breathing and sweating
- remove any heat from cooking and heating equipment
- provide air to replace that lost due to extraction from fans such as cooker hoods
- provide fresh air to make the environment healthy to live in.

There are several types of gas appliances and each has differing needs for ventilation:

- **Open-flued** appliances need air for combustion, and the movement of products up the flue will cause air to be taken from the room. This air needs to be replaced for the appliance to work safely.
- **Flueless** appliances such as cookers and some water heaters require a constant supply of fresh air to prevent the air in the room from becoming vitiated, which leads to incomplete combustion (see the section on 'Vitiation' on page 118). The smaller the room, the greater the problem.
- **Room-sealed** appliances take air directly from outside so do not require ventilation unless in a compartment.
- Appliances in **compartments** take air for combustion from outside but need ventilation to cool the compartment down and stop the appliance overheating.

Calculating the ventilation requirements for open-flued appliances

Where an open-flued appliance is installed with a rated input of more than 7 kW, the room it is in must have an air vent with a free area of 5 cm^2 for every kW input in excess of 7 kW (see the section on 'Adventitious ventilation' on page 167).

Example

To work out the ventilation requirement for a natural gas boiler rated at 15 kW gross heat input, first convert the gross kW to net kW:

$$\frac{15 \text{ kW gross}}{1.1} = 13.6 \text{ kW net}$$

To calculate the size of air vent required for this boiler, subtract the 7 kW (for adventitious air) and then multiply by 5 cm^2 for each kW:

(13.6 kW – 7 kW) × 5 cm^2 = 33 cm^2

This means that the boiler requires an air vent of 33 cm^2 free area.

It is worth noting that when a range-rated boiler of, say, 11 kW to 15 kW is used, the air-vent size should be calculated on the maximum setting (i.e. 15 kW).

Air-tight rooms

Ventilation calculations have relied on taking into account adventitious air in a room, however if that room has been draught proofed it would have no adventitious air and would require a different calculation. Air-tight rooms have an air permeability of less than 5.0 $m^3/hr/m^2$. Dwellings built after 2008 have a certificate showing air-tightness and this can be used to calculate the vent size required. If an older property has been draught proofed with double glazing and sealed floors then you should take this into account and seek further help and assistance if necessary.

Calculating the ventilation requirements for flueless appliances

The Building Regulations also state that all rooms with a flueless appliance must have an opening window or similar opening direct to outside. Table 6.01 (Table 4 from British Standard (BS) 5440) shows under what circumstances permanent openings are required.

Type of appliance	Maximum appliance rated input (net)	Room volume (m³)	Permanent vent size cm³	Openable window or see note (b)
Domestic oven, hotplate, grill or any combination thereof	None	< 5 5 to 10 >10	100 50 (a) see below Nil	Yes
Instantaneous water heater	11 kW	< 5 5 to 10 10 to 20 >20	Installation not allowed 100 50 Nil	Yes
Space heater in a room	45 W/m² of heated space		100 plus 55 for every kW (net) by which the appliance rated input exceeds 2.7 kW (net)	Yes
Space heater in an **internal space**	90 W/m² of heated space		100 plus 27.5 for every kW (net) by which the appliance rated input exceeds 5.4 kW (net)	Yes
Space heaters conforming to BS EN 449:1997 in a room	45 W/m² of heated space		50 plus 27.5 for every kW (net) by which the appliance rated input exceeds 1.8 kW (net)	Yes
Space heaters conforming to BS EN 449:1997 in an internal space	90 W/m² of heated space		50 plus 13.7 for every kW (net) by which the appliance rated input exceeds 3.6 kW (net)	Yes
Refrigerator	None		Nil	No
Boiling ring	None		Nil	No
Notes (a) If the room has a door direct to outside then no permanent vent is required. (b) Alternatives include adjustable louvres, hinged panel etc. that open directly to outside.				

Table 6.01 Permanent opening requirements (Table 4 from British Standard 5440)

You should take into account the following important considerations regarding the ventilation of flueless appliances:

- the vent must communicate directly to the outside air
- the appliance must not vent from a floor space
- the appliance must not vent through a loft space
- the vent must not pass through any other room or area unless ducted
- there must be sufficient room volume where the appliance is situated
- the appliance must not exceed the maximum heat input rating
- the room in which the appliance is situated must have an openable window or other means of ventilation direct to outside
- purpose-designed ventilation may be required for appliances in small rooms.

Example

If you are fitting a gas cooker to a kitchen which is less than 5 m³, with an openable window and door to outside, and has no other appliances in the room, what size should the vent be?

Using Table 6.01, you'll see that a 'Domestic oven, hotplate, grill or any combination thereof' with a door to outside, in a room less than 5 m³ requires a vent size 100 mm².

Domestic flueless space heaters, including catalytic combustion heaters

Appliances in this category, up to a maximum heat input (net) of 50 m³ of heated space, installed in a room volume of less than 15 m³ require 25cm²/kW, with a minimum free area of 50 cm² at high and low level, and an openable window or equivalent.

Appliances in this category, up to a maximum heat input (net) of 100 m³ of heated space, installed in an internal space of less than 15 m³, require 25 cm²/kW, with a minimum of 50 cm² at high and low level, and an openable window or equivalent.

Calculating the ventilation requirements for appliances located in compartments

Room-sealed or open-flued boilers may be installed in compartments. A compartment is an enclosure designed to house a gas appliance. It could also be a small room such as a coal house, outside WC, cupboard, lobby to a hallway etc. Essentially any small room that, because of its size, may be subject to significant heat build-up is classed as a compartment. This type of room therefore needs suitable air circulation provided by high- and low-level vents to outside air or to another room in the building.

Open-flued appliances in compartments

Open-flued appliances installed in compartments require ventilation to keep cool in addition to permanent ventilation for correct combustion. Room-sealed appliances do not require permanent ventilation for combustion but do require compartment ventilation. The size of compartment vents that must be provided is shown in Table 6.02 on page 172.

Vent position	Appliance compartment ventilated	
	To room or internal space (see note*)	Direct to outside air
	cm² per kW (net) of appliance Maximum rated input	cm² per kW (net) of appliance Maximum rated input
High-level	10	5
Low-level	20	10
*A room containing an appliance compartment for an open-flued appliance will also require ventilation		

Table 6.02 Compartment vents required for an open-flued appliance

IMPORTANT
FOR YOUR SAFETY

COMPARTMENT VENTILATION

Do not allow this compartment door to remain open. keep closed at all times.

Do not block or restrict any air vents or louvres in the walls, door, floor or ceiling.

Do not use this compartment as an airing cupboard, or for the storage of combustible material or chemicals.

Do not place anything close to the appliance or its flue.

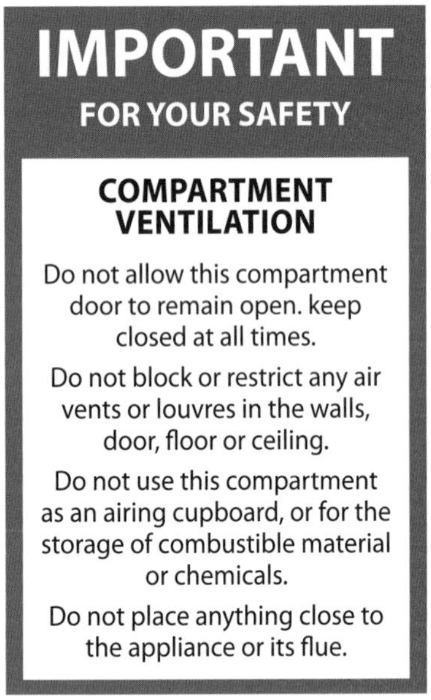

Figure 6.04 Warning label for appliance compartment

An appliance compartment with an open-flued appliance must be labelled to warn against blockage of the vents and advise against use for storage. The label should read – 'IMPORTANT – DO NOT BLOCK THIS VENT – Do not use for storage'.

Example

A 24 kW (gross) open-flued boiler is to be sited in a compartment. What is the compartment ventilation requirement if it is to be ventilated to outside air?

First, convert the boiler gross input to net input:

$$\frac{24 \text{ kW gross}}{1.1} = 21.6 \text{ kW}$$

Then, using Table 6.02, calculate the compartment vent size:

High-level: 5 cm² × 21.6 kW = 108 cm²
Low-level: 10 cm² × 21.6 kW = 216 cm²

Calculating the ventilation requirements for room-sealed appliances

To calculate the room-sealed appliance ventilation requirements, you'll need to consult the manufacturer's installation instructions. With the advances in boiler design, some room-sealed appliances do not need ventilation in compartments. If there is no guidance provided then ventilation must be provided in accordance with Table 6.03.

Vent position	Appliance compartment ventilated	
	To room or internal space	Direct to outside air
	cm² per kW (net) of appliance Maximum rated input	cm² per kW (net) of appliance Maximum rated input
High-level	10	5
Low-level	10	5

Table 6.03 Ventilation for room-sealed appliances

Example

A 24 kW (gross) room-sealed boiler is to be sited in a compartment. What is the compartment ventilation requirement if it is to be ventilated to outside air?

First, convert the boiler gross input to net input:

$$\frac{24 \text{ kW gross}}{1.1} = 21.6 \text{ kW}$$

Then, using Table 6.03, calculate the compartment vent size:

High-level: 5 cm^2 × 21.6 kW = 108 cm^2
Low-level: 5 cm^2 × 21.6 kW = 108 cm^2

Example

A 20 kW (gross) room-sealed boiler is to be sited in a compartment. What is the compartment ventilation requirement if it is to be ventilated to a room?

First convert the boiler gross input to net input:

$$\frac{20 \text{ kW gross}}{1.1} = 18.18 \text{ kW}$$

Then, using Table 6.03, calculate the compartment vent size:

High-level: 10 cm^2 × 18.18 kW = 181.8 cm^2
Low-level: 10 cm^2 × 18.18 kW = 181.8 cm^2

Where two or more appliances are installed in the same appliance compartment, whether or not they are supplied as a combined unit, the aggregate maximum rated input shall be used to determine the air vent free area, using Table 6.03 or the appliance manufacturer's instructions.

Calculating the ventilation requirements for decorative fuel-effect (DFE) space heaters

When calculating the ventilation requirements for DFE space heater, always refer to relevant section of the BS 5871-3 and/or manufacturer's instructions.

Single appliances

Normally, appliances will require a minimum of 100 cm^2 of purpose-provided ventilation. However, in the case of an appliance of not greater than 7 kW input which generates a clearance flue flow not greater than 70 m^3/h when tested in accordance with Annex A (BS 5871-3), an air vent may not be necessary.

A DFE without a throat in the flue, such as a canopy, should be treated for ventilation purposes as a solid fuel fire. You can determine the ventilation requirements for these from Table 6.04 (Table 1 Building Regulations J on page 174).

Type of appliance	Type and amount of ventilation*
Open appliance, such as an open fire with no throat, e.g. a fire under a canopy	Permanently open air vent(s) with a total equivalent of at least 50 % of the cross-sectional area of the flue
Open appliance, such as an open fire with a throat	Permanently open air vent(s) with a total equivalent of at least 50 % of the throat opening area**
Other appliances, such as a stove, cooker or boiler with a flue draught stabiliser	Permanently open vents as below: If design air permeability >5.0 $m^3/(h.m^2)$, then • 3000 mm^2/kW for first 5kW of appliance rated output • 850 mm^2/kW for balance of appliance rated output If design air permeability ≤5.0 $m^3/(h.m^2)$, then 850 mm^2/kW of appliance rated output***
Other appliance, such as a stove, cooker or boiler with no flue draught stabiliser	Permanently open vents as below: If design air permeability >5.0$m^3/(h.m^2)$, then • 550 mm^2/kW of appliance rated output above 5 kW If design air permeability ≤5.0 $m^3/(h.m^2)$, then 550 mm^2/kW of appliance rated output****

* = Equivalent area is as measured according to the method in BS EN 13141-1:2004 or estimated according to paragraph 1.14. Divide the area given in mm^2 by 100 to find the corresponding area in cm^2.

** = For simple open fires the requirement can be met with room ventilation areas as follows:

Nominal fire size (fireplace opening size)	500 mm	450 mm	400 mm	350 mm
Total equivalent area of permanently open air vents	20,500 mm^2	18,500 mm^2	16,500 mm^2	14,500 mm^2

*** = Example: an appliance with a flue draught stabiliser and a rater output of 7 kW would require and equivalent area of: [5 × 300] + [2 × 850] = 3200 mm^2

**** = It is unlikely that a dwelling constructed prior to 2008 will have an air permeability of less than 5.0 $m^3/(h.m^2)$ at 50 Pa unless extensive measures have been taken to improve air tightness

Table 6.04 Air supply to solid fuel appliances

Calculating the ventilation requirements for multiple appliances in the same room

You may encounter situations where more than one appliance is installed in the same room.

For one or more appliances totalling in excess of 7 kW (net) in a single room, internal space, through room or lounge/diner, ventilation is calculated at 5 cm^2 per kW (net) of total rated heat input above 7 kW.

For two or more gas fires in the same single space, up to a total rated heat input of 7 kW (net or gross) each (14 kW), ventilation is not normally required as adventitious ventilation will usually provide sufficient air for combustion. For a higher kW rating allow an additional 5 cm^2/kW (net) above 14 kW.

For two or more appliances, in a single room or internal space, you should calculate the total ventilation requirements of all the appliances based on which of the following has the greatest value:

- total rated heat input of flueless space heating appliances
- total rated heat input of open flue space heating appliances
- maximum rated heat input of any other type of appliance.

The permanent ventilation required for a multi-appliance installation should, wherever practicable, be sited between the appliances.

Example

A gas boiler rated 15 kW (net) is installed in the same room as a gas fire rated 3 kW (net).

The ventilation requirement is as follows:

15 kW + 3 kW = 18 kW
18 kW – 7 kW (adventitious air) = 11 kW (only deduct adventitious air for one appliance)
11 kW × 5 cm^2 = 55 cm^2 of free area of air vent.

The term 'space heating appliance' is taken to mean a central heating appliance, air heater, gas fire or convector. If permanent ventilation is required for a multi-appliance installation, wherever practicable, it should be sited between the two appliances. Where two or more chimneys use the same space, the pull of the stronger chimney can have a detrimental effect on the pull of the weaker one and cause spillage. This can happen with gas-fired appliances of different types, or if one of the chimneys serves a solid-fuel appliance. Where an interconnecting wall has been removed between two rooms and the resultant room contains two similar chimneys, each fitted with a gas fire or inset live fuel-effect fire, an air vent is not normally required, provided that the rated heat input of each of the appliances does not exceed 7 kW, and the installation instructions do not specify additional ventilation. For further information see BS 5871-1, BS 5871-2 and BS 5871-3.

Example

In addition to one DFE gas appliance, a room contains a gas cooker, and one open-flued instantaneous water heater of 25 kW heat input. The manufacturer's instructions for the DFE gas appliance specify that 100 cm^2 of purpose-provided ventilation is required for this appliance.

As there are no space heating appliances to consider, the overall ventilation requirement for the room is calculated as follows.

DFE gas appliance requirement plus *either* the gas cooker requirement or the open-flued water heater requirement, whichever is the greater.

DFE gas appliance requirement = 100 cm^2.
Gas cooker requirement = 0 cm^2 (plus openable window or equivalent opening).
Open-flued water heater requirement: 25 × 5 = 125 cm^2.

Therefore, the total ventilation requirement is **100 + 125 = 225 cm^2**.

Progress check

1 Complete the following statement: 'Where an open-flued appliance is installed with a rated input of more than 7 kW, the room it is in must have an air vent with a free area of 5 cm² for every …'.

2 What is natural ventilation through cracks in floorboards, windows and doors called?

3 If the width of 10 slots on an air vent is 65 mm and the depth is 10 mm, what is the free area?

Example

A room has one DFE, a cooker and an open-flued instantaneous water heater of 24 kW heat input. You find out from the manufacturer's instructions that the DFE requires a 100 cm² vent.

There are no space heating appliance considerations. The ventilation requirement of the room is calculated as follows.

DFE gas appliance requirement = 100 cm².

Gas cooker requires no purpose-made vent (plus openable window or equivalent opening).

Open-flued water heater requirement: $24 \times 5 = 120$ cm².

Therefore, the total ventilation requirement is **100 + 120 = 220 cm²**.

ACCEPTABLE LOCATIONS FOR APPLIANCE AIR VENTS

Installing a vent in the correct position is very important to ensure its satisfactory operation. The following information is for guidance only – always follow the manufacturer's instructions.

General principles of air vent installation

There is a wide variety of air vents in use, and their most important features are given below.

- They should be non-closable.
- No fly screen less than 5 mm² should be fitted.
- They should be corrosion-resistant and stable.
- The actual free area of the air vent is the size of slots or holes used (this applies to both sides of the ventilation arrangement, i.e. air vent and outside air brick).
- No air vent supplying air to an open flue appliance must communicate with any room or internal space that contains a bath or shower.
- No air vent should penetrate a protected shaft.
- An air vent must not communicate with a roof space or underfloor space if the space communicates with other premises.
- Where air vents are communicating with roof spaces you should take into account the problem of condensation and possible blockage by insulating material.
- Where an air vent incorporates a draught-reducing device or other restriction, this imposes a 25 per cent to 50 per cent reduction in equivalent area over that of an unrestricted air duct. The equivalent area should be obtained from the manufacturer to ensure compliance with the Gas Safety (Installation and Use) Regulations.

Air vents should not be positioned where they may easily be blocked by leaves or snow, or where they could cause the occupants discomfort through draughts. The inner and outer air vent must be connected by a liner to prevent anything from within the cavity interfering with the free air flow. It has been known for cavity foam fill to block the flow of air completely. When

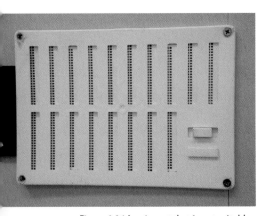

Figure 6.05 A cavity type vent

Figure 6.06 An air vent that is not suitable

used in internal walls the air vent should be placed no higher than 450 mm to the top of the vent above floor level to prevent the spread of smoke in the event of fire.

Restrictions to air vent locations

Air vents should be positioned so that there is a minimum separation distance from any flue terminal, as shown in Table 6.04. This is to ensure products of combustion cannot be drawn back into the dwelling through the vent, which might lead to carbon monoxide poisoning.

Air vent position	Appliance input (kW)	Room-sealed or non-sealed separation (mm)	
		Natural draught	Fanned draught
Above a terminal	0 – 7	300	300
	>7 – 14	600	300
	>14 – 32	1,500	300
	>32	2,000	300
Below a terminal	0 – 7	300	300
	>7 – 14	300	300
	>14 – 32	300	300
	>32	600	300
Horizontally to a terminal	0 – 7	300	300
	>7 – 14	400	300
	>14 – 32	600	300
	>32	600	300

Table 6.05 Positioning of air vents in relation to flue terminals (Table 3 BS 5440-2)

Table 6.05 gives the minimum requirements for the separation distances necessary between an air vent and an appliance terminal for the heat input of the appliance and the type of chimney to which the terminal is connected. The specified separation distances should be achieved upon installation of an appliance and/or air vent. In some cases, you may need to re-site an existing air vent.

These separation distances can also apply to room-extract fan outlets (e.g. cooker hood, window extraction fan – with or without a non-return flap), mechanical ventilation openings or any opening into a roof space. They do not apply to extract duct terminals fitted with non-return flaps that are connected via ducts to an appliance, such as a cooker hood or tumble dryer.

In addition, the air vent should not be located in a position where it is likely to be easily blocked (e.g. by leaves, snow or other debris), become flooded, or where contaminated air can be present at any time (e.g. in car ports). Location should prevent the air currents produced from passing through normally occupied areas of a room. This may be achieved by one or more of the following:

   ▪ siting the appliance in an appliance compartment with air vents communicating directly with outside air

- siting the air vent at a point adjacent to the appliance
- siting the air vent at or near ceiling height and orientating the internal louvers to direct incoming air away from the living or general accommodation area.

INSTALLATION OF DIFFERENT TYPES OF VENT

There are many types of vent so the correct installation is important to ensure their correct use.

Door vents

Vents can be fitted in external as well as internal doors. The door should have holes drilled into them of an adequate size so as to provide adequate free area. The low-level door vents should be no higher than 450 mm from the floor level to the top of the vent. The louver on both sides of the vents should point downwards, unless it is a high-level vent, in which case the vent louvre on the inside of the building should point upwards to prevent cold draughts.

Figure 6.07 Door vent

Case study

A gas engineer was called out to an open-flued boiler situated in a cellar that kept going out. Every time he came to fix it, on relighting it worked. The engineer left the building. Within two hours he was called back – it had gone out yet again. The owner asked him to shut the cellar door as it can be draughty with it open. This prompted the gas engineer to look for the air vent to the boiler – being in a cellar, the only air path route was the door.

What do you think was the fault?

How would you fix the problem?

Check your answers with your tutor.

Floor vents

Floor vents are designed to bring in air from under the suspended floor close to the appliance to prevent draughts. They are brass vents usually fitted with a plastic sleeve and are sized accordingly but adequate ventilation to the under floor area must be checked before installing.

Ceiling vents

An air vent should not communicate with a ventilated roof space if that space communicates with other premises/dwellings. This is to prevent effects from one dwelling on another when there's a possibility of high moisture levels, poor ventilation or excessive draught. Where an air vent draws air from a ventilated roof space, the total effective free area of the air vents or louvers into that space should be at least equal to the total effective free area of the air vents drawing air from that roof space. You should visually inspect ventilated roof space to confirm that it conforms to the minimum unrestricted ventilation requirements in BS 5250.

It is important that any ventilation provision is not blocked at the eaves and at the point where the ventilation passes through the ceiling insulation into the dwelling. This can be caused by poorly installed roof/cavity wall insulation, for example. Some modern construction methods use 'breathable' roofing felt designed to allow similar ventilation of the roof space. Where this has been used, do not consider the roof space to be outside air for the purpose of gas appliance air supply, because the membrane can deteriorate over time and be less effective.

It is also important that the requirements of Approved Document B regarding fire safety are taken into consideration and that any ventilation provision for a gas appliance provided through the roof space or loft does not compromise fire safety regulations. Further guidance on ventilation and fire safety can be found in BS 5250. A roof space not constructed to conform to BS 5250 can only be regarded as another internal space and therefore should be treated as such when calculating air vent free areas.

In all circumstances air vents leading into roof spaces should be designed or located to avoid blockages from, for example, insulation material. Air vents supplying air to the appliance from the roof space should be sited not less than 300 mm above joists or 150 mm above the level of insulation, whichever is the greater (measured up to the bottom openings of the air vent), in order to avoid blockage by insulation material. It is also preferable that terminal guards or bird guards are fitted to air vents to prevent blockages in general.

Circular wall vent

A circular vent can be used and allows for a neat cut into walls using a diamond tipped core drill. This makes them ideal for installation in flats. The vent has baffles fitted to stop water and wind from coming in. They should be fitted on the outside of the tube, while still allowing air flow through it.

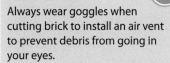

Safe working

Always wear goggles when cutting brick to install an air vent to prevent debris from going in your eyes.

Figure 6.08 A circular vent

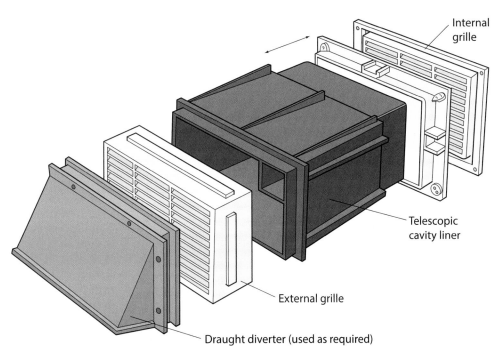

Internal grille

Telescopic cavity liner

External grille

Draught diverter (used as required)

Figure 6.09 Air vent through cavity wall liner

Intumescent air vents in fire-rated compartments

An intumescent air vent is one which allows air at room temperature to flow through it but closes itself by filling with fire-resistant foam when exposed to fire. Intumescent air vents may be used if the manufacturer of the air vent, in calculating the free area, has accounted for a space between the intumescent block and metal louvre/air vent.

The air vent should have no means of being closed, other than by an intumescent device.

When carrying out work on an intumescent air vent you should advise the end user that it is essential that an intumescent air vent that has been triggered by a high temperature, such as a fire, is replaced before any gas appliance is operated.

WHERE TO POSITION AIR VENTS

The positioning of air vents is very important to ensure they work as intended by the manufacturer.

In rooms for open-flued appliances

Figures 6.10 and 6.11 show air vent positioning requirements within rooms for open-flued appliances. They have the same size free area through the internal and external walls so there is no restriction of air flow.

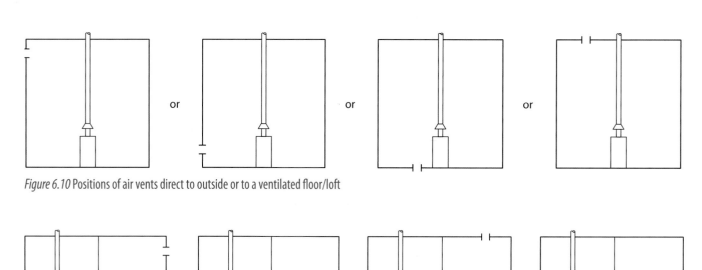

Figure 6.10 Positions of air vents direct to outside or to a ventilated floor/loft

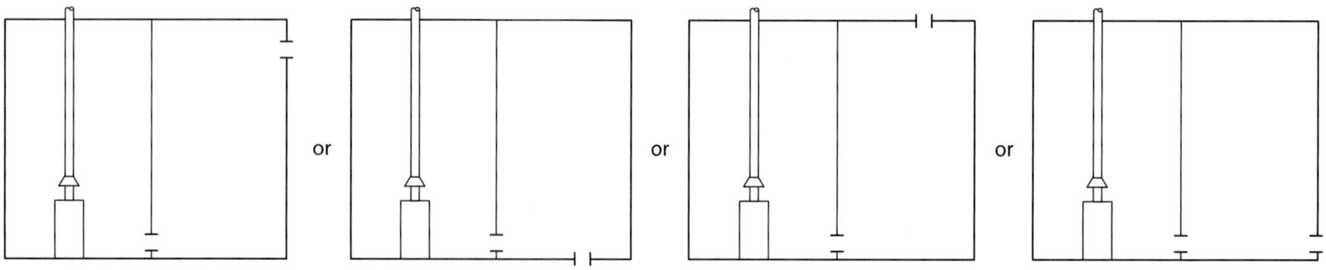

Figure 6.11 Vent positions via another room or space

Via another room (in series)

Where air cannot be taken directly from the outside, you need to position air vents to connect to an air vent through another room. This is known as positioning the vents 'in series'. There are some general rules regarding the sizing of air vents that are in series as they are usually required to be at least 50 per cent above the requirements of a single vent, as shown in Figures 6.12, 6.13 and 6.14.

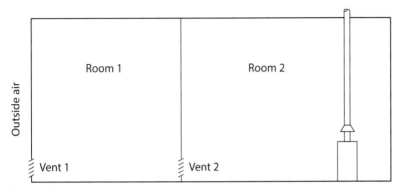

Figure 6.12 Two vents in series

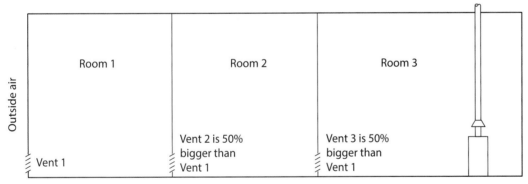

Figure 6.13 Three or more vents in series

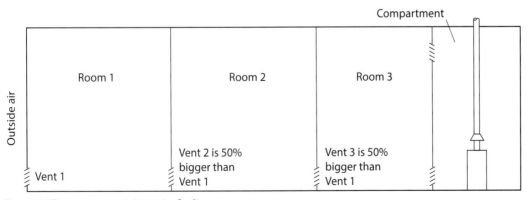

Figure 6.14 Three or more vents in a series feeding a compartment

When vents are in series going through more than two walls then extra ventilation has to be provided by increasing the vent size to the internal walls by 50 per cent.

Figure 6.13 shows three vents in series. The outside wall vent is 50 cm^2 so you need to increase vent two and three by 50 per cent by following the calculation below.

50 per cent of the outside wall vent is $\dfrac{50 \text{ cm}^2}{2} = 25 \text{ cm}^2$

Added to the original vent size: $\mathbf{50cm^2 + 25cm^2 = 75cm^2}$

Therefore we need the two internal vents to be 75 cm^2. Figure 6.14 shows three or more vents feeding a compartment that would require the same increase in vent size – by 50 per cent.

In compartments

Air vents should be provided at both the lowest and highest practicable levels in the appliance compartment. Both high and low level air vents should communicate either with the same room or internal space, or with the outside air through the same wall.

Ventilation from two differing external walls, for example, can create a cross-flow of ventilation, leading to unsatisfactory burner and/or flue performance.

Open-flued appliances

Figures 6.15 and 6.16 are examples of ventilator positions for open-flued boilers in compartments.

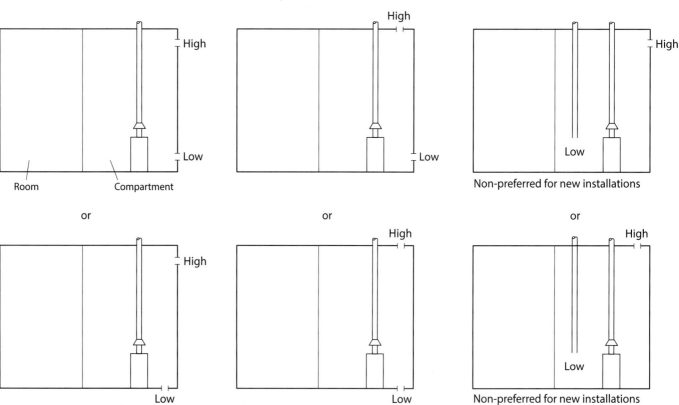

Figure 6.15 Examples of high and low vents direct to outside air or ventilated floor/loft

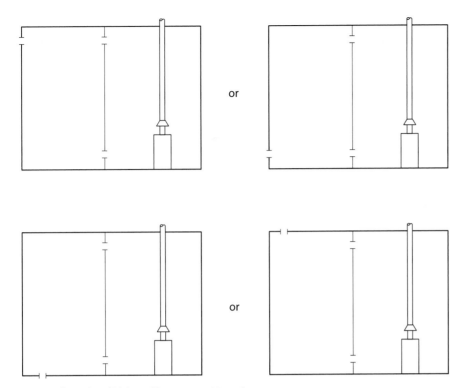

Figure 6.16 Examples of high- and low-vent positions via a room or space

If an air duct is used for compartment ventilation it should be no longer than 3 m. The vertical distance between low- and high-level vents should be as great as possible.

The appliance should have a clearance all the way around of 75 mm unless manufacturer's instructions say differently.

Where two or more appliances are installed in the same appliance compartment, whether or not they are supplied as a combined unit, the aggregate maximum rated input shall be used to determine the air vent free area, using Table 6.02 (page 172) or the appliance manufacturer's instructions.

Room-sealed appliances

Figures 6.17 and 6.18 are examples of ventilator positions for room-sealed boilers in compartments.

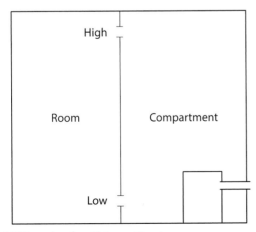
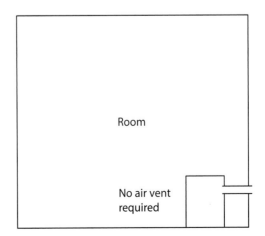

Figure 6.17 Examples of ventilator position via a room

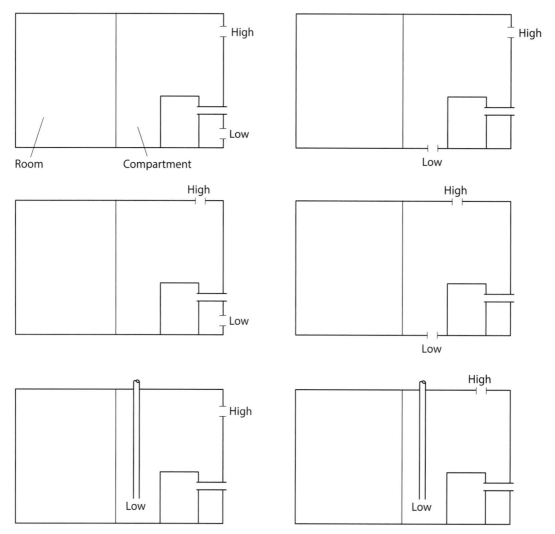

Figure 6.18 Examples of compartments

When installing air vents always allow space for maintenance access. For example, the siting of air vents behind radiators can cause problems of accessibility for inspection and therefore might not be considered to be good practice.

The installation of vents through doors, windows and ducts is illustrated in Figure 6.19. This should be used for guidance when installing and inspecting the ventilation requirements for a dwelling.

For flueless appliances

The ventilation for flueless appliances must communicate directly to outside air otherwise it has to be ducted through any other space. The room must also have an openable window, a hinged panel, adjustable louvre or other means of ventilation that opens directly to outside. The layout in Figure 6.20 shows the typical paths the ventilation is required to follow. To comply with the building regulations all ducts should only communicate with one room.

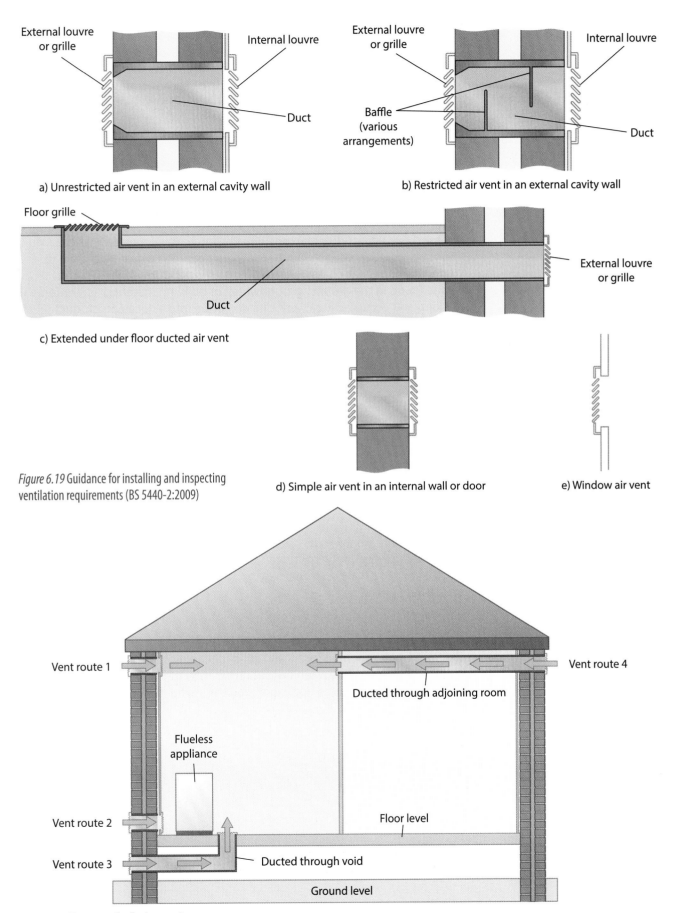

External louvre or grille

Internal louvre

Duct

a) Unrestricted air vent in an external cavity wall

External louvre or grille

Internal louvre

Baffle (various arrangements)

Duct

b) Restricted air vent in an external cavity wall

Floor grille

External louvre or grille

Duct

c) Extended under floor ducted air vent

d) Simple air vent in an internal wall or door

e) Window air vent

Figure 6.19 Guidance for installing and inspecting ventilation requirements (BS 5440-2:2009)

Vent route 1

Vent route 4

Ducted through adjoining room

Flueless appliance

Floor level

Vent route 2

Vent route 3

Ducted through void

Ground level

Figure 6.20 Vent route for flueless appliances

185

Location of air vent

Where a builder's opening, fireplace recess or inglenook installation is served by an existing underfloor air supply, the air vent should be sealed to avoid draughts etc. from interfering with the correct operation of the flue and the appliance burner. Outside the hearth area, the use of a floor vent communicating with a ventilated underfloor void is permitted.

Venting from under the floor – dangers of radon gas

Radon can cause cancer, you cannot see it or smell it and has no taste, it may be a danger in some homes. If householders smoke and their house has higher than normal radon levels, they have an increased risk of getting lung disease. Scientific studies of exposure to radon have suggested that children might be more sensitive to radon. The main pathways of potential human contamination to radon gas is through ingestion and inhalation. Radon can be found in the ground, groundwater, or building materials as it enters working and occupied areas as it disintegrating into a decaying product.

If an area has identified radon problems, then in no circumstances can ventilation be taken from under the building.

THE EFFECTS OF OTHER HEAT-PRODUCING APPLIANCES AND TYPES OF EXTRACTION

It is very important to take careful consideration of other appliances in the room and air extraction equipment when working on ventilation.

Oil and solid fuel appliances

If a room/internal space also has a solid fuel or oil-burning appliance in it, any ventilation requirements need to be calculated accordingly to prevent vitiation or excessive pull from these appliances having a detrimental effect on the gas appliances.

A gas appliance situated in the same room as a solid fuel appliance will require a separate vent.

The ventilation provision for oil and solid fuel appliances is given in Building Regulations Part J. If the rated output rather than the rated input is shown on an oil or solid fuel burning appliance, you can calculate the rated input using the following formula:

$$\text{Input} = \frac{(\text{output} \times 10)}{6}$$

Example

The rated input of a closed, multi-fuel stove with a declared output of 8 kW is calculated as follows.

$$\frac{(8 \times 10)}{6} = 13.33 \text{ kW}$$

	Open flued	Room sealed
Appliance in a room or space	Open flued appliance 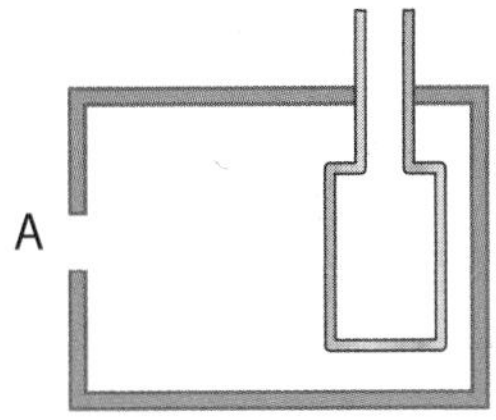 A A = 550mm² per kw output	Room sealed appliance 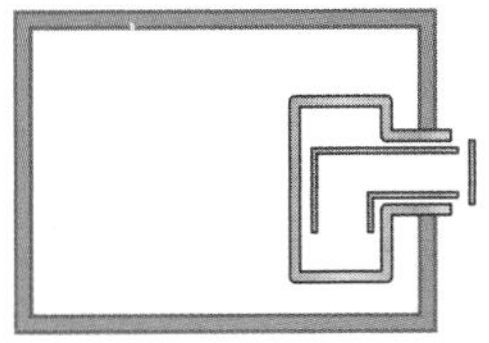 No vent needed
Appliance in an appliance compartment ventilated via an adjoining room or space	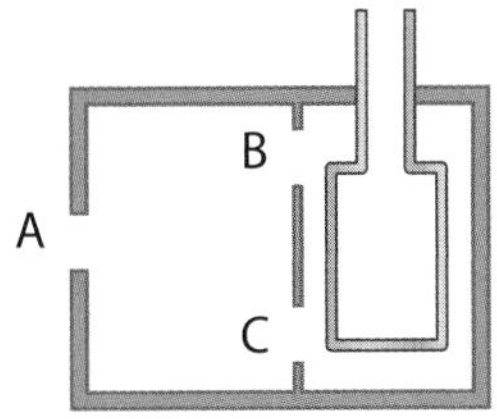 B A C A = 550mm² per kw output B = 1100mm² per kw output C = 1650mm² per kw output	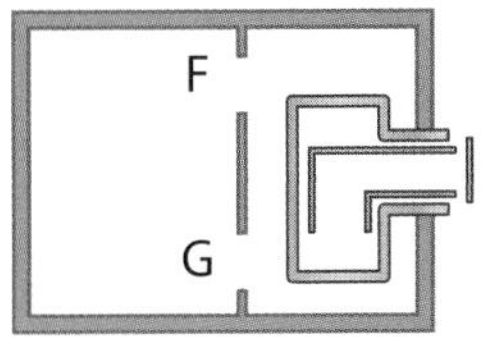 F G F = 1100mm² per kw output G = F
Appliance in an appliance compartment ventilated direct to outside	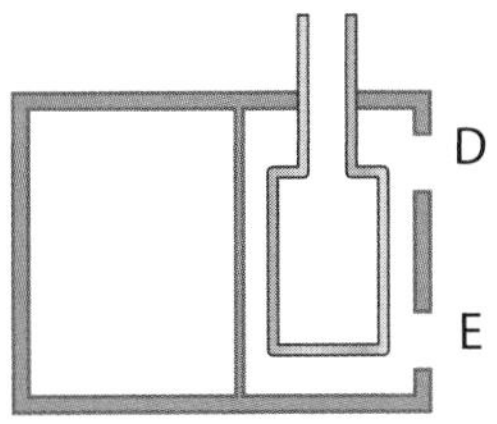 D E D = 550mm² per kw output E = 1100mm² per kw output	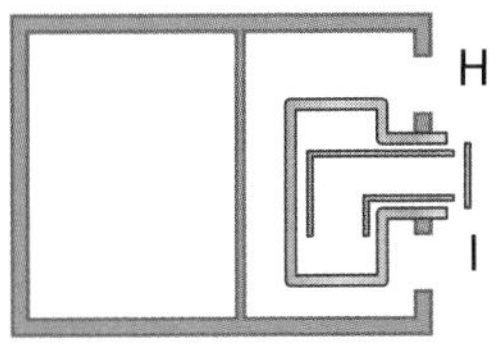 H I H = 550mm² per kw output I = H

Figure 6.21 Free areas of permanently open air vents for oil-fired appliance installations

For a solid fuel open fire or small closed stove, add on 2,500 mm² for the gas appliance requirement where the heat input is unknown to ensure an adequate supply of air. Oil appliances should be calculated as per Building Regulations Part J (see Figure 6.21).

Passive stack ventilation

Passive stack ventilation (PSV) should be considered independently from the provision of ventilation for gas appliances. PSV is a ventilation system that uses ducts that pass from the ceiling of a room to terminals situated on the roof. This acts as a natural stack effect, caused by the movement of air due to a temperature difference inside and outside the building, and also the effect of wind pressure passing over the roof of the property. In order to avoid the outlets for a PSV system and a chimney being positioned in different pressure zones, both the outlets should be located on the same face of the building. The chimney outlet should be at the same level or higher than the PSV outlet, so as to avoid the possibility of setting up a pressure gradient across the two systems. This could lead to appliance flue reversal; in other words, the ventilation system working backwards, leading to products of combustion entering the building.

Extractor fans, cooker hoods and tumble driers

You must check to see if any extractor fans are fitted that may affect the performance of an open-flued appliance. Set the fan to its maximum setting and check for spillage (see the section on 'Spillage test' page 448). Additional air vents may be necessary to prevent any spillage from taking place. If the spillage test fails, a larger vent will need to be fitted until the spillage test passes.

Extractor fans include room-extractor fans (which remove condensation/smells), warm air heater fans, external ducted tumble dryers, cooker hoods, or any other type of air extraction fan.

This is not an exhaustive list as any piece of equipment that contains a fan designed to move air can adversely affect the safe operation of an open-flued gas appliance and it is recommended that these are checked in all modes of operation.

As a general guide, an extra 50 cm^2 of air vent free area should be sufficient for most situations. However, the spillage test should be repeated after extra air vents have been fitted.

SERVICING AND MAINTAINING AIR VENTS

When carrying out a service or maintenance work on an appliance, make sure you examine the provisions for air supply. You should check existing air vents for internal and external obstructions. Gauze, fly screens and/or adjustable controls should be removed to render the ventilation permanent and non-adjustable. The size of air vent apertures, the free area and the position of an air vent should all be inspected for conformity to current regulations.

When checking an existing air vent, the vent and/or louvres should be examined visually for signs of obstruction. Where practicable, a screwdriver or similar tool should be inserted into the openings to check for blockages such as clear plastic sheets or fly screens. Where the internal parts of an air vent or ducting are inaccessible, try as much as is practicable to check for and clear any obstructions. These checks are in addition to the formal inspections for correct operation of the appliance. Where any defects are identified with an existing installation or appliance, refer to the current Gas Industry Unsafe Situations Procedure (GIUSP).

How to check a vent

Checklist

PPE	Tools and equipment	Source information
• Overalls • Protective footwear • Barrier cream	• Electric drill • Screwdriver • Measurer	• Manufacturer's instructions

1 When checking installed vents, ensure they are not installed more than 450 mm above the floor to help prevent the spread of smoke.

2 Check, using a screwdriver, that the vent passes through the wall and has not been blocked by cellophane etc.

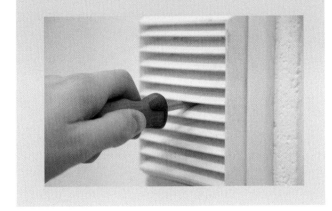

3 Check the vent is large enough for the appliance installation in the room. Measure the depth of the slot at the correct angle otherwise you will get an incorrect measurement.

4 Measure the length of each slot. Work out the area of one slot and multiply by the number of slots – check this against the actual requirements for the appliances. If incorrect, enlarge as necessary by putting in a larger vent or adding another vent.

Knowledge check

1 Complete the following statement: 'Room-sealed appliances take air directly from outside so do not require ventilation unless in a ...'.

a Room
b Compartment
c Shed
d Loft space

2 If the gross heat input of a boiler is 18 kW, what is the net heat input?

a 18 kW
b 16.36 kW
c 10.11 kW
d 9 kW

3 What is the upper height limit for a low-level door vent?

a 250 mm
b 100 mm
c 500 mm
d 450 mm

4 At least how high above the joists should air vents supplying air to an appliance from the roof space be sited?

a 300 mm
b 500 mm
c 900 mm
d 25 mm

5 Complete the following statement: 'An air vent should not be located in a position where it is likely to be easily ...'.

a Seen
b Painted
c Blocked
d Removed

6 What is the longest an air duct used for compartment ventilation should be?

a 3 m
b 7 m
c 10 m
d 1 m

7 What is the permanent vent size of any space or room containing a DFE?

a 50 mm^2
b 100 cm^2
c 150 cm^2
d 50 cm^2

8 Where must flueless appliance ventilation communicate?

a Into the room
b To a compartment
c Directly to the outside air
d Indirectly to a flue

9 What is the air permeability of an air-tight room?

a Less than 5.0 m^3/hr/m^2
b Less than 3.0 m^3/hr/m^2
c More than 9.0 m^3/hr/m^2
d More than 1.0 m^3/hr/m^2

10 To calculate the rated input (in kW) from the rated output, which equation should be used?

a Input = (output x 10) ÷ 5
b Input = (output x 10) ÷ 6
c Input = (output x 12) ÷ 4
d Input = (input x 10) ÷ 6

11 What should an air vent not communicate directly with?

a Builder's opening or fireplace recess
b Outside wall
c Door
d Internal wall

12 What should you check existing air vents for?

a Internal and external obstructions
b Colour match with wall
c Conductivity
d Country of origin

13 What should you check inaccessible internal parts of an air vent or ducting for?

a Dampness
b Noise
c Obstructions
d Spiders

Flues and chimney systems

Key terms

Flue
A flue is a passage for conveying the products of combustion to the outside atmosphere.

Chimney
A chimney is a structure consisting of a wall or walls enclosing a flue or flues.

INTRODUCTION

In this chapter you will be instructed on the standards of chimneys and flue systems that can be used with gas appliances. Flues and chimney systems are an integral part of an appliance's installation. It is important that you, as a gas operative, understand the need for them in the effective removal of the products of combustion, how they are constructed and the materials that they can be manufactured from.

REGULATIONS AND STANDARDS ON FLUES AND CHIMNEY SYSTEMS

As with all aspects of gas installation work, there are certain standards and regulations which must be adhered to. With respect to working on **flues** and **chimney** systems, there are rules laid out in British Standards (BS) and in the Gas Safety (Installation and Use) Regulations (GSIUR). There are rules set out for the designer, supplier and installer of flue and chimney systems and for landlords with regard to their maintenance.

The standard relevant to this area of work is BS 5400 Flueing and ventilation for gas appliances of rated input not exceeding 70 kW net (1st, 2nd and 3rd family gases) – Part 1 Specification for installation of gas appliances to chimneys and for maintenance of chimney and Part 2 Installation and maintenance of flues and ventilation for gas appliances. As with other work on gas, it is essential that persons carrying out work on the flues for gas appliances are competent to do so, and any work that is subject to the GSIUR must comply with these requirements.

The building regulation which applies to gas appliances is Approved Document J (Combustion Appliances and Fuel Storage Systems). This document has been updated and came into force on the 1 October 2010. Section 1 sets out the general provisions which apply to combustion installations. For the safe accommodation of combustion appliances, you must ensure:

- there is sufficient air for combustion purposes and where necessary for the cooling of the appliance
- that appliances operate normally without the products of combustion (POC) causing a hazard to health (spillage)
- that a device is fitted to warn of carbon monoxide where a fixed appliance is installed
- that the appliance operates without causing damage to the fabric of the building through heat exposure
- that the appliance and chimney/flue have been inspected and are fit for the purpose intended
- that the chimney/flue has been labelled to indicate its performance capabilities.

Exchange of information and planning

The designer or installer of the chimney, and the provider or installer of the gas appliance should agree and document the important compatibility details with the customer as appropriate. When erecting a new chimney or chimney configuration or modifying an existing one, these important details include:

- the type, size and route of the chimney
- the type and size/heat input of the gas appliance that is intended to be connected to it.

This is particularly important when different trades are involved in the erection of the chimney or chimney configuration and the fitting of the gas appliance.

When you are fitting a gas appliance to an existing open-flue chimney or room-sealed chimney configuration, it is essential that you confirm that the chimney is suitable for the appliance. When the chimney is provided as part of the appliance, for example a room-sealed configuration (including balanced flue), you should agree and document, with the customer, that the chimney configuration is suitable for the application. When you are installing either a new or replacement appliance to an existing chimney/flue configuration you are responsible for checking that the installation is suitable for the appliance being installed.

Maintenance of flues

The responsible person (e.g. landlord) should be advised that, for continued efficient and safe operation of the appliance and its chimney, it is important that adequate and regular maintenance is carried out by a competent person (i.e. a Gas Safe registered gas installer) in accordance with the appliance manufacturer's recommendations.

The GSIUR impose a general obligation on landlords who provide appliances in tenanted premises to have them maintained and checked for safety every 12 months.

CLASSIFICATION OF CHIMNEYS AND APPLIANCES

Under the Accredited Certification Scheme (ACS) convention you will be required to identify types of appliances and category of the flue system. It is important that you, as a gas operative, are able to undertake this task. It is your responsibility to know if the appliance is correct for a given situation. This section informs you of the classification types.

Classification of chimneys

Chimneys are classified according to BS EN 1443:2003, according to the following performance characteristics:

- temperature class
- pressure class

- resistance to condensate class
- corrosion resistance class
- soot-fire resistance class (G or O), followed by a distance to combustibles.

Chimney products are specified in the European chimney standards according to the materials which are being used, i.e. concrete, clay/ceramic, metal or plastic.

Classification of appliances

All appliances are now classified by PD CR 1749:2005; this is the new European standard for the method of evacuation of the POC. It means that the classification of appliances burning combustible gases is the same across the European Community.

There are three main types of appliance, grouped according to how they discharge their POC:

- **Type A Flueless** – This type of appliance is not intended for connection to a flue or any device for evacuating the POC to the outside of the room in which the appliance is installed. Products of combustion are released into the room in which the appliance is installed. The air for combustion is taken from the room.
- **Type B Open-flued** – This type of appliance is intended to be connected to a flue that evacuates the POC to the outside of the room containing the appliance. The air for combustion is taken from the room.
- **Type C Room-sealed** – The air supply, combustion chamber, heat exchanger and evacuation of POC (i.e. the combustion circuit) for this type of appliance is sealed with respect to the room in which the appliance is installed.

These types of appliance are then further classified according to flue type, as shown in Table 7.01 opposite.

Progress check

1. Which of the Building Regulations would you refer to for combustion appliances?
2. Complete the following description: 'Type A Flueless appliances are. . .'
3. Type C appliances are classified as which type of appliance?
   a Flueless
   b Open-flued
   c Room-sealed
   d All of the above
4. Which type of appliance would you install in SE duct or 'U' duct systems in the UK?
5. Complete the following requirements: 'When erecting a new chimney or chimney configuration or modifying an existing one, the important details shall include . . .'

Letter classification and type	Classification and first digit	Classification and second digit		
		Natural draught	Fan downstream of heat exchanger	Fan upstream of heat exchanger
A – Flueless		A1*	A2	A2
B – Open-flued	B1 – With draught diverter	B11*	B12* B14	B13*
	B2 – Without draught diverter	B21	B22*	B23
C – Room-sealed	C1 – Horizontal balanced flue/inlet air ducts to outside air	C11	C12	C13
	**C2 – Inlet and outlet ducts connected to common duct system for multi-appliance connections	C21	C22	C23
	C3 – Vertical balanced flue/inlet air ducts to outside air	C31	C32	C33
	C4 – Inlet and outlet appliance connection ducts connected to a U-shaped duct for multi-appliance system	C41	C42	C43
	C5 – Non-balanced flue/inlet air-ducted system	C51	C52	C53
	C6 – Appliance sold without flue/air-inlet ducts	C61	C62	C63
	C7 – Vertical flue to outside air with air-supply ducts in loft. Draught diverter in loft above air inlet	C71	C72* (Vertex)	C73* (Vertex)
	C8 – Non-balanced system with air-supply from outside and flue into a common duct system	C81	C82	C83

* Common types of flue in the UK.

** Used for SE ducts and 'U' ducts systems in the UK.

Table 7.01 Classification of gas appliances according to flue type

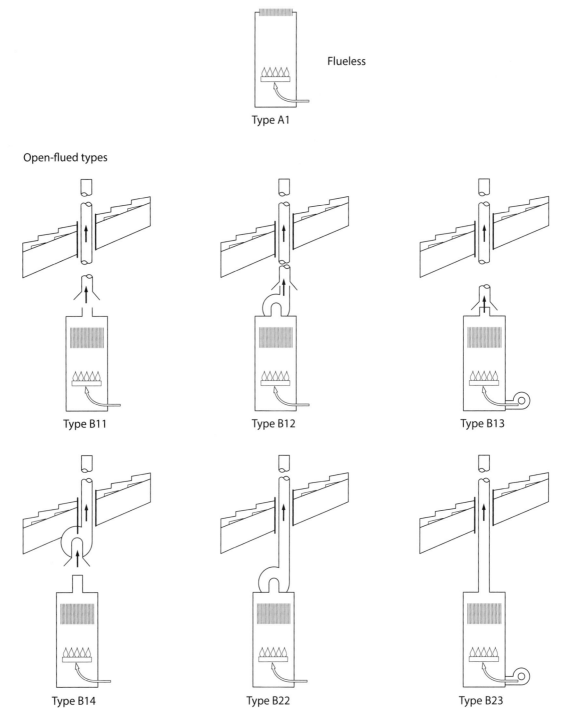

Figure 7.01 Typical appliances of types A and B

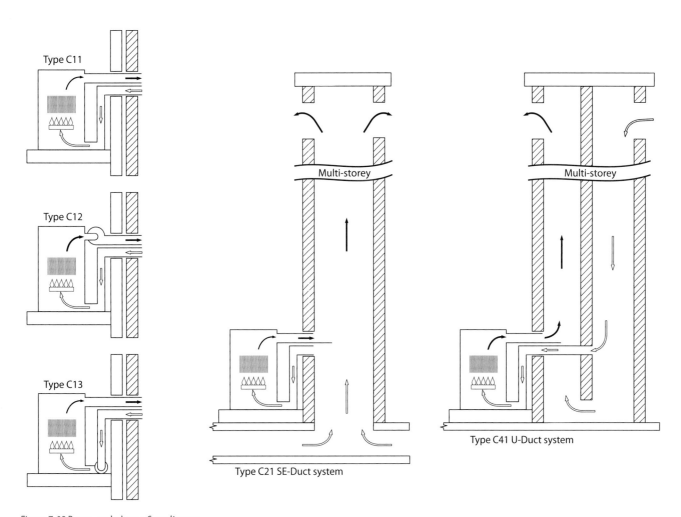

Figure 7.02 Room-sealed type C appliances

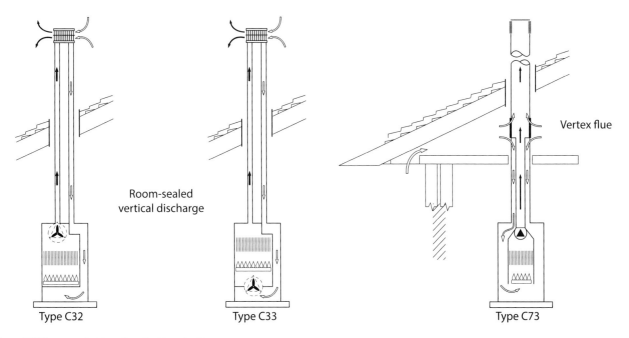

Figure 7.03 Room-sealed type C vertical terminations

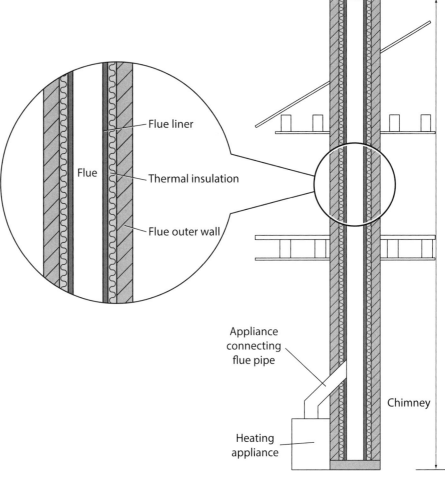

Figure 7.04 Chimney components

Progress check

1 Which of the Building Regulations would you refer to for combustion appliances?

2 Complete the following description: 'Type A Flueless appliances are...'

3 Type 'C' appliances are classified as which type of appliance:
   a Flueless
   b Open-flued
   c Room-sealed
   d All of the above

4 Which type of appliance would you install in SE duct or 'U' duct systems in the UK?

5 Complete the following requirements: 'When erecting a new chimney or chimney configuration or modifying an existing one, the important details shall include...'

WORKING PRINCIPLES AND FEATURES OF OPEN-FLUED SYSTEMS

Open flues are still an integral part of appliance installation, in particular when fitting space heating appliances (gas fires). You need to understand the importance of ensuring that the flue is correct for the installation.

Type B Open-flued (natural draught)

Natural draught systems take combustion air from the room and the POC travel up the flue by natural draught or 'flue pull'. This is caused by the difference in the densities of hot flue gases and the cold air outside.

The strength of the flue pull or draught is increased when the flue gases are hotter or if the flue height is increased. Factors that will slow down the flue pull are 90 ° bends and horizontal flue runs, so these must be avoided.

This flue draught is created by natural means and is quite slight, so it is important to design/install a flue carefully to allow for the necessary up-draught. Fans can be fitted in flues to overcome problems and allow more flexibility. See the section on 'Type B Open-flued forced (fanned draught)' on page 201.

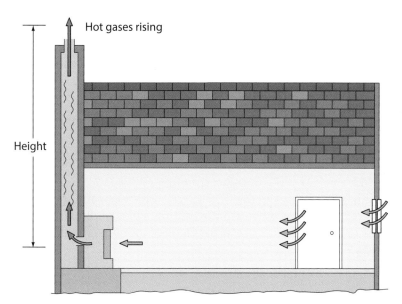

Figure 7.05 Flue draught

Open flues are sometimes referred to as 'conventional flues' and have four main parts:

- primary flue
- draught diverter
- secondary flue
- terminal.

Both the primary flue and draught diverter are normally part of the appliance, while the secondary flue and terminal are installed on the job to suit the particular position of the appliance.

The **primary flue** creates the initial flue pull to clear the POC from the combustion chamber.

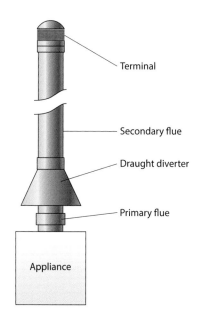

Figure 7.06 Four main parts of a flue

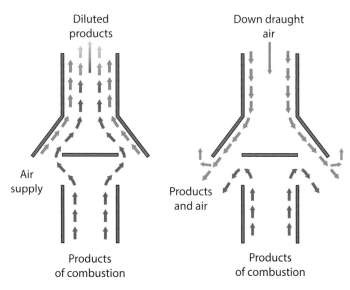

Figure 7.07 Section through a draught diverter

The **draught diverter**:

- diverts any downdraught from the secondary flue from the combustion chamber of the appliance, as this can interfere with the combustion process
- allows dilution of flue products with air
- breaks any excessive pull on the flue (i.e. in windy weather), as this can also interfere with the combustion process.

Where a draught diverter is fitted it should be installed in the same space as the appliance.

The **secondary flue** passes all the POC up to the terminal and should be constructed in such a way as to give the best possible conditions for the flue to work efficiently. Resistance of the installed components should be kept to a minimum by:

- avoiding horizontal/shallow runs
- keeping bends to a minimum of 45 °
- keeping flues internal where possible (warm)
- providing a 600 mm vertical rise from the appliance to the first bend
- fitting the correctly sized flue – at least equal size to the appliance outlet and as identified by the manufacturer.

The **terminal** is fitted on top of the secondary flue. Its purpose is to:

- help the flue gases discharge from the flue
- prevent rain, birds and leaves etc. from entering the flue
- minimise downdraught.

You should fit terminals to flues with a cross-sectional area of 170 mm or less. The terminal needs to be suitable for the appliance type fitted to the flue.

Terracotta chimney rain inserts are *not* suitable for use with gas appliances. Only use approved terminals, as these have been checked for satisfactory

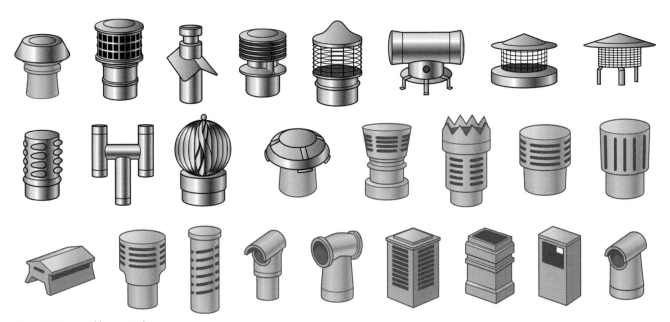

Figure 7.08 Acceptable terminals

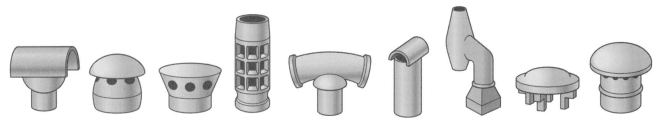

Figure 7.09 Unacceptable flue terminals

performance and have limited openings of not less than 6 mm but not more than 16 mm (except for incinerators, which are allowed 25 mm). Figures 7.08 and 7.09 show examples of acceptable and unacceptable types of terminals for use with certain flue systems.

Type B Open-flued forced (fanned draught)

Fanned draught flues allow for greater flexibility in the positioning of the appliance. There are two types of fanned draught flue systems:

- where the fan is an integral part of the appliance (positive pressure)
- where a fan is located in the outlet to a chimney or flue system (negative pressure) and has been specified or supplied by the appliance manufacturer.

Where the fan is not factory fitted, always connect it to the appliance in accordance with the appliance manufacturer's instructions. Do not make any modifications to the appliance without the agreement of the appliance manufacturer.

When fitting a fan in the chimney/flue system, ensure the fan size allows for full clearance of POC against adverse wind pressures. This also includes the route which the chimney/flue must take and requires you to calculate the resistance to flow (including the specified adverse pressure) at the design flow rate and to compare it with the pressure available from the chosen fan. The responsibility for safe installation lies with the installer and the appliance manufacturer.

Proprietary fan kits are available from manufacturers and include fail safe features to prevent the appliance from operating should the fan fail.

Minimum flow rates for fanned flues		
Appliances	**Maximum CO_2 concentration (%)**	**Minimum flue flow rate (m^3/h per kW input*)**
Gas fire	1	10.7
Fire/back boiler	2	5.4
All other appliances	4	2.6
* These figures refer to natural gas.		

Table 7.02 Minimum flow rates for fanned flues

Table 7.02 may be used to calculate flue velocity although care is required to relate a specific measured velocity to a mean volumetric flue flow rate. The final test for correct operation of a chimney is a spillage test at the appliance. For decorative fuel effect gas appliances you should refer to BS 5871-3.

Safety control

Where fans are fitted in secondary chimneys/flues they should incorporate a safety control in the secondary flue which is external to the appliance. The safety device should be capable of cutting off the flow of gas to the main burner if the flow in the secondary flue becomes insufficient for more than 6 secs.

The safety control means the flue flow sensor must be in the 'no flow' position before the fan can be set in operation. Should the safety control be activated then manual intervention is required to re-establish the gas supply to the main burner, unless the appliance incorporates a flame supervision device (FSD) and the correct flue flow is re-established.

TYPES OF OPEN FLUES AND CHIMNEYS

There are certain considerations that must be taken into account when fitting new chimneys and open flues, including the construction materials of the flue/chimney and their suitability for the particular appliance and circumstances.

Always read the manufacturer's instructions to check if the appliance is suitable for the flue it is being installed with.

Chimney construction materials

Where new open-flued appliances are fitted, the chimney should be designed to comply with BS EN 15287-1 Chimneys – Design, installation and commissioning of chimneys (Part 1: Chimneys for non-room-sealed heating appliances). These chimneys are classified as being allowed by the manufacturer. New chimneys can be classed as system chimneys or custom-built chimneys.

Chimney/flue construction materials must now be capable of removing condensed combustion products which are mildly acidic. Materials such as copper, mild steel and lower grades of stainless steel are not suitable for this type of application.

Where a new chimney is being installed, the chimney should be constructed from either brick (or other masonry) or flue blocks. Brick/masonry chimneys should be lined with clay liners conforming to BS EN 1457 or concrete liners conforming to BS EN 1857. Poured concrete linings are *not* acceptable as a method of lining new masonry chimneys. Flue block chimneys should be lined with clay conforming to BS EN 1806 or concrete conforming to BS EN 1858.

Rigid flues

Where factory-made insulated metal chimneys are used, they must conform to BS EN 1856-1. If they are single walled then they must not be used externally. When chimneys are used externally, they must be twin walled to BS EN 1856-1 and installed to manufacturers' instructions.

Rigid metallic flues

Twin-wall metal flues

Twin-walled metal flues are available in a variety of lengths and diameters, with a vast range of fittings and brackets to suit every installation. You should consult the manufacturers' information booklets to familiarise yourself with the range of products available before deciding which to fit.

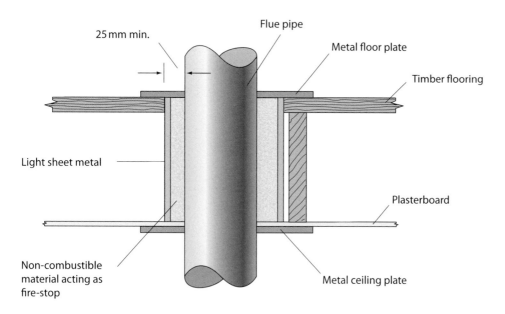

Figure 7.10 Flue passing through combustible material

There are two types of twin-walled flues: fully insulated or with an air gap. Twin-walled flues with an air gap are only suitable for use internally but can be used externally for lengths less than 3 m. For all other external situations fully insulated twin-walled pipe should be used.

The joints are designed to be fitted with the 'male' or spigot end uppermost. Where a pipe passes through a combustible material like a floor/ceiling, a sleeve must be provided to give a minimum circular space of 25 mm (see Figure 7.10).

Where a flue pipe passes through a tiled sloping roof (see Figure 7.11 on page 204), a purpose-made weathering slate is required with an upstand of 150 mm minimum at the rear of the slate. Aluminium weathering slates are also available to purchase. You should always consult the manufacturer's instructions before fitting the weathering slate.

Vitreous-enamelled steel flues

A vitreous-enamelled steel flue is a single-skin pipe available in many lengths and sizes, although it can be cut to any length. It is often used as the connection between an appliance and the main flue, and may include a disconnecting collar to allow appliance removal. The socket on single-wall pipe is fitted uppermost, unlike the twin-wall.

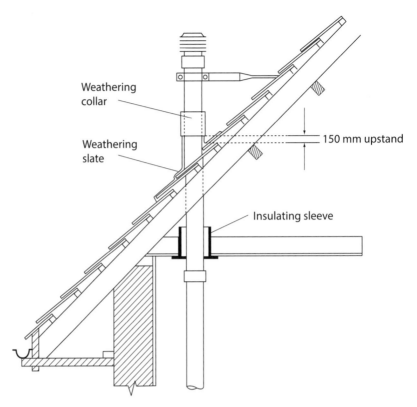

Figure 7.11 Flue passing through sloping roof

Asbestos flues

Under no circumstances should asbestos cement materials be used for new flue pipes. Where an existing chimney is suspected to be made of asbestos cement or contain asbestos then a risk assessment must be conducted prior to carrying out any work. See the section on 'Risk assessments' (page 62).

You should only reuse an existing asbestos cement chimney or chimney component if it is sound and does not require cutting or machining. More information regarding asbestos-related products can be found on the HSE website.

Pre-cast concrete flue blocks

Pre-cast concrete flue blocks are the same size and shape as a house brick and can therefore be built into (or 'bonded' with) the walls of a new property during construction. Non-bonded blocks are available and are more suited to existing properties.

You must use flue blocks certified to BS EN 1858 and fit them to the manufacturer's instructions. Excess cement should be carefully removed from the block during construction and no air gaps should be left.

All flue blocks must be laid spigot end up with a 3mm thick complete and gas-tight joint. The most convenient method of jointing is to apply cartridges of ready-mixed high-temperature mortar with a cartridge gun. Always use the nozzle and cut it 35 mm from the end to give an 8 mm bead. Before jointing, ensure the upper face of the block is dry and clear of debris.

Particular attention should be given to the connection from the flue blocks to the ridge terminal; the flue pipe installed should be of twin-wall insulated

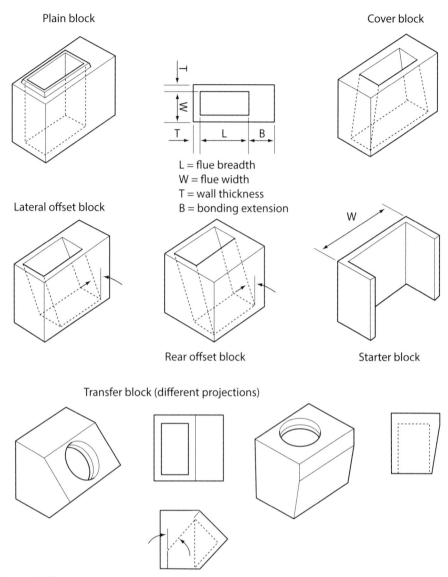

Plain block

Cover block

L = flue breadth
W = flue width
T = wall thickness
B = bonding extension

Lateral offset block

Rear offset block

Starter block

Transfer block (different projections)

Figure 7.12 Flue blocks

pipe construction using the correct fittings. When a metal chimney is connected to a flue block system then a transfer block must be used. When making the connection, be sure not to let the metal flue component project into the flue as this would cause a restriction of the cross-sectional area of the flue system.

Flue block chimneys should not be directly faced with plaster, as the heat will cause the plaster to crack. They should be faced with either concrete (or similar material) blocks or with plasterboard. Where plasterboard is used as dry lining, the dabs or batons should not be in direct contact with the flue blocks. You must also ensure that no fixing devices penetrate the blocks and that the joint between the facing and the blocks is sealed around the fireplace opening. An example is where plasterboard has been fitted and the gap between the blocks and the plasterboard must be sealed to prevent POC from escaping between the gap.

Where a new chimney is constructed using flue blocks, the minimum cross-sectional area should not be less than 16,500 mm^2 with no dimension less than 90 mm. There are, however, appliances which cannot be connected to flue blocks which have a cross-sectional area between 12,000 mm^2 and 13,000 mm^2 or a minor dimension of 63 mm or less. These are:

- drying cabinets
- appliances having a flue duct outlet area greater than 13,000 mm^2
- gas fires and combined appliances incorporating a gas fire, unless a special starter block/adapter has been designed for the purpose, tested and supplied by the appliance manufacturer or the appliance manufacturer's instructions specifically state that this is acceptable.

Some manufacturers state that their appliances are not suitable for connection to flue block systems of certain sizes constructed to BS 1289:1975. Further guidance is given in BS 5440 – Part 1:2008.

Figure 7.13 below shows a detail of a completed installation with flue blocks.

Pre-case flue blocks can have a bad reputation for poor flue performance and spillage of the POC. This is because they have been badly installed in the past. As with any open flue, always ensure that a thorough visual inspection of the flue system is conducted. You should be particularly careful when measuring the cross-sectional area to ensure accuracy, and always conduct adequate spillage tests prior to handover.

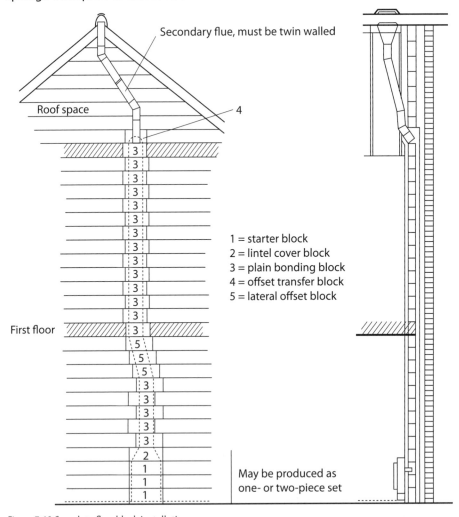

Figure 7.13 Complete flue-block installation

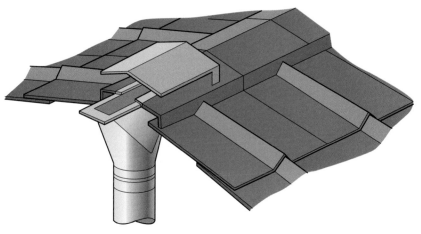

Figure 7.14 Terminal ridge tile

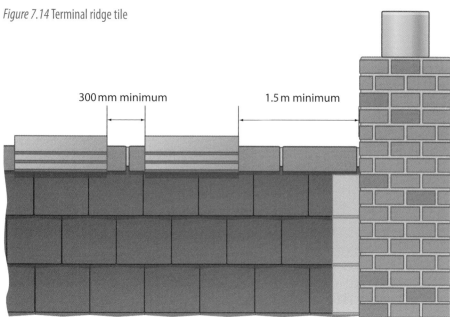

300 mm minimum

1.5 m minimum

Figure 7.15 Positioning of ridge vents

Other types of open flue

Flexible stainless steel flues – liners

Flexible stainless steel flues must comply with BS EN 1856-2. They are used internally to line existing flues that do not have a suitable clay lining as part of the original building construction. Liners are also used when the existing chimney or flue has given unsatisfactory performance in the past.

Liners must be installed in one continuous length and *not* be joined to reach the required length. Any bends in the liner must be of a maximum of 45 ° and there should be no kinks or tears present.

It is essential to secure liners with a clamp plate and to seal the top and the base of the chimney. A sealing plate must be included at the base of the flue system to prevent debris from falling into the appliance opening and onto the appliance. Where the diameter of the flue is less than 170 mm, use an approved terminal to protect the end of the liner.

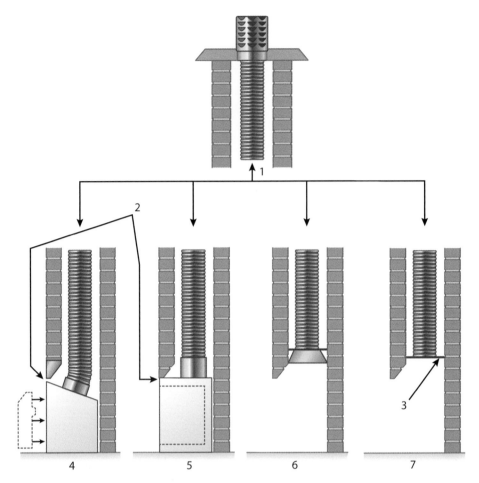

Key

1 Flue liner conforming to BS EN 1856-2
2 Joints to be well made where the closure plate or the flue box is sealed to the face of the opening or fire surround
3 Debris or register plate
4 Flue liner connected to a proprietary flue gas collector. For use with an appliance with a closure plate
5 Flue liner conforming to BS EN 1856-2 connected to a gas flue box conforming to BS 715
6 Flue liner secured and sealed into a proprietary gather above the builder's opening
7 Flue liner mechanically secured and sealed with a clamp to a debris or register plate above a builder's opening

Figure 7.16 Connections for flexible stainless steel liner

The liner should not project more than a nominal 25 mm below the plate. Where gas supplies are to be made through the wall of a gas flue box (4 and 5, Figure 7.16) it should be routed as close as practicable to the bottom of the box and sealed with non-setting sealant.

Where flexible liners are connected to the tops of flue boxes they should rise as near vertical as possible no angle should be any greater than 45 °. The correct method of connecting flexible flues to a gas fire back boiler is shown in Figure 7.17 – note the disconnection socket and the sealing plate.

A typical way of sealing the annular space between the chimney and the flexible flue liner would be with mineral wool. For larger openings, it might be necessary to use, for example, a register plate to hold the mineral wool in place. Follow the advice given by the liner manufacturer, particularly in relation to the location of the liner where it passes around bends in the chimney.

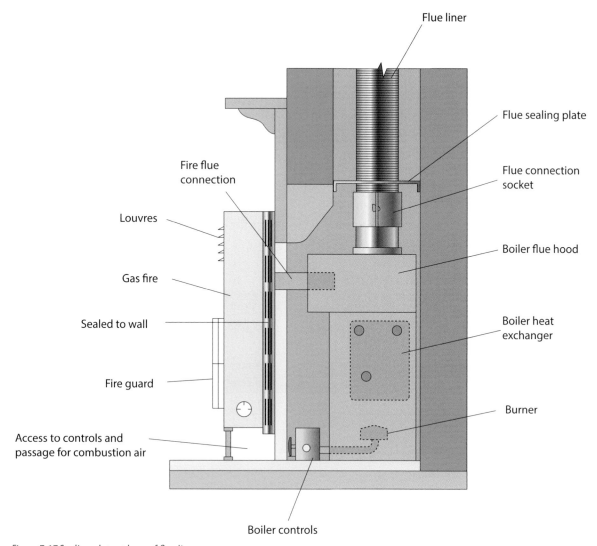

Figure 7.17 Sealing plate at base of flue liner

Unless otherwise stated by the manufacturer, decorative fuel effect fires must be installed with a minimum flue size of 175 mm diameter.

When chimneys exceed certain lengths, they may need to be lined, depending on the type of appliance fitted, as shown in Table 7.03.

Appliance type	Flue length
Gas fire	> 10 m (external wall)
	> 12 m (internal wall)
Gas fire with back boiler	Any length
Gas fire with circulator	> 10 m (external wall)
	> 12 m (internal wall)
Circulator	> 6 m (external wall)
	> 1.5 m (external length and total length > 9m)
Other appliance	Flue lengths greater than in Table 7.05 (see page 213)

Table 7.03 Required flue lining for different appliances

Shared flues

In special cases you are allowed to connect two or more appliances into the same flue, but special rules apply. BS 5440 Part 1:2008 identifies six important rules:

- Each appliance shall have a draught diverter.
- Each appliance shall have a flame-supervision device.
- Each appliance shall have a safety control (an atmospheric sensing device) to shut down the appliance before there is a dangerous quantity of POC in the room.
- The flue must be sized to ensure complete removal of the products of the whole installation.
- The chimney must have access for inspection and maintenance.
- All appliances shall be in the same room or space or on different floors as below.

Type of appliance	Nominal cross-section area of main flue			
	Greater than 40,000 mm² but less than 62,000 mm²		62,000 mm² or more	
	Maximum number of appliances	Total input rating (kW)	Maximum number of appliances	Total input rating (kW)
Gas fire	5	30	7	45
Instantaneous water heater	10	300	10	450
Storage water heater, central heating unit or air heater	10	120	10	180

Table 7.04 Appliances discharging by way of subsidiary flue into a main flue

Where appliances are to be installed on different floors using the same shared flue then the following rules apply:

- The main chimney must not to be part of an external wall.
- The nominal cross-sectional area of the main flue serving two or more appliances should be no less than 40,000 mm² and should be sized as detailed in Table 7.05 on page 213.
- Each appliance is to discharge into the main flue by way of a subsidiary flue
  - the connecting pipe is to be not less than 1.2 m above the outlet of the appliance
  - the connection shall be a minimum of 3 m above the outlet of the appliance where it is a gas fire.
- Where chimneys are newly built all appliances are to be the same type.
- When connecting to an existing chimney replacements are to be of the same type and not greater in input to the original appliance.
- Fanned-flued appliances of Types B_{14}, B_{22}, B_{23} should not be used.
- All appliances connected to the flue must be labelled to indicate that they are on a shared flue system.

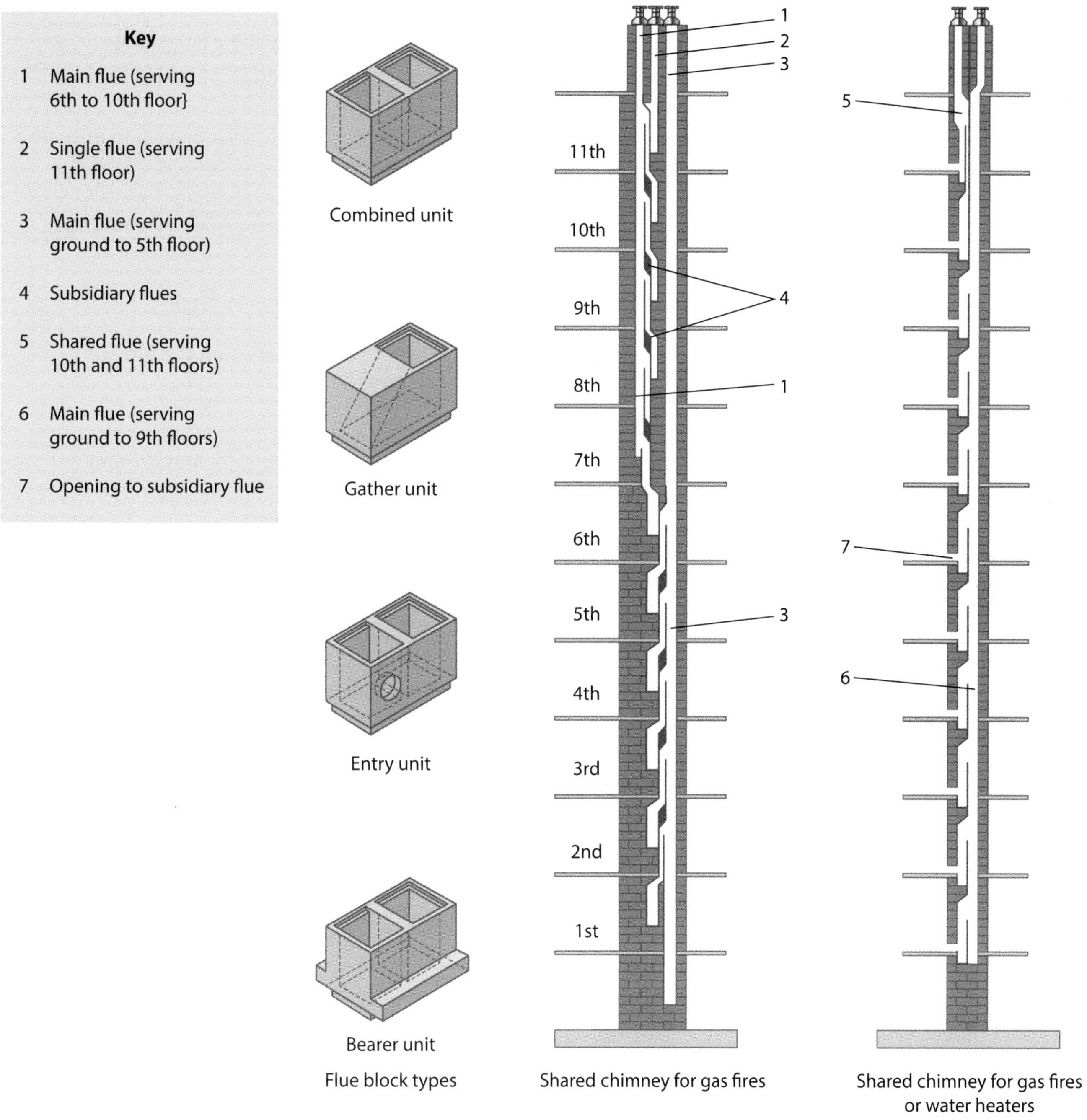
Key
1 Main flue (serving 6th to 10th floor}
2 Single flue (serving 11th floor)
3 Main flue (serving ground to 5th floor)
4 Subsidiary flues
5 Shared flue (serving 10th and 11th floors)
6 Main flue (serving ground to 9th floors)
7 Opening to subsidiary flue
Combined unit
Gather unit
Entry unit
Bearer unit
Flue block types
11th
10th
9th
8th
7th
6th
5th
4th
3rd
2nd
1st
1
2
3
4
1
3
Shared chimney for gas fires
5
7
6
Shared chimney for gas fires or water heaters

Figure 7.18 Shared flues

Flue boxes

Flue boxes funnel the POC into the flue system and can be fitted to the back of:

- radiant convector gas fires
- insert live fuel effect gas fires
- decorative fuel effect gas fires
- combined gas fire and back boilers
- gas heating stoves.

Flue boxes *must not* be installed in solid fuel appliances. Check with manufacturer's instructions if the gas appliance can be used with gas flue box systems. Flue boxes are used in builder's openings or in a purpose-built chimney without bricks or masonry.

Flue boxes are manufactured to BS 715 and can be of single-wall or insulated twin-wall construction. All joints are sealed to prevent leakage of POC.

The type of flues which connect to these can be either metallic flexible flue liner to BS EN 1856-2 or double-walled chimney/flue system to BS EN 1856-1.

Considerations when installing open flues

When an open flue is installed you must take into account the type of material, the length of the flue and the exposure of the flue.

Condensation

An open flue should be installed to keep flue gases at their maximum temperature and avoid problems of excessive condensation forming in the

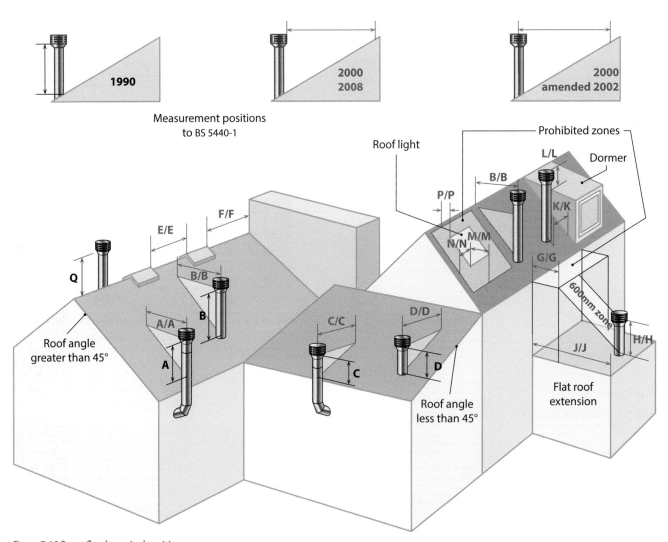

Figure 7.19 Open-flued terminal positions

flue. This is why single-wall flues are not allowed to be installed externally except where they project through a roof, and why twin-wall flues, with only an air gap for insulation, are only allowed up to 3 m in length when used externally.

Table 7.05 shows the maximum lengths of open flue to be used with a gas fire in order to avoid condensation.

Flue exposure	Condensate-free length		
	225 mm² brick chimney or pre-cast concrete block flue of 1300 mm² *	125 mm flue pipe	
		Single wall	Double wall
Internal	12 m	20 m	33 m
External	10 m	Not allowed	28 m

* See BS 5440-1 for more details.

Table 7.05 Maximum lengths of open flue used with a gas fire in order to avoid condensation

Terminal positions

It is important that you understand the positioning of open-flued appliance terminal positions. This is important when you are servicing or maintaining appliances which have been in a number of years. Table 7.06 indicates the positioning requirements from earlier editions of the standards (BS 5440-1).

Roof position, angle or situation		Minimum (mm)			
		BS 5440–1 Version			
		2008	2000		1990
A/A/A	Roof angle greater than 45 °	1,500	1,500	1,500	1000
B/B/B	Roof angle greater than 45 °	1,500	1,500	1,500	1000
C/C/C	Roof angle less than 45 °	1,500	1,500	1,500	600
D/D/D	Roof angle less than 45 °	1,500	1,500	1,500	600
E/E	Between ridge terminals	1,500	1,500	1,500	600
F/F	Ridge terminal to a higher structure	300	300	300	
G/G	Prohibited zone	2,300	2,300	2,300	
H/H	Flat roof extension	600	600	600	
J/J	Between edge of flat roof structure and dwelling. Either 10,000 mm along flat roof or to edge of structure which is ever the:	least	least	greater	
K/K	Where distance 'K' between flue and dormer is less than	1,500	1,500	1,500	1,500
L/L	Above dormer	600	600	600	600
M/M	Prohibited zone from a roof light downwards	2,000	2,000	2,000	
N/N	Prohibited zone from a roof light either side	600	600	600	
P/P	Prohibited zone from a roof light upwards	600	600	600	
Q	Above the ridge of a roof	600			

Table 7.06 Open-flued terminal position measurements

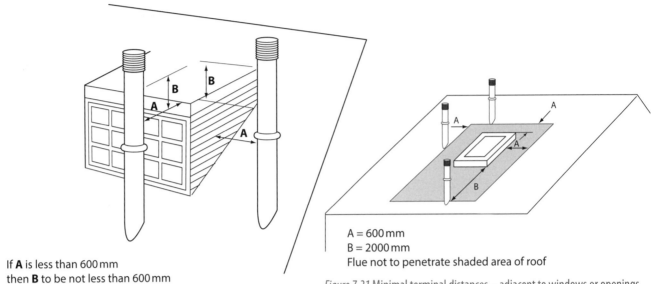

If **A** is less than 600 mm
then **B** to be not less than 600 mm

Figure 7.20 Locations of terminals for pitched roof with structures

A = 600 mm
B = 2000 mm
Flue not to penetrate shaded area of roof

Figure 7.21 Minimal terminal distances – adjacent to windows or openings on pitched roof

Figures 7.20 and 7.21 illustrate acceptable open-flued terminal positions. Please note that measurements are taken from the bottom of the terminal.

Prohibited zones

There are prohibited zones on or adjacent to buildings where open flues must not terminate, primarily to prevent downdraught and reversal of the flue-pipe operation, which would spill the POC into the building. Figures 7.22 and 7.23 show where flues must not be sited.

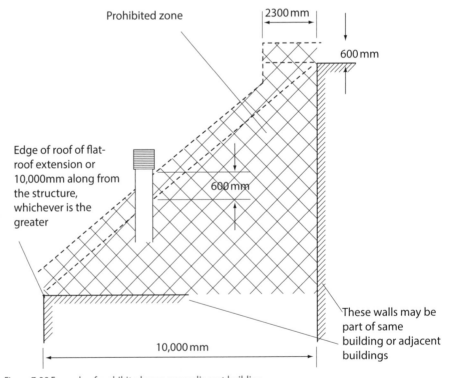

Prohibited zone

2300 mm

600 mm

Edge of roof of flat-roof extension or 10,000 mm along from the structure, whichever is the greater

600 mm

These walls may be part of same building or adjacent buildings

10,000 mm

Figure 7.22 Example of prohibited zone near adjacent building

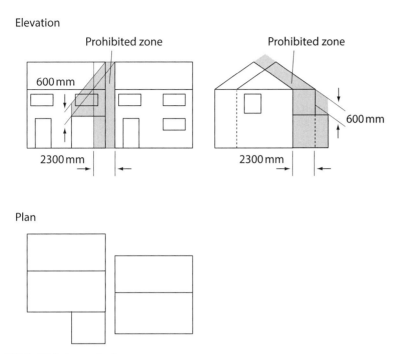

Figure 7.23 Prohibited zones for flues

To protect against the problems of downdraught where appliances are fitted to flues that exit through a steeply pitched roof, it is recommended that the route of the chimney is diverted to an outlet at the highest point on the roof, rather than terminating on the slope of the roof.

WORKING PRINCIPLES OF ROOM-SEALED SYSTEMS

While open-flued appliances can be difficult, as the route and terminal position are critical to ensure that safe dispersal of the products occurs and vents can easily be blocked or restrict the provision of air for combustion, room-sealed or 'balanced flue' appliances don't have such difficulties.

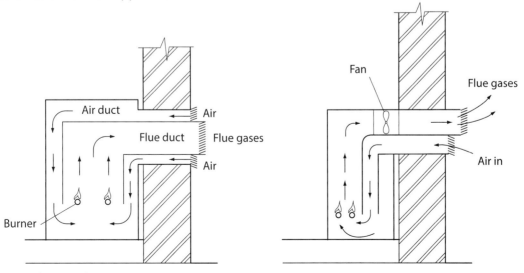

Figure 7.24 Principles of balanced flue operation

215

Air for combustion in a room-sealed or balanced-flue appliance is taken directly from outside and, as the name implies, the appliance is 'room-sealed', so there is no danger of POC entering the room.

Room-sealed appliances are therefore preferable to open flues. The flueing options are increased greatly with these types of appliance where a fanned flue system is chosen.

Terminating room-sealed systems

Room-sealed systems must be fitted within the vicinity of an external wall or roof termination. It can be seen in Figure 7.24 that the POC outlet and the air intake are at the same point and are therefore at equal pressure, whatever the wind conditions. This is why it is called 'balanced' flue. The special terminal that is part of the appliance must be fitted in such a position so as to:

- prevent products from re-entering the building
- allow free air movement
- prevent any nearby obstacles causing imbalance around the terminal.

BS 5440 Part 1 details acceptable positions for flue terminals on buildings, as shown in Figure 7.25 (larger appliances need greater distances).

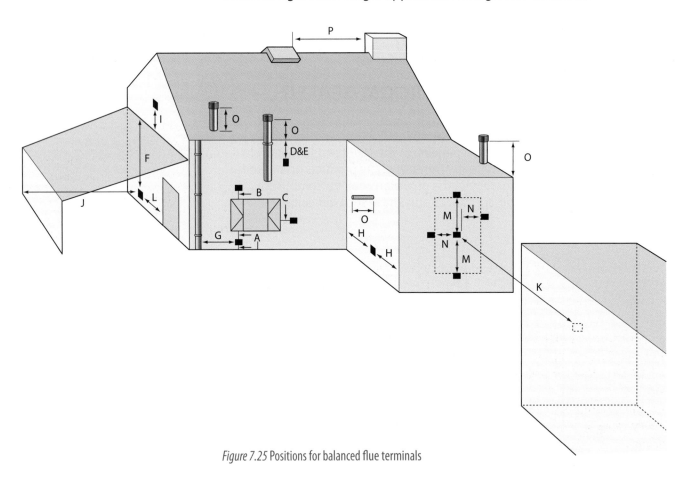

Figure 7.25 Positions for balanced flue terminals

Dimension	Terminal position	Heat input (kW net)	Natural draught	Fanned draught
A – see note 1	Directly below an opening, air brick, opening window, door etc.	0–7 kW >7–14 kW >14–32 kW >32–70 kW	300 mm 600 mm 1,500 mm 2,000 m	300 mm 300 mm 300 mm 300 mm
B – see note 1	Above an opening, air brick, opening window, door etc.	0–7 kW >7–14 kW >14–32 kW >32–70 kW	300 mm 300 mm 300 mm 600 mm	300 mm 300 mm 300 mm 300 mm
C – see note 1	Horizontally to an opening, air brick, opening window, door etc.	0–7 kW >7–14 kW >14–32 kW >32–70 kW	300 mm 400 mm 600 mm 600 mm	300 mm 300 mm 300 mm 300 mm
D	Below gutters, drain pipes or soil pipes	0–70 kW	300 mm	75 mm
E	Below eaves	0–70 kW	300 mm	200 mm
F	Below balconies or car-port roofs	0–70 kW	600 mm	200 mm
G	From a vertical drain pipe or soil pipe	0–70 kW		1,500 mm – see note 4
H – see note 2	From an internal or external corner	0–70 kW	600 mm	300 mm
I	Above ground, roof or balcony	0–70 kW	300 mm	300 mm
J	From a surface facing a terminal – see note 3	0–70 kW	600 mm	600 mm
K	From a terminal facing a terminal	0–70 kW	600 mm	1,200 mm
L	From an opening in the car-port into the dwelling	0–70 kW	1,200 mm	1,200 mm
M	Vertically from a terminal on the same wall	0–70 kW	1,500 mm	1,500 mm
N	Horizontally from a terminal on the same wall	0–70 kW	300 mm	300 mm
O	Above intersection with the roof	0–70 kW	N/A	Manufacturer's instructions
P	Between a chimney and a ridge terminal		1,500 mm (300 mm between similar ridge terminals)	

Note 1

In addition the terminal should not be closer than 150 mm (fanned draught) or 300 mm (natural draught) from an opening in the building fabric for the purpose of accommodating a built-in element such as a window frame.

Note 2

This does not apply to building protrusions less than 450 mm, e.g. a chimney or an external wall, for the following appliance types: fanned draught, natural draught up to 7 kW, or if detailed in the manufacturer's instructions.

Note 3

Fanned-flue terminal should be at least 2 m from any opening in a building that is directly opposite and should not discharge POCs across an adjoining boundary.

Note 4

This dimension may be reduced to 75 mm for appliances up to 5 kW (net) input.

Table 7.07 Balanced flue terminals

Note that the outlet part of the terminal can become quite hot, and therefore a guard must be fitted if the terminal is within 2 m of ground level or if persons have access to touch it, e.g. on a balcony.

Take special care when fitting room-sealed flues through walls, particularly in timber-framed buildings to protect against fire. As always, follow the manufacturer's instructions carefully.

Terminals for room-sealed flues or fanned draught open flues must be positioned to ensure the safe dispersal of flue gases. In general this means that no terminal should be located more than 1 m below the top level of a basement area, light well or retaining wall. The products must discharge into free open air. Further guidance is given in BS 5440.

Balanced compartments

The balanced compartment is a method of installing an open-flued appliance in a room-sealed situation and arranging the chimney and ventilation so that a balanced flue effect is achieved.

This method is useful where higher rated appliances are installed. This method can also be an alternative to longer external flues, or can be used where a boiler is housed in an adjacent boiler house to a tall building.

The chimney outlet location in Figure 7.26 is only suitable for balanced compartment applications and is not to be used for other open-flue installations.

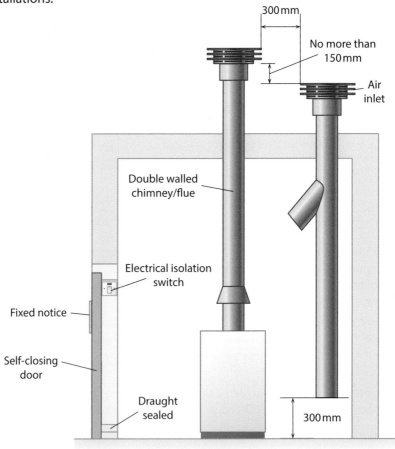

Figure 7.26 Non-proprietary balanced compartment

When a proprietary system is used, as in Figure 7.27, then the systems manufacturer's instructions should be followed.

Balanced compartments must have:

- a self-closing flush door with a draught-sealing strip
- a door that should not open into a room containing a shower or bath, or where the appliance has a rated input greater than 12.7 kW (14 kW gross) and the door is opening into a room intended for sleeping accommodation
- a notice attached to the door stating that the door must be kept closed
- no other ventilation opening in the compartment other than those designed
- a door opening fitted with a switch that acts as an electrical isolator, shutting down the appliance when the door is opened
- twin-wall insulated pipes flue pipes to BS EN 1856-1
- insulated any exposed hot water pipework to a minimum thickness of 19 mm in order to minimise heat transfer within the compartment.

In addition, the chimney and ventilation should ensure the full clearance of the POC. The supply duct (air inlet) terminal should be no more than 150 mm from the base of the chimney outlet.

Where the air is ducted to a low level, i.e. 300 mm or less from the floor (see Figure 7.26) within the balanced compartment, the cross-sectional area of the air supply duct should be not less than 7.5 cm^2 per kW (net) of the appliance maximum rated input. Maximum rated input is calculated from 1.5 times the allowance for the maximum air vent area for a high level, direct to outside air opening specified for open-flued appliances in BS 5440-2.

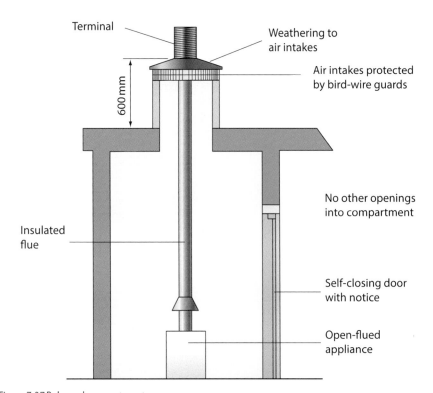

Figure 7.27 Balanced compartment

Where the air is ducted to a high level only (see Figure 7.27) within the balanced compartment, the cross-sectional area of the air supply duct shall be not less than 12.5 cm^2 per kW (net) of the appliance maximum rated input. This is calculated from 2.5 times the allowance for the maximum air vent area for a high level, direct to outside air opening specified for open-flued appliances in BS 5440-2.

TYPES OF ROOM-SEALED FLUES

Shared flues

Shared flues are mainly for use in multi-storey buildings, but since you may work on an appliance in a domestic flat, it is important that you recognise the main features.

The two types of system are the SE duct and the 'U' duct, as shown below in Figure 7.28.

Appliances used in these type of ducts are specially adapted versions of room-sealed flue appliances which are fitted into the vertical flue and air duct on each floor. Only types C_2 and C_4 appliances are suitable for use. Replacement appliances must be of the same type and suitably labeled, stating that they are fitted to a shared flue.

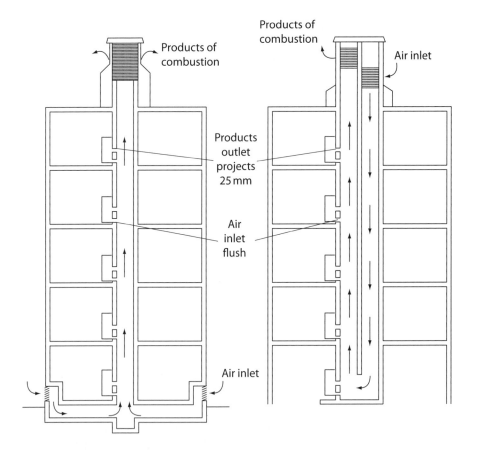

Figure 7.28 SE duct (on left) and 'U' duct (on right)

The responsibility for the shared flue itself is that of the landlord or the person responsible for the building. Annual checks need to be carried out on the shared flue.

Only type C_2 appliances should be connected to a SE duct or 'U' duct chimney (where the flue duct and air supply duct of the appliance are connected into the same common duct of the chimney) in accordance with the appliance manufacturer's installation instructions.

Only a C_4 appliance should be used with the shared chimney where the flue duct and the air supply duct of the appliance are connected to separate common ducts of the shared chimney.

Fanned flue

Condensing boilers are normally of the fanned-flue type and are becoming more popular, owing to their increased efficiency. These appliances have a tendency to form a plume of vapour from the flue terminal – you should take this into account when siting the appliance. You should also consider the disposal of condensate and follow manufacturer's instructions. Typical condensate drainage is discharged in plastics which are to be solvent welded (MUPVC) to avoid corrosion problems, and the position of termination may be to internal or external discharge stacks with a trap fitted, and where no stack or drain gulley is available to soak away.

Vertex flues (type C7)

Vertex flues are unusual, with the air supply being taken from the roof space. The secondary flue is connected to a draught break in the attic which should be ventilated to the standard of current Building Regulations. The draught break must be at least 300 mm above the level of any insulation, and the flue above the break should be vertical for at least 600 mm before any bend is used.

Vertex type appliances

Consider the following guidance where a chimney for a Type C_7 ('Vertex') appliance is used (the primary flue and draught break are both parts of the appliance):

- install in accordance with the instructions provided by the appliance manufacturer
- install the secondary flue connected to the appliance draught break in the roof space in accordance with the instructions provided by the chimney manufacturer
- make provision for an adequate unobstructed air supply to the roof space in which the draught break is located
- make sure the secondary flue connected to the appliance draught break is constructed of a non-corrosive material such as stainless steel and is vertical
- if a change of direction (offset) is unavoidable, the first section of the chimney above the draught break should rise vertically by a minimum of 600 mm before it changes direction

Progress check

1 At what height above a dormer window should an open-flued appliance terminal finish?

   a 300 mm

   b 600 mm

   c 1,100 mm

   d 1,500 mm

2 By how much is the prohibited zone to the side of a roof light?

   a 300 mm

   b 600 mm

   c 900 mm

   d 1,200 mm

3 Complete the following statement: 'To prevent downdraught where appliances are fitted to flues that exit through steeply pitched roofs it is recommended that. . .'

4 If an 11 kW fanned flue boiler terminates above an air brick, what is the minimum measurement above the air brick that the flue should terminate?

5 How far from an opening directly opposite a fanned flue should the outlet be terminated?

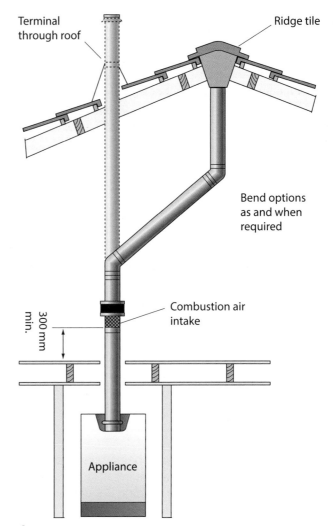

Figure 7.29 Vertex flue

- make provision to collect and remove any condensate that may form above the draught break
- ensure the draught break is at least 300 mm above the level of the insulation in the roof space
- conduct a spillage test in accordance with the manufacturer's instructions at the draught break in the roof space (see the section on 'Checks with appliance connected (spillage test)', page 225).

The ventilation design parameters should be considered and any roof ventilators should be checked to ensure they are unobstructed by insulation, etc.

TESTING GAS APPLIANCE FLUE SYSTEMS

Regulation 26(1) of the GSIUR states:

26(1) No person shall install a gas appliance unless it can be used without constituting a danger to any person.

Approved Code of Practice 26(1) also gives guidance that, as a gas operative, you should ensure that any appliance you install, or flue to which you connect an appliance, is safe for use. You should ensure requirements in Appendix 1 of the GSIUR are met, as applicable, and refer to the appropriate standards.

It is essential, then, that you inspect and test any flue for gas appliances, not just at installation but each time the appliance is worked on, including service/maintenance. It is necessary to carry out checks on the complete flue system, as follows.

Inspection and tests for open-flued systems

Building Regulation Document J (Appendix E) gives guidance on the testing of natural draught flues, both existing and new. These procedures only apply to open-flued appliances and are only used to assess whether the flue in the chimney, the connecting flue pipe and the flue gas passage in the appliance

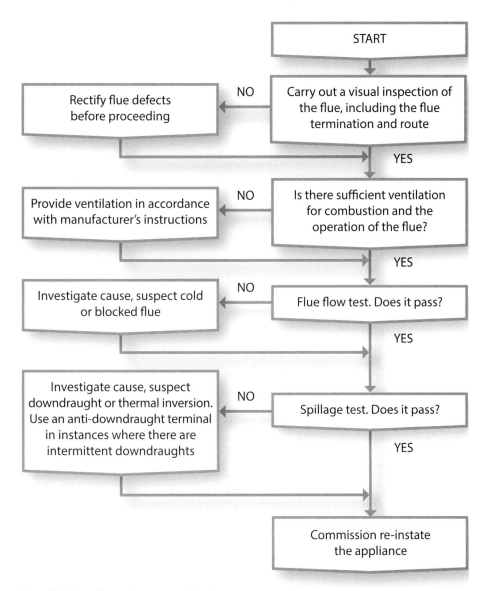

Figure 7.30 Open-flue testing sequence flow chart

are free from obstruction and are acceptably gas tight. The procedure for flue testing is also outlined in BS 5440 Part 1. Where possible, test flues at the most appropriate time during the building work and before any finishing coverings have been applied e.g. plaster or dry-lining boards.

Methods of testing open-flued systems involve a visual check, and spillage and flow tests.

Visual inspection

Visual inspections are covered in 6.3.2.1 of BS 5440 Part 1.

The chimney, whether existing, newly erected, adapted or altered, should be visually checked before you fit an appliance. You should ensure it is fit for the intended use with the intended appliance, and:

- is unobstructed, complete and continuous throughout its length
- serves only one room or appliance
- has the terminal correctly sited and a weather-tight joint between the terminal and the chimney
- any dampers or restrictor plates have been removed or permanently fixed in the open position to leave the main part of the flue unobstructed
- any catchment space is the correct size, free of any debris and any gaps into the catchment space are sealed from the surrounding structure
- where an existing chimney has been used, any signs of spillage should be investigated and faults rectified.

In addition to the above:

- Inspect the loft space to ensure that any chimney passing through it is complete, continuous and not damaged; that all joints are properly made; and that it is properly supported using suitable brackets, especially non-vertical sections.
- Where the flue passes through or is connected to an adjoining property, inspect the adjoining property so far as is practicable.
- As far as practicable, inspect masonry chimneys to ensure they are free from debris and soundly constructed. Remove any debris. If a masonry chimney is in poor condition it should be renovated to ensure safe operation. One solution might be to fit a correctly sized liner.

Flue flow testing (smoke test)

The smoke test is covered in 6.3.2.2 of BS 5440 Part 1. On satisfactory completion of the visual check, you should then inspect the flue flow as follows:

- Having established that an adequate air supply for combustion has been provided in accordance with the appliance requirements, close all doors and windows in the room in which the appliance is to be installed.
- Carry out a flow visualisation check using a smoke pellet that generates at least 5 m^3 of smoke in 30 secs burn time at the intended location for the appliance. Ensure that there is discharge of smoke from the correct terminal only and no leakage into the room.

Where gas fires are fitted that require a closure plate, the flue flow test should be carried out with the closure plate in situ.

Where the chimney is reluctant to draw and there is smoke spillage, introduce some heat into the chimney for a minimum of 10 mins using a blow torch or other means and repeat the test. The pre-heating process might require as much as half an hour before the chimney behaves as intended, as a blow torch does not provide a representative volume of heat into the chimney consistent with normal appliance operation.

When an adequate air supply and correct flow have been confirmed, there should be:

- no significant escape of smoke from the appliance position
- no seepage of smoke over the length of the chimney
- a discharge of smoke from only the correct terminal.

If smoke comes out of a chimney outlet other than the correct one, or the downdraught or 'no flow' condition indicates that the chimney has failed the test, see Figure 7.32.

Where the chimney has failed the test, you should undertake a thorough examination of the chimney to identify any obvious cause of failure. The appliance should not be connected until any defect has been found and rectified.

A smoke test is very subjective and is only intended to establish that the chimney serving the appliance is of sufficient integrity that it can safely remove the POC when the appliance is alight.

Weather conditions and the temperature of the chimney at the time of testing can influence the results of the test. Also, the material the chimney has been constructed from may determine the outcome of the test.

If the chimney has been correctly applied and constructed, check it for adequate and safe performance while connected and lit, and then re-test until satisfied that the chimney is functioning properly.

If the chimney continues to fail after a longer preheating period, and there is no obvious reason for this, it might be necessary to have the appliance installed in position but not connected to the gas supply, so that the smoke test can be carried out with representative flue flow conditions.

Checks with appliance connected (spillage test)

The spillage test is covered in 6.3.2.3 of BS 5440 Part 1.

Do not install new or used appliances unless the appliance manufacturer's instructions are available. Where the appliance manufacturer's instructions are not available, the appliance manufacturer shall be consulted.

On satisfactory completion of the flue flow test, check the chimney, with the appliance connected, as follows:

- In the room:
  - close all doors and windows
  - close all adjustable vents
  - switch off any mechanical ventilation supply to the room other than any that provides combustion air to an appliance
  - operate any fan and open any passive stack ventilation (PSV) (see the section on 'Passive stack ventilation' page 235).

Figure 7.31 Flue flow test passed

Figure 7.32 Flue flow test failed

■ With the appliance in operation at its set input setting, check that the appliance clears its POC using the method described in the appliance manufacturer's instructions. If spillage is detected, switch off the appliance, disconnect and rectify the fault.

■ Close any PSV and repeat the test. If spillage is detected, switch off the appliance, disconnect and rectify the fault.

Where the installation instructions do not contain specific instructions for checking spillage, proceed as follows:

■ In the room:
  — close all doors and windows
  — close all adjustable vents
  — switch off any mechanical ventilation supply to the room other than any that provides combustion air to an appliance
  — operate any fan and open any PSV.

■ With the appliance in operation carry out a flow visualization check by applying a smoke-producing device such as a smoke match to the edge of the draught diverter or gas fire canopy within 5 mins of lighting the appliance.

■ Apart from an occasional wisp, which may be discounted, all the smoke should be drawn into the chimney and evacuated to the outside air.

■ Close any PSV and repeat the test.

■ If spillage occurs, leave the appliance operating for a further 10 mins and then re-check. If spillage still occurs, switch off, disconnect the appliance and rectify the fault.

■ If there are fans elsewhere in the building, the tests should be repeated with all internal doors open, all windows, external doors and adjustable vents closed and all fans in operation.

Examples of fans which might affect the performance of the chimney by reducing the ambient pressure near to the appliance are:

■ fans in cooker hoods

■ wall- or window-mounted room extract fans

■ fans in the chimneys of open-flued appliances, including tumble dryers

■ circulating fans of warm air heating or air conditioning systems (whether gas fired or not)

■ ceiling (paddle) fans – these could particularly affect inset live fuel effect fires.

All fans within the appliance room and adjoining rooms should be operated at the same time. In addition, if a control exists on any such fan, then the fan should be operated at its maximum extract setting when the spillage test is carried out.

Any appliance found to be spilling POC is classed as 'immediately dangerous' (ID), and must be disconnected. See Chapter 11 for information on responding to unsafe situations.

Do not leave the appliance connected to the gas supply unless it has successfully passed these spillage tests. Disconnect the gas supply to the appliance, inform the user/owner or responsible person and attach a label to

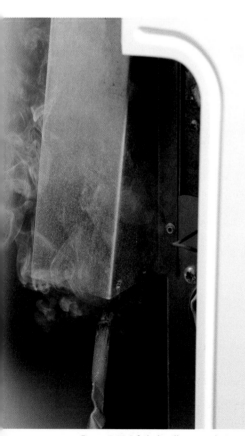

Figure 7.33 A failed spillage test being carried out

the appliance to warn that it should not be used until the fault is remedied in accordance with the Gas Industry Unsafe Situations Procedure [15].

Where radon gas extraction systems are installed, test the spillage performance of every open-flued appliance in the building in accordance with 6.3.2.3 with the radon gas extractive system in operation. Do not leave the appliance connected to the gas supply unless it has successfully passed the spillage test specified in 6.3.2.3.

Case study

Angus has been instructed to undertake a flue flow test on a chimney which is to have a gas fire fitted. Answer the following questions and check your answers with your tutor.

1 What visual checks should Angus make prior to the flue flow test?

2 How should Angus perform the flue flow test and what should he check for?

3 What should Angus do should the test fail on the first test?

Case study

Leigh has been called to a customer who complains that they always feel tired when they have the gas fire turned on.

List the procedures Leigh should undertake on entry to the property.

Inspection and tests for room-sealed flue systems

For room-sealed flue systems (natural draught and fan assisted), you should visually examine the appliance and the chimney configuration before leaving the appliance connected to the gas supply. You should check to ensure that:

■ the sealing method used by the appliance manufacturer to provide the room seal between the combustion chamber and the room is intact and in good condition – this includes checking that any sight glasses are properly fitted and ensuring that the appliance back plate or case has not been distorted such as to make any seal ineffective

■ the flue duct or combined chimney configuration are either continuous throughout the wall or, if they are telescopic, then any sealing tape required by the appliance manufacturer's instructions has been fitted

■ no debris is contained within the room-sealed chimney configuration

■ the joint between the terminal and the wall is weatherproof

■ internal equipment such as thermocouples and wires are securely held or positioned to ensure that they cannot interfere with the sealing of the combustion circuit, and all grommets etc. are in place.

Testing room-sealed positive pressure fanned-draught flue installations

BS 5440-1 gives guidance for the checking/testing of **positive pressure** case appliances. When a fan is installed at the inlet of the flue, the boiler combustion chamber operates at a positive pressure. Special precautions must be taken to prevent POC escaping. When testing room-sealed fanned-draught flue installations, you need to follow the four steps: checking case seals and integrity before fitting; checking the case fits and the appliance can be operated properly; lighting and operating the appliance; and checking flue pipe air inlet connections for leakage.

Key term

Positive pressure
This is when pressure within a system is greater than that of the environment surrounding it.

Case seal checks

Before the case is put back on the appliance, the following checks should be carried out:

- Are any water leaks evident?
- Is the backplate or case corroded?
- Where corrosion is evident, is it likely to affect the integrity of the case, backplate or seal? The extent of the corrosion should be carefully checked with a sharp instrument e.g. a screwdriver. If the instrument does not perforate the corroded area, this should be deemed acceptable, but be sure to advise the gas user of the problem and potential consequences if a repair is not made.
- Are the combustion chamber insulation linings intact?
- Is the backplate or the case distorted or damaged? Pay particular attention to the area where the case and seal meet. Distortion or damage here may have been caused by explosive ignition of the main burner.
- Is the case sealing material intact and in good condition (e.g. pliable, free from discoloration, trapped debris, etc.)?
- Will the case seal continue to form an adequate seal between the case and the backplate?
- Is anything trapped or likely to be trapped when the case is put back on (e.g. wires, thermocouple capillaries, tubes, etc.)?
- Are other gaskets and seals intact?
- Is the pilot inspection glass undamaged?
- Are the case fastenings and fixings (including fixing lugs) in good condition (e.g. screws/nuts stripped)?
- Are there any signs of discoloration on or around the appliance, which may have been caused by leaks of POC from the appliance?

You must rectify any defects identified in these checks as necessary before refitting the case.

Where defects are identified they should be classified using the following criteria in accordance with the current Gas Safe Gas Industry Unsafe Situations Procedure. Where there are inappropriate or missing case fittings or defective seals which cannot be remedied, but there is no evidence of leakage, the appliance should be classified as 'at risk' (AR). If there is evidence of actual leakage, then the appliance should be deemed ID. Where suitable replacement seals are no longer available the appliance should be classed as ID and regarded as obsolete.

Case fitting checks

When the case has been put back on the appliance, carry out the following checks:

- Is the case fitted correctly?
- Is a 'mark' visible, showing that the case had previously been fitted closer to the backplate?
- Are all the case screws adequately tightened?
- Is a bright area visible on the screw thread of any of the case-securing screws, indicating that the screw was previously secured more tightly?
- Is anything trapped and showing through the case seal?

Any defects identified when refitting the case should be rectified before lighting and operating the appliance.

Lighting and operating the appliance

Once the case is refitted satisfactorily, you can light the appliance:

- ensure the main burner remains lit (i.e. set the appliance and room thermostats to their highest settings)
- check for possible leakage – initially this can be done by running your hands around the boiler casing and backplate.
- then check the appliance thoroughly for leakage, as described below.

Leakage checks

Where joints have been disturbed, check with leak detection fluid to confirm that there are no gas escapes. To check for possible leakage of POC from the appliance:

- Use a taper (for less accessible locations), an ordinary match, or similar. Whilst smoke tubes and smoke matches can be used, the results may require further interpretation and these methods are currently being validated.
- Light the taper/match and allow the flame to establish. Position the flame very close to the case seal or any possible leakage point (e.g. back panel).
- The flame will be blown quite easily by the draught caused by a leak. Move the taper around the entire seal, using fresh tapers as required.
- To investigate the seal at the bottom of the case, hold the lit taper between the bottom of the case and the appliance control panel. Does the flame flicker slowly or is it disturbed by leakage flowing from the case? Try the taper in several positions.

Be careful not to confuse natural convection with leakage.

If you find any defects, rectify them as necessary and repeat the checks. If you are still unsure, seek expert advice.

Safe working

Do not look for a gas escape with a naked flame, e.g. matches or lighter.

Safe working

When checking for POC leakage with lit tapers/matches, be careful not to set fire to surrounding fixtures/furnishings.

Flues in voids

Gas Safe issued Technical Bulletin 008 (Edition 3) 'Existing concealed room-sealed fanned draught boiler chimney/flue systems in domestic premises' on 1 April 2013 to replace TB 008 (ed. 2.1).

The aim of this bulletin is to provide guidance to gas engineers to assist in meeting the requirements of the Regulations when working on existing concealed room-sealed fanned-draught boiler chimney/flue systems when working in domestic premises.

Note: The advice in this bulletin is for guidance only and engineers can take other action if they wish. However if the engineers follow this guidance they will be complying with the law.

Regulation 26(9) of the Gas Safety (Installation and Use) Regulations (GSIUR) places a legal duty on registered engineers to immediately examine and confirm the effectiveness of a flue whenever they undertake work on flued appliance(s). This becomes difficult to fulfil where a concealed chimney/flue system is encountered, e.g. where they are concealed in ceiling voids etc., which occurs in many existing developments. These mainly occur in flats or apartments built between 2000 and 2007.

From 1 January 2013 any concealed room-sealed fan-draught boiler chimney/flue system installation being worked on, where the effectiveness of the chimney/flue system cannot be confirmed, should be classified as 'At Risk' with the responsible person's permission and turned off in accordance with the Gas Industry Unsafe Situations Procedure (GIUSP). Where a customer/responsible person does not give permission to turn off their 'At Risk' boiler, they should be asked to sign paperwork to confirm they accept responsibility for a situation which could result in a serious incident. In the case of an 'Immediately Dangerous' situation where permission to disconnect has not been given, the registered engineer should contact the Gas Emergency Contact Centre

The term 'Concealed' when used in the context of situations of chimney/flue system passing through refers to ceiling and floor voids and behind false walls.

However it does **not** apply to chimney/flue system incorporating:

- vertical condensing flexible room-sealed fanned-draught chimney/flue systems installed in enclosures, such as constructional chimneys etc., which are sealed so that any leakage of the products of combustion cannot pass from the enclosure to any room or internal space
- short chimney/flue systems connected directly from an appliance to the outside air through an external wall
- air inlet pipes of twin chimney/flue systems.

Where the entire chimney/flue system can be examined, for example where the system does not pass through a concealed void, then this information is not relevant and the chimney/flue system carries no risk.

Note: A 'constructional chimney' is considered to be an existing construction designed and built to operate as a chimney, in accordance with the Building Regulations that were in place at the time of the dwelling construction.

Preferred industry options

The gas industry guidance explains how to deal with existing/concealed room-sealed fanned draught boiler chimney/flue systems, in order of preference as follows.

Installation of appropriately specified and located inspection hatches, room monitoring carbon monoxide (CO) alarms and regular service and maintenance by a registered engineer

Where systems are concealed and cannot be visually confirmed as being complete/ intact and effective, appropriately located and installed inspection hatches are considered the most effective method for the inspection of chimney/flue systems. This form of examination and other operational safety checks are necessary to confirm the safe operation of the boiler as specified by GSIUR 26(9). The installation of room monitoring CO alarms throughout the length of the route will ensure as is reasonably practicable the system is safe for continual use.

The following installation defects may contribute to an increased risk of chimney/flue system failure to joints or supports:

- For condensing boilers, inadequate gradient/fall of the flue system back to the boiler may trap condensate, putting excessive strain on the chimney/ flue joints or supports.
- Incorrect or inadequate flue system support may cause significant risk of chimney/flue system failure.

- Signs of condensate/water leakage at the chimney/flue system joints.
- The use of incorrect materials/joints other than those specified by the manufacturer.

Installation of a CO void monitoring safety shut-off system (COSSVM) and regular service/maintenance by a registered engineer

Where instances restrict the installation of hatches for inspection, for example:

- the enclosure around chimney/flue system may be too small
- where the installation of hatches may affect the integrity of fire protection for buildings.

The installation of a system which monitors the presence of CO in the void and will automatically switch off the boiler should an incident occur may be considered to ensure the boiler and chimney/flue system are safe for continual use.

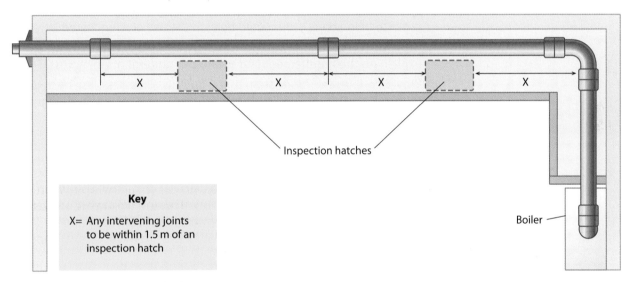

Key

X= Any intervening joints to be within 1.5 m of an inspection hatch

Inspection hatches

Boiler

Figure 7.34 Positioning of inspection hatches

Note: This system is not acceptable for new or replacement installations and will need to meet the relevant Building Regulations/standards. The customer needs to understand the need for hatches and why they will have to be fitted at the same time. Void monitoring systems alone will not satisfy the Building Regulations/standards.

Exceptions

Where you have identified that a short chimney/flue system is concealed in a void which has no means of access to allow inspection the following factors will need to be confirmed when completing your safety check documentation:

- There are no changes in the chimney/flue direction.
- There are no signs of distress, this would indicate that a chimney/flue issue exists.
- The chimney/flue length does not exceed the maximum single chimney/flue system component length supplied by the manufacturer.
- Documentation exists from the installation that no chimney/flue flue joint are within the concealed part of the building.

Once these are confirmed, the installation can then be considered acceptable. Copies will be needed for the customer/responsible person and yourself for future reference.

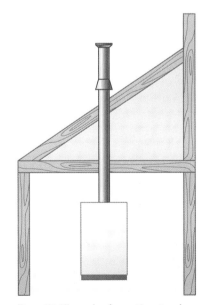

Figure 7.35 Example of exceptions to rules

Flues passing through adjacent property

On some older flue/chimney systems you may encounter flues which have been routed through an adjacent property. This is no longer allowed on new flue/chimney systems. Where these situations occur, and on the basis of checks of the boiler and flue/chimney system in the property are satisfactory, you must take reasonable steps to ensure the overall flue/chimney systems integrity (see Appendix 1 of GSIUR Approved Code of Practice L56). Gaining access to adjacent properties will normally require the full co-operation of others e.g. Housing Associations, landlords and neighbours. You can demonstrate that you have taken reasonable steps by taking the following actions:

- Make enquiries with all parties and request to see evidence of reports of examinations made by them, or on their behalf.
- Make enquiries with the occupants of those other adjacent properties in order to gain access – this could be with the use of a registered letter.
- Leave documentation with the occupier of the adjacent property, explaining the requirement and seeking arrangements for access.
- Providing relevant information to the flues in voids database hosted by the Gas Safe Register and accessed at https://engineers.gassaferegister.co.uk/FluesInVoids.aspx.

Where access to the property cannot be achieved, despite taking all reasonable steps and there is no evidence to indicate that any chimney/flueing problems exist (based on evidence from checks undertaken in the property where the boiler is located), the boiler and chimney/flue system can be left operational. However if there is good reason to suspect problems with the system, it is essential that the complete length of the system is checked through the adjacent property. Until access to the adjacent property is gained, then the boiler should be classed as 'At Risk' and actions should be applied to this classification.

Boiler operational safety and other checks

When you work on a boiler with a concealed chimney/flue system, as well as confirming the effectiveness of the flue, the supply of combustion air, operating pressure and/or heat input (gas rate), engineers must verify the following requirements in order to ensure the boilers' safe functioning:

- that the combustion performance is correct in all modes of operation.
- where manufacturers provide an air sampling point, that any specified O_2 levels are in accordance with the manufacturer's instructions.
- the plume/heat discharge is evident from the chimney/flue termination with the boiler in operation.
- there is no evidence of distress on the enclosure or ceiling along the complete length of the chimney/flue system, likely to arise due to system integrity issues.
- there is no knowledge of previous history issues relating to the property, or other properties in the same development that could be related to concealed systems issues that have not been corrected/rectified before, e.g. enquire from the responsible person.

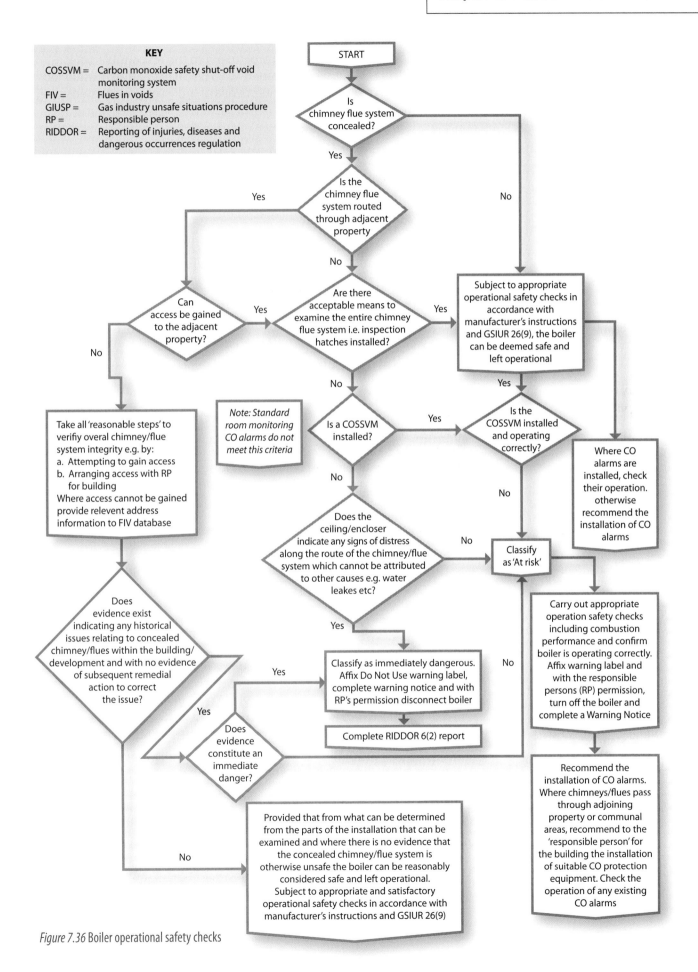

Figure 7.36 Boiler operational safety checks

The registered engineer should also recommend to the responsible person that they install CO alarms which conform to BS EN 50291 and are positioned in accordance with manufacturer's instructions or industry guidance, ensuring that they work when tested.

Factors affecting flue system performance

Below are the situations you may encounter when you are investigating why a flue is not performing as it should.

Downdraught conditions

Downdraught is a condition of the flue where the POC are forced back down the flue and into the room where the appliance is sited. This condition is normally not permanent and can be caused by wind conditions acting on a poorly sited flue terminal, ventilation opening or chimney/flue route configuration.

Wind effects at the appliance termination

When wind blows across a building, it is most likely to produce a pressure differential between the bottom of the flue and the flue terminal. This will depend upon the:

- wind speed
- direction of the wind
- position of flue outlet in relation to the building
- location of neighbouring buildings and structures
- geographical features (e.g. hills and valleys).

These may cause or prevent a natural draught to occur in the flue. Because of this it may cause any one of the following to occur:

- increased flow up the flue
- reduced flow up the flue
- intermittent downdraught.

Because of its nature, wind pressure on occasions can be greater than the flue draught from open flues. To minimise this effect it is important that you position the flue terminal where the effect of the wind is minimised. This can be done by checking the position as in Figure 7.19 (page 212).

Figure 7.37 shows that the zone with the greatest pressure is on the windward side of a dwelling, i.e. the direction the wind is blowing into the dwelling. As you can see from Figure 7.37, the flue is on the windward side of the dwelling which could cause downdraught and spillage of the POC into the dwelling.

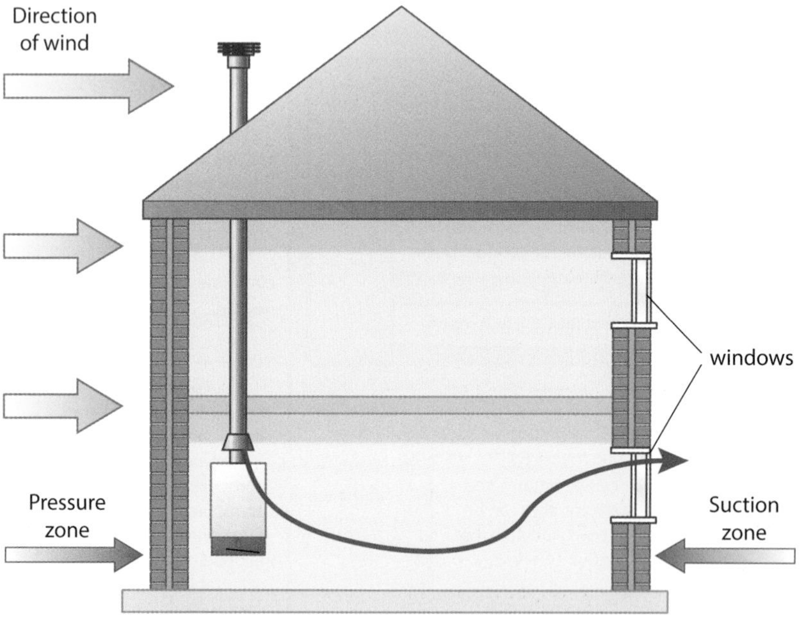

Figure 7.37 Adverse effects of wind on open flues

Passive stack ventilation

Passive stack ventilation (PSV) is a system to extract air from rooms in which it is installed and it can be built into modern homes. These rooms are mainly bathrooms, kitchens and WCs. The system is made up of a series of ducts, one to each of the rooms that is to be ventilated. These ducts rise vertically through the building and terminate above the roof level of the dwelling. They work by natural ventilation: the difference in the temperature between the air outside and inside, and also the effects of wind passing over the top of the stack.

These PSV's have to be positioned so that it does not adversely affect the operation of the open-flued system. The ideal positioning of the two stacks is on the same face of the building. The open flue should terminate at the same height or higher than the PSV.

Extraction fans

Extraction systems which are within the vicinity of or in an adjacent room to an open-flued appliance can cause problems with the safe removal of the POC.

If this problem is found to be where an appliance is fitted in the same or adjacent room, then additional ventilation may need to be installed to overcome the problem by depressurising the effect of the fan. Some examples of fans that may cause problems to open-flued appliances are listed in the section on 'Checks with appliance connected (spillage test)' (page 225).

COMMISSIONING FANNED-DRAUGHT OPEN-FLUE SYSTEMS

During the commissioning process for fanned-draught open-flue systems it is important that as well as following manufacturer's instructions you commission using the guidance below.

Fans integral to an appliance

For fans integral to an appliance, be sure to carry out the commissioning in accordance with the appliance manufacturer's instructions.

Fans installed on site

The commissioning of open flues with a fan installed on site in the secondary flue should conform to section 6.3 of BS 5400-1 and the following:

- set the fan speed in accordance with the fan manufacturer's commissioning instructions
- check all safety controls specified for safe operation
- check the safety control to ensure it shuts off the gas supply to the main burner within 6 seconds of any spillage occurring from the draught diverter or any other flue break in accordance with the appliance manufacturer's instructions
- check the clearance of POC from any other open-flued appliance in the room, adjoining room or space, with the fan-powered chimney in operation and all external doors and windows closed, and with the interconnecting door open.

The minimum flow rates for fanned flues should be as given in Table 7.02 (page 201).

ANALYSING FLUE COMBUSTION PRODUCTS

With the introduction of BS 7967 Parts 1 to 4, gas operatives have had to undertake Combustion Performance Analysis (CPA) as part of the ACS scheme. The competency is a requirement for all installers who work with central heating appliances and space heating appliances.

This has meant that gas operatives have had to purchase combustion gas analysers, the cost of which can range from £300 to £700.

The current standard with which the analysers should comply is BS 7927:1998, includes Amendment 1:1999, for the measurement of CO and CO_2. The standards require that the analyser has an accuracy that should be better than + or – 3 parts per million (ppm) below 20 ppm and + or – 5 per cent or better above 20 ppm. The analyser should be certificated by the manufacturer or supplier to say that it complies with these regulations. The equipment must also carry a current calibration certificate.

Because hard copy evidence is required as part of the analysis work, you will also need to purchase a portable printer that is compatible with the equipment. These printers connect directly to the analyser as they are either in-built or connect via infra-red.

Because of its greater toxicity, instructions for the determination of POC levels are mainly aimed at the measurement of CO only. However, in some circumstances the measurement of CO_2 can give a better indication of the presence of POC. This is particularly true where the volume of combustion products released is small, the concentration of CO in the POC is low or access to the appliance is restricted, for example when appliances are fitted in small appliance compartments. Where CO_2 is used to check for the presence of POC, as a precaution it is essential that CO is monitored as well.

This procedure is not intended to measure the absolute levels of CO within a space. It is intended to measure an increase in CO levels, which, for example, could be emitted from a faulty appliance installation, above the background CO level.

These measurements are normally only made to provide reassurance to the customer, or where there is still a cause for concern of an appliance which has been installed correctly, is in apparently good working order, and is operating in accordance with the manufacturer's instructions.

If an absolute CO level is required it will be necessary to set the analyser to zero with a CO-free sample. If the analyser measures absolute CO levels it will be necessary to subtract the outdoor background CO level from the level recorded in the building.

When forming the final conclusion of the test, bear in mind that busy roads and car parks can generate significantly high levels of outdoor background CO.

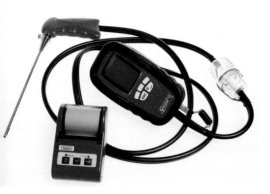

Figure 7.38 Combustion analyser with infra-red printer and flue sampling probe

Types of portable combustion gas analysers

BS 7967-2 gives guidance on the types and use of portable gas analysers. Generally analysers should be treated with care, and used and maintained in accordance with the manufacturer's instructions. Before the analyser is used it is essential that you read the operator manuals and, as appropriate, ensure:

- the batteries are correctly inserted, charged and not leaking
- the analyser has a current proof of calibration
- the display is functioning correctly
- the analyser is zeroed and purged in accordance with the manufacturer's instructions
- the pump is working
- filters and water traps are clean and dry
- probe tubing is free from leaks or damage.

These analysers should not be used to identify leakage of POC from the appliance combustion circuit except where their use for this purpose is detailed in the appliance manufacturer's instructions.

The levels they read are very low – near to zero. They can fail to give any reading if the sensor has failed, so check the sensor before use by, for example, measuring POC from a cooker grill burner or around a cold saucepan on a hotplate. A small bottle of calibration gas can also be used to check the sensor.

Electronic portable combustion gas analysers conforming to BS EN 50379-3

- Identified by a durable label on the analyser.
- Hand held or portable, with a display and keypad for information and control.
- Battery and/or mains powered and have an automatic pump.
- Use an external probe to extract POC from a sample point for measurement and calculation within the analyser.
- Use a filter/water trap to remove particles and water vapour from the POC sample.
- Measure the oxygen (O_2) and CO concentration in a sample of POC drawn from the chimney/flue system.
- Measure or calculate the CO_2 concentration and CO/CO_2 ratio of the POC sample.

Electronic portable combustion gas analysers conforming to BS 8494

- Identified by a durable label on the analyser.
- Hand held or portable, with a display using characters not less than 8 mm high (unless the display is backlit).
- Battery powered and have an automatic pump.
- Use probes and sensors constructed from materials that are not affected by substances found in the environment of its intended use.
- Constructed to prevent damage to sensors and pumps from particulate matter and liquids that might be expected in its application.
- Measure concentrations of CO_2 in indoor ambient air.

Electronic portable combustion analysers conforming to BS 8494 that read CO_2 *in ambient air only* do not fully meet the complete instrumentation requirements outlined in BS 7967-1, BS 7967-2, BS 7967-3 and BS 7967-4 where measurements of CO_2 in flue gases are required.

Safe working

You must only operate an analyser if you are:
- competent in their use
- have an understanding of the results obtained
- have an awareness of the necessary safety actions and appropriate regulations, such as GSIUR.

Key term

Electronic portable combustion gas analyser Electronic apparatus that will detect and measure the presence of combustion gases and clearly display the result.

Flue gas samples to be taken during the commissioning process

Sampling is essential nowadays, in particular when flueless appliances such as water heaters and space heaters. This is to ensure that prior to signing off the job the appliance(s) are safe to use and will cause no danger to the user.

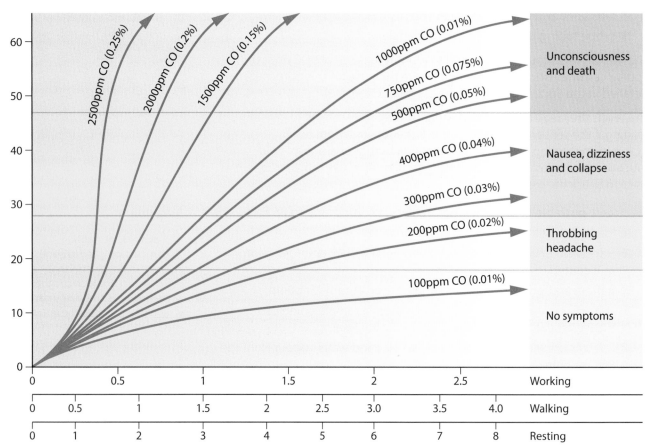

Figure 7.39 CO absorption into the bloodstream

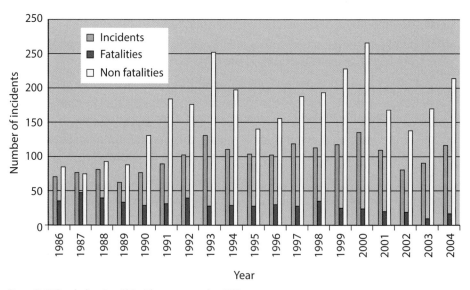

Figure 7.40 Graph showing CO incidents reported to HSE

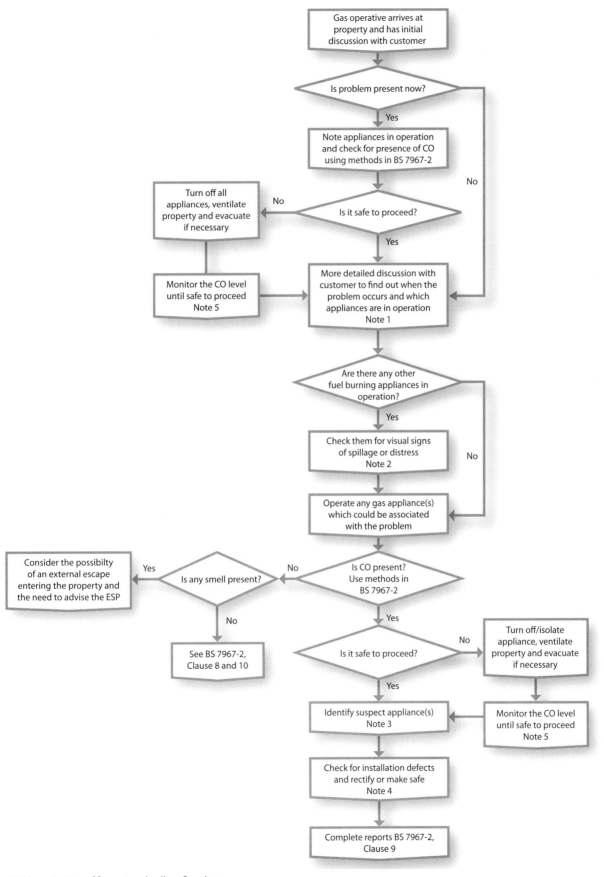

Figure 7.41 Investigation of fumes in a dwelling flowchart

Ambient levels of CO in dwellings

For the measurement of ambient CO levels in dwellings the gas analyser can be:

- of a type that measures CO
- of a type that measures CO and O_2, and calculates the CO_2 levels from the O_2 measured.

The accuracy of gas analysers used in the measurement of CO in dwellings should be guaranteed by the manufacturer. Gas analysers should be of a type that measures and displays CO in ppm. For readings at or below 20 ppm the accuracy should be ±3 ppm or better, and for above 20 ppm the accuracy should be ±5 per cent or better of the instrument reading. Regardless of which gas analyser is used, the manufacturer's written specification should be checked to confirm that the analyser is accurate to these levels.

For the detection of CO_2 in dwellings the analysers should conform to BS 7927:1998 including Amendment 1:1999. Where the analysers are being used for the determination of CO and CO_2 products from gas-fired appliances, the analyser should also conform to BS 7927:1998 including Amendment 1:1999.

If the analyser is not labelled as conforming to the above regulations then the manufacturer of the analyser should be asked to provide written confirmation as to its suitability for the intended purpose.

Preparation for testing atmospheric CO levels

Initial checks to undertake for the absence of CO are essential. The key here is to ensure that you do not put yourself or the customer under any danger whilst undertaking sampling tests of the atmosphere or an appliance test.

It is recommended that only persons required to carry out the test work are in the room under test and that these persons do not smoke before or during the tests. Smoking in other parts of the building adjacent to the area being tested should be discouraged during the testing procedure as this can also adversely affect results.

The initial test is taken with all appliances and/or gas supply turned off. With them off any trace of CO that are above the permitted background levels will be indicated on the analyser.

Position an open ended sampling probe approximately 2 m above floor level in the centre of the room and at least 1 m away from any suspect appliance installation (see Figure 7.43 on page 241).

An initial check for CO in a dwelling involves:

- closing all external doors, windows and customer adjustable ventilation
- recording the level of CO over a 15-min period.
- if the indoor level of CO starts to rise during this period, checking for CO migration from other sources
- if there is no rise, proceeding to the appropriate test in General Testing Procedures.

After completion of the test in Figure 7.43 the results can confirm only one of two outcomes: (1) an absence of CO, in which case carry out a gas tightness test; or (2) the presence of CO, in which case the CO discharge is likely to be coming from another fuel-burning appliance within the property. The fuel-burning appliance does not necessarily have to be natural gas, for example it could be a wood-burning stove in the dwelling that is causing the CO discharge.

Figure 7.42 Angled sampling probe

Normally an open-ended sampling probe should be used, positioned in the POC stream of the secondary flue via the draught diverter and its position adjusted until the highest steady value of CO_2 or lowest steady value of O_2 is obtained.

Where there is no access, it will not be possible to measure the POC. See the section on 'Checks to make where it has not been possible to check CO/CO_2 ratio' (page 249).

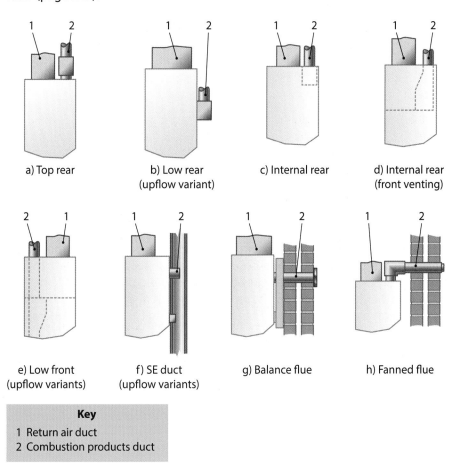

a) Top rear

b) Low rear
(upflow variant)

c) Internal rear

d) Internal rear
(front venting)

e) Low front
(upflow variants)

f) SE duct
(upflow variants)

g) Balance flue

h) Fanned flue

Key
1 Return air duct
2 Combustion products duct

Figure 7.48 Flue/draught diverter configurations for warm air heaters

For **gas fires**, a sampling probe as shown in Figure 7.38 (p. 236) should be used where practical. If not, an open-ended probe can be used, this may have to be purpose made to gain access into the flue behind the fire: see Figure 7.49 for more details.

The sampling probe should penetrate the flue by at least 200 mm in the POC stream and as far away from the burning gas as practicable. Adjust the position of the sampling probe until the highest steady value of CO_2 or lowest steady value of O_2 is obtained.

If the chosen probe cannot be positioned in the POC stream without first removing the fire (e.g. a box-radiant fire), then it will not be possible to obtain a reliable combustion sample for the fire as found. Consequently, if the probe cannot be positioned without removing the fire, this test should not be carried out. See the section on 'Checks to make where it has not been possible to check CO/CO2 ratio' (page 249).

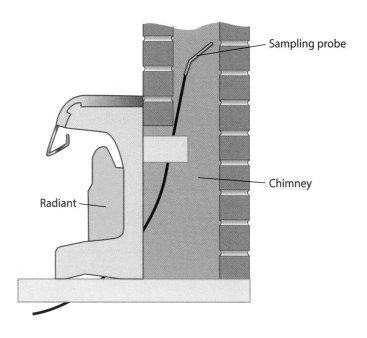

Figure 7.49 Open-flued gas fire sampling

For **combined gas fire/back boiler units** (BBUs), an open-ended sampling probe should be inserted into the POC stream in the secondary flue via the draught diverter. The connecting tube to the sampling probe needs to be routed so that the fire front can be replaced without displacing the probe. The POC samples should be taken for both the boiler and the fire, when operating separately and together.

You may find it necessary to remove the outer case of the fire unit to gain access. If it is still not possible to position the sampling probe correctly when the case of the fire unit has been removed then this test cannot be carried out. See the section on 'Checks to make where it has not been possible to check CO/CO_2 ratio' (opposite).

Type C – Room-sealed appliances

An open-ended sampling probe should be positioned 200 mm inside the POC outlet duct and its position adjusted until the highest steady value of CO_2 or lowest steady value of O_2 is obtained.

Without a custom-made probe it will not be possible to gain access to the POC stream for installations of **warm air heater** types shown in Figure 7.48.1f) and where, it is not advisable to attempt to do so. See the section on 'Checks to make where it has not been possible to check CO/CO_2 ratio' (opposite).

For installations of appliance type shown in Figure 7.48.1g) access can be achieved via the terminal but only at a safe working height. This type of appliance has not been manufactured since about 1990. Room-sealed appliance types shown in Figure 7.48.1h) may have a sampling position fitted.

Checks to make where it has not been possible to check CO/CO$_2$ ratio

When it has not been possible to check the combustion performance ratio (CO/CO$_2$), then the following checks should be made:

- Check all disturbed gas connections. Test for gas tightness where a gas escape was previously identified/repaired.

- Confirm that the burner operating pressure and/or gas rate is correct.

- Where possible, carry out a final combustion performance test in accordance with the methods in BS 7967-2 to confirm that the combustion performance is within the limits allowed in BS 7967-3:2005, Clause 5.

- Check the operation of any flame supervision device using the manufacturer's approved method.

- Ensure all seals or fastenings are present, in good condition and secured in accordance with the manufacturer's instructions. Be especially vigilant with room-sealed positive fan pressure appliances. Refer to Gas Safe Register Technical Bulletin 006 of 19 August 2010, *Industry guidance for the checking of case seals and the general integrity of room-sealed fan assisted positive pressure gas appliances* (14).

- Where appropriate for the appliance design, carry out a spillage test.

- Ensure any warning labels necessary to ensure safe use of the appliance are present and correct.

Knowledge check

1 Which of the following materials would *not* be suitable for use in constructing new open-flue chimneys?

a Concrete liners
b Clay liners
c Metallic pipes
d Pumped or poured concrete liners

2 When installing a gas fire, the maximum length of flue permitted in order to prevent condensation within an internal 225 mm x 225 mm brick chimney is:

a 10 m
b 12 m
c 28 m
d 33 m

3 When installing a gas fire to an unlined brick chimney a terminal is required when the flue size is less than:

a 120 mm
b 150 mm
c 170 mm
d 300 mm

4 A suitable method of sealing the annular (circular) space around a flexible stainless steel flue liner is to use:

a Any flexible material
b Mineral wool, clamping and/or register plate
c Sand and cement
d All of the above

5 When installing a gas fire and back boiler unit the chimney shall be lined if the length of the chimney exceeds:

a 12 m
b 15 m
c 18 m
d Any length

6 Single wall flue pipe with socket joints shall be:

a Sealed with the socket of each section fitted uppermost
b Sealed using mastic jointing compound
c Sealed and the socket of each section shall be facing downwards
d Fitted with the socket of each section clipped to the wall

7 The temperature effects on pre-cast flue blocks (caused by heat from the products of combustion) should be avoided by:

a Plastering directly onto the flue blocks
b Dry lining with an air gap in front of the flue blocks
c Using plasterboard with the supporting dabs in direct contact with the flue block
d Screwing the plasterboard to the flue block

8 Which of the following might lead to incomplete combustion?

a Poor flueing
b A blocked heat exchanger
c A blocked primary air port
d All of the above

9 What are the symptoms associated with carbon monoxide poisoning?

a Severe abdominal pains
b Headaches, nausea and dizziness
c Soreness of joints and muscles
d All of the above

10 Products of incomplete combustion, including CO leaking from an appliance, might be found:

a In the location of the appliance
b In adjoining rooms
c Migrating into other properties
d In all of the above

11 When testing the combustion performance of an open-flue boiler the analyser sensing probe should be:

a 200 mm into the primary flue
b 200 mm into the secondary flue
c 200 mm into the draught diverter
d Inside of the terminal

12 What immediate advice should be given to someone suffering from the symptoms of carbon monoxide poisoning?

a Leave the room and close all internal doors
b Call a service engineer
c Get out of the building and into fresh air
d Turn off the appliance and take headache tablets

13 In accordance with BS7967 Part 3 what is the maximum CO/CO_2 ratio permitted for a flueless space heater?

a 0.004
b 0.008
c 0.001
d 0.01

14 In accordance with BS7967 Part 3 what is the maximum CO/CO_2 ratio permitted for a combination boiler?

a 0.004
b 0.008
c 0.001
d 0.01

15 When a flueless space heater is tested and the CO/CO_2 ratio is exceeded, what action should you take?

a None
b Classify the appliance as 'at risk' (AR)
c Classify the appliance as 'not to current standards' (NCS)
d Classify the appliance as 'immediately dangerous' (ID)

Installation

INTRODUCTION

It is essential that the gas installation systems you install are safe and gas tight. You need to be able to install pipework using tools and methods which are acceptable within the industry. For this, you will need to know which tools to use and the basic principles of electricity and domestic circuits.

SURVEYING THE WORK SITE PRIOR TO COMMENCING WORK

The responsible person would need to undertake a site survey to ensure that all services, i.e. gas, water and electrics, are in place. An example of a job would be that you are changing the heating and hot water to a combination boiler. Where this occurs, consider the position of the boiler in relation to the services that are already installed; an example of this is to check the cold supply for the boiler is in close proximity. Which of the services have to be altered or removed to suit the new boiler; an example of this is the removal of a cold water storage cistern and feed, and expansion cistern and associated pipework. One situation which could also determine whether a combination boiler can be fitted would be if there was insufficient flow and pressure in the cold main to serve the system. All of these considerations should be taken into account when surveying prior to starting the job.

TOOLS NEEDED TO COMPLETE INSTALLATION, DECOMMISSION AND COMMISSION WORK

There is a wide range of tools used to install appliances and pipework. There will be specialist tools and equipment that you will need to purchase and use over your time as a gas operative. Some specialist tools, such as fusion fittings for installing polyethylene gas pipework underground, can be hired when necessary from specialist hire companies. When carrying out this type of one-off job, it can be more cost-effective and practical to hire the specialist equipment required.

Each of the sections which follow gives a brief description of the types of tool you are likely to come across in your installation and commissioning work, including fixing and tightening tools; cutting, marking and measuring tools; preparation and power tools; and tools for patching up.

Fixing and tightening tools

When using fixing and tightening tools, ensure the teeth are free from jointing compounds. If they are clogged up, the tool may slip and cause damage or injury. Once the teeth become worn, the tool should be replaced. Check for wear on the ratchet mechanism when using pump pliers. These often slip when under pressure. Be careful when loosening a joint or pipe that is difficult to move. It might give suddenly, and you could damage your hands, or even pull a muscle.

PROTECTING FURNISHINGS

Before starting a job, be sure to survey your work space for any damage. Where any damage is found, including chips, stains, cracks and scratches, then bring this to the attention of the client and, if possible, get them to sign a document which identifies what has been found. This protects you from possible damage claims that weren't your fault.

When working in occupied dwellings, it is important to protect carpets and furnishings.

For stair protection, use a stair protector. This comes in a roll and must be non-slip. Nowadays they're made from non-permeable membrane.

Carpets can be protected by dust sheets. Ensure that they are secured all around. They don't provide much protection from spilled fluids so if there is a risk of fluid spillage, use plastic carpet protector.

Even though laminate floors are hard wearing you should protect them from possible damage from scratches, certain spills etc. Because of their smoothness, do *not* use a cloth dustsheet. Instead use the 'sticky' carpet protector, as this adheres itself to the floor and helps to prevent trips and falls.

Cover furniture with dust sheets, especially when fitting a gas fire if the chimney needs sweeping. Ask the occupier to remove any delicate ornaments from the work area.

For existing systems and appliances, ensure they are safely isolated from the gas and electrical supply. If undertaking work inside a boiler (e.g. changing a pump in a combination boiler), ensure all electrical connections are protected from water which could gain entry into the electrical control box.

BRITISH STANDARDS FOR INSTALLATION WORK

Before commencing installation work, you will require any current or updated information from normative documents.

A list of the many normative documents that are required under Gas Safe and the regulations are within TB999 (see example in Table 8.01 opposite). This is available from Gas Safe to registered engineers. It is important that, as a gas engineer, you keep up to date with the latest information from Gas Safe.

BASIC ELECTRICS

Any operative who works with electrics, whether undertaking maintenance work, fault finding or installation needs a basic understanding of electricity. This is a requirement of COP20 (Standards of training in safe gas installation, Approved Code of Practice) this is available to download from the HSE website: www.hse.gov.uk/pubns/priced/cop20.pdf

Principles of electricity

Modern-day society relies heavily on **electricity** for everything from heating and washing to entertainment systems and traffic controls. But if you misuse it or touch it, it can hurt or even kill. You must, therefore, understand how it works so that you can work with it safely.

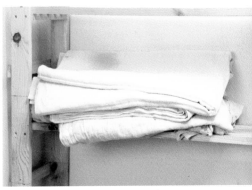

Figure 8.33 Cloth dust sheets are used to protect customer property

Progress check

1 What type of job would you use a pair of pump pliers for?
2 When would you use pincers?
3 List the tools that can be used for cutting pipe and other materials.
4 What device/tool would you use to locate pipes and cables that are hidden in walls?
5 How often should a PAT be undertaken?

Key term

Electricity
Electricity is a flow of electrons through a conductor.

Molecules and atoms

Every known substance is made up of molecules. A molecule is a very tiny part of matter which can only be seen using special microscopes. Molecules are always in a state of rapid motion, and the ease with which they move around determines the form of the substance they make up:

- When molecules are densely packed together, their movement is restricted and they form a **solid** substance.
- When the molecules are less densely bound together and move more freely, the substance they form is a **liquid.**
- A substance that allows the molecules almost unrestricted movement is a **gas.**

These three molecular conditions – solid, liquid and gas – are sometimes known as the 'three states of matter'.

Molecules are made up of **atoms**, which are the smallest parts of matter that can be subdivided. They are not solid. They have a **nucleus** at their centre, made up of very tiny particles, known as **protons** and **neutrons**. Protons hold a positive charge (+) and neutrons have no charge – they are electrically neutral. We can think of neutrons as the glue that holds the nucleus together. Around the nucleus orbits a third type of particle – **electrons** – which hold a negative charge (−).

Like charges repel each other (+ repels +, − repels −) and unlike charges attract each other (+ and −). All atoms contain equal numbers of protons and electrons, and, in this unaltered state, the matter is electrically neutral, i.e. no electricity is flowing. In some cases, it is possible to add or remove electrons, leaving the atom with a positive or negative charge.

In each atom, electrons orbit the nucleus containing protons and neutrons. Sometimes these electrons 'break free' and flow to a neighbouring atom. It is these free or 'wandering' electrons, moving through the material structure, that give rise to electricity.

Conductors

In some materials, it is very easy to get the electrons to move; these materials are called **conductors**. In other materials, it is very hard to get the electrons to move; these materials are called **insulators**. Examples of good conductors of electrons are copper and aluminium. Gold is probably the best conductor, but it is too expensive to use in everyday installations. Typical insulators are wood, plastic and rubber.

Measuring electricity

There are three necessary elements that make up an electrical circuit:

- current
- voltage
- resistance.

Current

Electricity is the movement of electrons through a conductor. The current is the flow of electric charge. To create a practical electric circuit, you need to know *how much* electricity will flow in a given circuit in a given time and to control how much electricity is flowing. A single electron is much too small

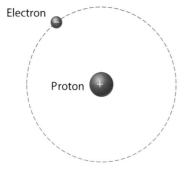

Figure 8.34 A hydrogen atom

to have any practical use in this measurement. Instead, millions and millions of electrons are grouped together into useful amounts. These groups of electrons are known as **coulombs**.

The flow of electricity or 'current' is measured in **amperes** (usually abbreviated to amps – A) which is the rate of coulombs per second. Electric **current** is given the symbol I.

Voltage

When a coulomb of electrons leaves a battery or generator it has a potential energy, but as it travels round the circuit this energy is used up. The amount of energy used up by one coulomb passing between two points in a circuit is known as the **potential difference**. This is measured in volts (V).

Using a water analogy, voltage can be thought of as the pressure that pushes the water (electricity) around the system. By applying a voltage to the end of a conductor, we provide an electrical pressure that causes a current to flow. The voltage may be supplied by a battery or a mains supply. A current that is produced by connecting a battery to a circuit is called **direct current (DC)**. The electricity that we use in our homes and which is produced by power stations is called **alternating current (AC)**. See the sections on 'Direct current (DC)' (page 268) and 'Alternating current (AC)' (page 269).

Resistance

In the water analogy, voltage can be thought of as the head of water in a cistern, providing the pressure. The electrical current can be thought of as the rate per second at which the water will flow with the tap open. The final principle of basic electrics is resistance. To extend the analogy, if the size of the pipe and tap are increased, the natural resistance to the flow of water will decrease and the bath will fill more quickly. If the size of the pipe and taps is decreased, the resistance to the flow will increase and the bath will fill more slowly. In electrical terms, the bigger the conductor, the lower the resistance to current flow. Electrical resistance (R) is measured in units called ohms (sometimes denoted by the Greek letter omega – Ω).

Gravity drives the water through the system, but for this to happen there must be a difference in level between the tank (the electricity supply) and the bath (the **load**). This gives the equivalent of potential difference. The tap performs the same function as an electric switch.

Key terms

Coulomb
A coulomb consists of approximately 6,240,000,000,000,000,000 electrons.

Ampere
A flow of one coulomb per second. In other words, the quantity or amount of electricity that flows every second.

Load
Load refers to the power consumed by a circuit or appliance.

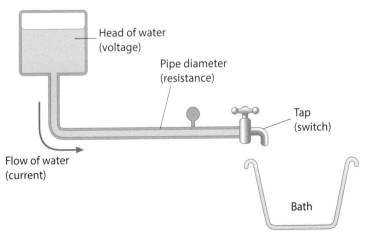

Figure 8.35 Analogy of a water system as an electrical circuit

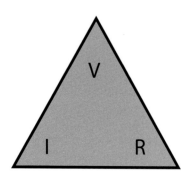

Figure 8.36 Ohm's Law Triangle

Safe working

The ohmmeter has its own internal power supply from a battery. It must *never* be connected to a live circuit as it could damage the meter.

Ohm's Law Triangle

Ammeters and voltmeters are connected into live electrical circuits to measure current and voltage. They pick up the electricity that they need to operate from the live circuit.

The three major components that make up an electric circuit – voltage, current and resistance – are interrelated. If any two of the quantities are known, the third can be calculated by using a basic rule known as 'Ohm's Law'. Ohm's Law is named after a German physicist, George Ohm, who defined electrical resistance in 1827.

Ohm's Law states the relationship between the electrical quantities as:

$V = I \times R$

Voltage = Current x Resistance

It can also be shown in a simple form, often called the 'Ohm's Law Triangle' as shown in Figure 8.36.

Provided you know two of the three values, you can work out the third by using the calculations below.

$I = V - R$

Current = Voltage – Resistance

$R = \dfrac{V}{I}$

Resistance = Voltage ÷ Current

For example:

A circuit has a current of 2 A and resistance of 120 Ω. What would be the voltage applied to the circuit?

Using the calculation $V = I \times R$ you get:

$2 \times 120 = 240\,V$

How an electric circuit works

If you were to bend a conductor, such as a copper wire, into a loop and connect a battery across the two ends, you would have a complete electric circuit and a current would flow. Unfortunately this would not create a practical circuit that you could use. A single length of copper wire has a very low resistance on its own – depending on the length and cross-sectional area it may have only a few millionths of an ohm of resistance. If you were to connect this across a normal domestic voltage supply of 240 V, using Ohm's Law you could show that, as $I = V - R$ and R is very low, thousands of amps of current will flow, causing serious damage to the circuit and risk of fire (see the section on 'Why fuses are needed' below).

Even if you could have the circuit described above, it would still not be practical because there is no load to limit the current, no way to switch the current off and no protection for the circuit from the high current flowing due to the very low cable resistance.

Checks during decommissioning

Once you have tested the existing system and made sure all is ok, it should be safe to proceed.

When disconnecting/decommissioning appliances, you need to be aware that the appliance may have an electrical connection. Where there is an electrical connection you need to follow the safe isolation procedure (see pages 9–11). Failure to follow this can, in some circumstances, be fatal.

Always make sure before commencing a decommission task that you have the appropriate tools and fittings available. In some instances it would be advisable to prepare for the unexpected and have spares to do an emergency cap-off. If you have not got the appropriate fittings to safely decommission an appliance or installation, *do not* proceed until you have the fittings. If you need to leave the job at any time unattended then you *must* leave the installation in a safe condition. Failure to do so can lead to criminal convictions.

When having to 'cut back' or cut into pipework, you need to ensure the continuity of the earthing (see page 12). When undertaking any task on a gas installation, you must always ensure that equipotential bonding is fitted within 600 mm of the meter and must be a minimum 10 mm^2 earthing cable. Where this is not installed, or connected in the wrong place, inform the responsible person that protective equipotential bonding work should be carried out by an electrically competent person.

Case study

Joe has been given the job of decommissioning an Ideal Mexico balanced flue boiler in a domestic dwelling, which feeds a 'Y' plan system of heating. The system is an open-vented system connected to a double-feed indirect cylinder. The new boiler is a condensing combination boiler and is being installed in the same position as the existing boiler.

Consider the needs of the customer and, referring to a copy of the manufacturer's instructions, write out a plan of the works to be undertaken.

List the procedure to take when decommissioning the gas pipework to the boiler.

Check your results with your tutor.

Knowledge check

1 The supply earthing system which has a combined neutral and earth in part of the system is designated as what?

a TT
b IT
c TN-S
d TN-C-S

2 What is the name of the unit which contains separate circuit breakers for all types of different electrical appliances within a dwelling?

a Fuse box
b Consuming unit
c Consumer unit
d Residual current device

3 Where should a domestic consumer unit be installed?

a Not less than 1,000 mm from the main supply entry to the dwelling
b Immediately after the supplier's meter
c No higher than 500 mm from floor level
d Immediately before the supplier's sealed fuse unit

4 Equipotential bonding is intended to increase safety by the bonding of?

a The electrical supply and the incoming water pipework
b Any two metal sections of pipework which display evidence of discontinuity
c The hot and cold water pipework only
d All exposed metalwork within a building

5 A consumer unit contains what information about the electrical circuit you are working on?

a Wiring requirements
b Electricity reading
c Voltage rating
d Current rating

6 It is important that you keep your cutting tools

a Clean
b Sharp
c In a set
d All of the above

7 Which of the following practices should be adopted to protect a customer's stair carpet when a boiler is to be installed upstairs?

a Wrap the boiler in polythene
b Cover the steps with polythene or insulating material
c Insist that the customer removes the carpet before the work commences
d Cover the carpet with non-slip dust sheets at the start of the job and before taking the boiler upstairs

8 Working in houses or other occupied areas is potentially problematic for a gas operative. Which of the following would you recommend as the best course of action before work commences?

a Carry out an inspection of the work area and report any existing damage or defects to the customer
b No special precautions are required
c Only commence the work when the area has been fully sheeted
d Simply complete the work as quickly as possible with no particular precautions taken

9 Who should you notify before turning off the electrical supply to a heating system?

a Building Control Officer
b Customer
c HSE
d Power company

10 Which of the following tools is most likely to be affected by mushrooming?

a Screwdriver
b Hammer
c Cold chisel
d Hack saw

Operating principles
of controls

INTRODUCTION

Controls are a major part of any appliance and system. As a gas operative you will be working with old and new types of controls. You need to be able to identify what each control is, what its purpose is and its function on the appliance or system.

LEGISLATION FOR GAS SAFETY DEVICES AND CONTROLS

Regulation 26(1–10) of the Gas Safety (Installation and Use) Regulations (GSIUR) deals with the requirements of gas safety controls on appliances. These controls and devices play an important part in preventing accidents in the use of appliances. Regulation 26(1) states that:

> 'No person shall install a gas appliance unless it can be used without constituting a danger to any person.'

As a gas operative you are required to never leave an appliance which may cause injury or death to the user. You must also ensure that the safety controls and devices are working correctly. Check that they have not been modified in any way, such as being wedged open to prevent them shutting off the gas supply to an appliance. Such negligence constitutes a contravention of the regulations and can lead to a criminal prosecution, as the operative or person who modified the safety control will be charged under the GSIUR and taken to court by the Health and Safety Executive.

OPERATING PRINCIPLES OF GAS CONTROL DEVICES

Natural gas and liquefied petroleum gas (LPG) are utilised for heating, cooking and, in some circumstances, lighting and must be able to be controlled. There are several reasons for the need to control the flow of gas:

- to help ensure safety
- to control and regulate temperature
- to ensure that an appliance is receiving the correct gas input rate.

The types of controls are varied; they range from those controlling the flow of gas into systems such as gas cocks and emergency control valves (ECVs). They also include non-electrical devices such as bi-metallic strips and electrical devices such as multifunction valves.

Pressure regulators

There is a need to control the pressure of gas in systems, whether it is natural gas or LPG. The purpose of the regulator is to reduce the pressure, ensuring

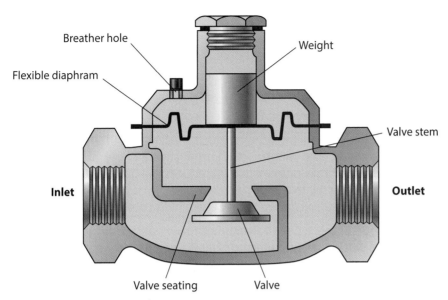

Breather hole

Weight

Flexible diaphram

Valve stem

Inlet

Outlet

Valve seating   Valve

Figure 9.01 Constant pressure regulator

that the lower pressure is constant throughout its operation. On low pressure installations this is normally kept at 21 mbar + or – 2 mbar. Installations like these are able to compensate as the demand for gas changes within the system. There are occasions where at peak demand the pressure may drop to 19 mbar at the emergency control, but these are in extreme circumstances.

How do the regulators work?

Constant pressure meter regulators are simple devices that ensure the pressure within a system is kept at constant. The breather hole in the top of the regulator allows the upper compartment above the diaphragm to vent to atmosphere thus allowing the diaphragm to move up and down. Gas flows into the inlet through the regulator to the outlet, the gas passes a weighted or spring-loaded diaphragm (as in Figure 9.01).

Figure 9.02 shows how the position of the valve seating is dependent on the rate of gas flow required to feed appliances. The left-hand regulator in Figure 9.02 shows the valve seating partly open when there is one appliance/burner demanding gas. As more appliances are used or turned off then the position of the valve seating changes to compensate for this demand. This is the basic principle of how the governor works to keep a constant pressure.

Flow through governor

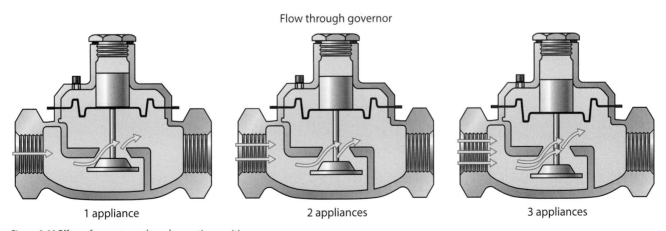

1 appliance   2 appliances   3 appliances

Figure 9.02 Effect of gas rate on the valve seating position

Air/gas ratio valves

Alternative names for an air/gas ratio valve are a zero governor, air/gas valve and 1:1 valve.

They are used mainly in high-efficiency condensing boilers to maintain gas and air flow to the burner in the correct proportions for high combustion efficiency.

The appliance circuit board controls the combustion air flow rate by varying the fan speed. The air/gas ratio valve then reduces the gas pressure to equal the air pressure (i.e. a relative 'zero' gas pressure). The gasways to the burner are sized to produce an air/gas mixture of very high combustion efficiency.

The introduction of this type of valve brings new complexities to the gas engineer's work. Each manufacturer has its own procedures for commissioning appliances with air/gas ratio valves. Adjustments must not be attempted without careful adherence to the specific appliance manufacturer's instructions.

Typically, setting and adjustments are made using a combustion analyser rather than a manometer. A small change in setting may produce a large change in combustion products including carbon monoxide (CO) levels.

Air/gas ratio valves have two adjustment parameters, adding further to the level of complexity involved in correctly setting them. Some manufacturers seal one or both adjustment screws to prevent alteration. These seals must not be broken.

You should test these valves in accordance with the manufacturer's instructions. Manufacturers may provide a test mode in which the boiler temporarily operates under certain conditions (e.g. at maximum rate) for combustion analysis.

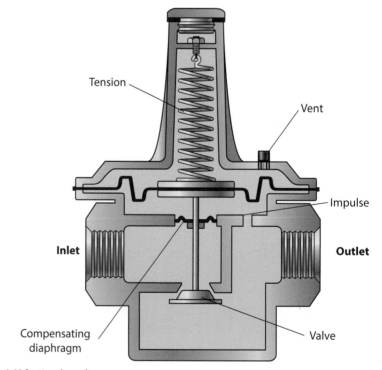

Figure 9.03 Section through a zero governor

Only correct a fault that requires you to adjust the valve if the manufacturer's instructions permit it. If you are not permitted to do so, contact the manufacturer for specific guidance.

Cut-off valves

A typical cut-off valve is shown in Figure 9.04. The valve is a fitting with a spring-loaded plunger which is held in place by a ring of low melting solder which melts at a temperature of 95 °C. When the solder melts the spring forces the closure plug down into the valve and cuts off the gas supply to the meter and dwelling's connections.

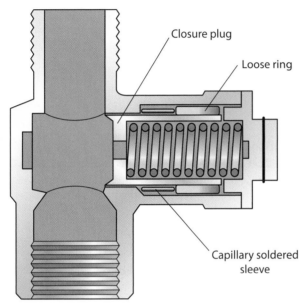

Closure plug

Loose ring

Capillary soldered sleeve

Figure 9.04 Section through a thermal cut-off valve

Low pressure cut-off valves

GSIUR Regulation 14(2) part b states:

> That there is adequate automatic means for preventing the installation pipework and gas fittings downstream of the regulator from being subjected to a pressure different from that for which they were designed.

So, when the gas is supplied through a pre-payment meter, LPG bottles or bulk storage tank, in the event of a failure in the supply, the user may open a gas outlet which causes the pressure in the downstream side to decrease. When this occurs the cut-off valve closes automatically, thus preventing gas passing through the downstream side of the pipework.

Low pressure cut-off valves come in two types: automatic reset and manual reset. In a domestic dwelling low pressure cut-off valves normally have a manual reset. To reset a low pressure cut-off valve manually:

- turn off all appliance outlets
- open the upstream valve slowly
- allow pressure to build up

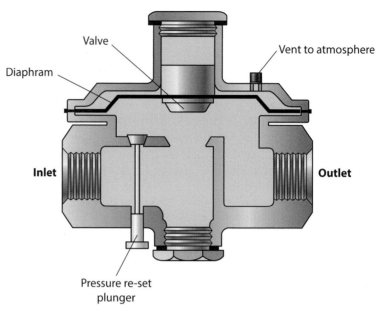

Figure 9.05 Section through low pressure cut-off valve

- depending upon model of valve, pull or push the reset valve until pressure is restored in the system
- check that the appliances light.

Thermal cut-off valves

Regulation 12(1) of the GSIUR states:

> No person shall install a meter in any premises unless the site where it is to be installed is such as to ensure so far as is reasonably practicable that the means of escape from those premises in the event of fire is not adversely affected.

Where buildings have two or more floors above the ground floor and the stairway or other means of exit is the only possible escape route in the event of a fire, then the meters should not be sited under the stairway.

Where premises have fewer than two floors above the ground floor then meters need to be installed where reasonably practicable in accordance with Regulation 12(1).

Where it is necessary to install new or replacement meters in or under a stairway, or where any other part of the premises forms an escape route where the meter is sited, then at least one of the following must be adhered to:

- the meter should be fire-resistant
- the meter should be housed in a fire-resistant compartment with automatic self-closing doors
- the pipe immediately upstream of the meter, or regulator if fitted, should be provided with a thermal cut-off device which is designed to automatically cut off the gas supply if the temperature of the device exceeds 95 °C.

Gas cocks/valves

Valves that are used for gas need to be manufactured from materials that are not going to be affected by the gas itself. These materials include the 'O' rings in particular where they are being used for gate valves. The most common form of isolation valve for gas is the simple gas cock. This is a tapered plug with a hole which is lined up with the gasway by rotating the plug 90 ° to allow gas to pass.

The plug is turned by a thumb piece called a 'fan' or a square top to which a lever can be fitted, as on an emergency meter control. The square top has a groove which is lined up with the supply when it is in the 'on' position.

Ball-o-fix valves must be of a type suitable for gas and gate valves may be used on larger installations. Remember Gas Regulation 5(1) states a gas fitting cannot be installed unless it is of good construction and sound material.

Cooker hotplate lid control valves

Some cookers come with a lid that drops down over the hotplates to provide extra workspace in kitchens when the cooker hotplates are not in use. The lids are fitted with a safety shut-off valve (SSOV) device which prevents gas from passing to the burners when the lid is in the down position (closed) (see Figure 9.06).

It is important when undertaking tightness testing that the cooker lid is in the open position, as stated in Institute of Gas Engineers' (IGE) Utilisation Procedure 1B 5.3.1 for new installations tightness testing procedures and 5.3.2 for existing installations. This is so that all controls on the cooker are included within the test.

> **Key term**
>
> **Ball-o-fix valve**
> A valve with a spherical disc – the part of the valve which controls the flow through it. The sphere has a hole, or port, through the middle so that when the port is in line with both ends of the valve, the gas will flow.

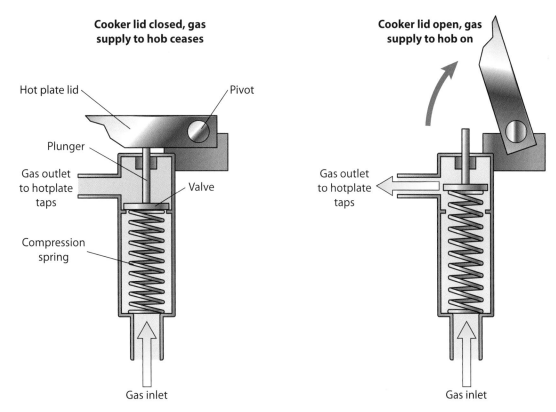

Figure 9.06 Section through cooker lid valve when open and closed

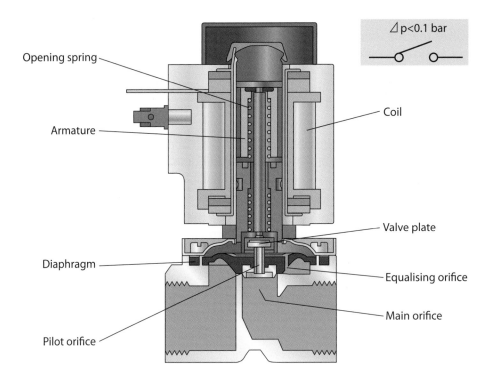

Figure 9.07 Section through a solenoid valve

Electric solenoid valves

Although not a thermostat, an electric solenoid valve works in conjunction with electrical thermostats/switches to control the flow of gas to a burner – it's essentially an on/off valve. They are built into multifunctional valves, for use on boilers.

Solenoid valves will open when energised, as the current flowing into the coil will cause the soft iron core to be magnetised, lifting it open against the spring. Loss of current allows the valve to spring to the closed position.

Excess flow valves

Excess flow valves can be installed on both LPG and natural gas installations. This is a requirement now of IGE/G/5 Gas in flats and multi-dwelling buildings. The valve works when there is a pre-determined flow rate in the gas pipework above the normal flow rate for a dwelling, for example following a breakage in the pipework or connections to an appliance. The valve effectively shuts off the gas supply to the dwelling. After the repair is made the valve will automatically reset when the system is re-pressurised.

The location for the installation of this valve in domestic dwellings is on the downstream side of the ECV.

TYPES AND OPERATING PRINCIPLES OF FLAME SUPERVISION DEVICE

The purpose of any flame protection device is to prevent gas from passing into an appliance in the event of a fault condition. Examples of such faults are:

- the pilot or main burner failing to light
- the pilot or main burner is extinguished.

This device was commonly known as a flame failure device (FFD) but the new term now often used is **flame supervision device (FSD)**. If the presence of a flame (usually a pilot flame) is not detected, then the device shuts off the gas supply to the burner. The presence of the flame can be sensed in several ways:

- the use of a flame
- the use of flame conductance and rectification, which converts an AC current to a DC current (for more information see the section on 'Flame conduction and rectification systems' on page 296)
- the use of photo-electric cells which can detect ultra-violet light and infra-red.

The devices in Table 9.01 will use one of these methods. FSDs have to comply with regulations for maximum shut-down time – the longest the FSD on any of the appliances in Table 9.01 should take to shut off the gas supply is 60 seconds.

Suggested maximum times for FSD shut down	
Appliance	**Time (secs)**
Boilers	60
Circulators	60
Cooker hotplates and grills	60
Cookers and ovens	90
Decorative gas fires	60
Gas fire and back boiler	60
Instantaneous water heaters	60
Radiant gas fires (space heaters)	180
Storage water heaters (≤35kW(net)) (35< – ≤70kW (net))	60 45
Warm air unit	60

Table 9.01 Suggested maximum times for FSD shut down

Vapour pressure devices

Vapour pressure devices are used on room heaters, water heaters and cooker ovens. When the liquid (volatile alcohol or water) temperature is raised it turns into a vapour, which has a much greater volume. All liquids when heated expand and with the volume increased in this device it has the capacity to push the end of the device against a lever which then opens the valve allowing gas into the burner. The closed position is shown in Figure 9.08.

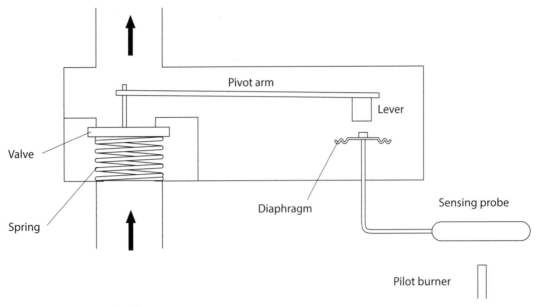

Figure 9.08 Vapour pressure device in closed position

The advantage of this type of device is that if there is a leakage of the liquid from the device or if it is not positioned correctly then the valve returns to failsafe condition (turns off the gas supply to the burner).

If the sealed system containing the liquid fails, you will need to change the device. The lever system and valve may also require cleaning and lubricating from time to time.

The main source of faults that occur from incorrect siting of the probe, relative to the pilot flame. The vapour pressure principle is also used for other devices including appliance thermostats (see the section on 'Types and operating principles of thermostat/thermistor' on page 302).

Older types of vapour pressure valves may contain mercury in the phial, so extra care needs to be taken when disposing of these.

Thermoelectric valves

These are FSDs and work on the principle of heat applied to a **thermocouple** producing a small amount of electrical energy. The thermocouple is simply a loop of two dissimilar metals joined at one end, which is heated by a pilot flame. The other ends are connected to a magnet in a spring-loaded gas valve, which holds the valve open as long as the pilot produces heat. If the pilot light fails and the thermocouple goes cold, the magnet is de-energised and the valve closes.

It is therefore essential that the tip of the thermocouple is properly positioned in the pilot assembly to work correctly. You can check whether or not the thermocouple is operating correctly by measuring the electrical current generated when it is heated between its tip and gas valve connection point using a multimeter. The reading should generally be between 10 mA and 30 mA.

The valve in Figure 9.09 is in the closed position; there is no gas passing through the valve at this stage. The pilot is not activated. Lighting the pilot

Key term

Thermocouple
A device that consists of two dissimilar conductors in contact, which produce a voltage when heated. The size of the voltage is dependent on the difference of temperature of the junction to other parts of the circuit.

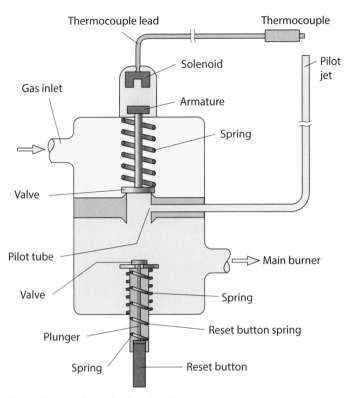

Figure 9.09 Thermoelectric valve in the closed position

is done manually on this type of valve. Depress the plunger (see Figure 9.10) and this pushes up the valve in the bottom chamber, which closes off the gas supply to the main burner. The plunger also pushes the top valve away from its seating allowing gas to enter the pilot tube. You should attempt to light the pilot using the Piezo-ignition device at this point.

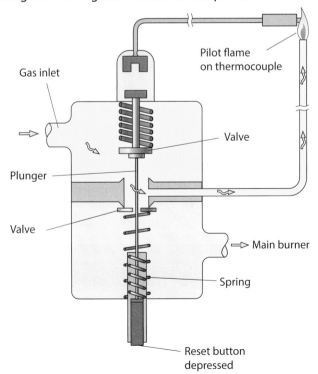

Figure 9.10 Thermoelectric valve in pilot position

With the pilot now lit, keep the plunger depressed for approximately 30 seconds longer. This time would enable the thermocouple to generate heat, which in turn generates a small DC voltage. The voltage is needed to activate the solenoid in the valve to hold the valve open.

Release the plunger and, if it is operating correctly, then the solenoid will hold the top valve off the seating and the bottom valve will return to its original position. Full gas flow now goes to the main burner, as shown in Figure 9.11.

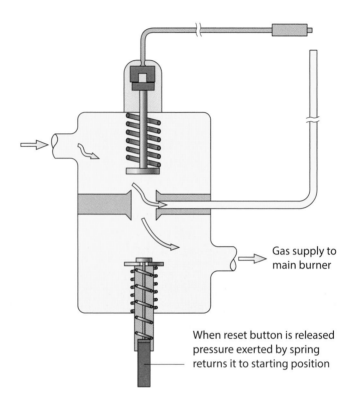

Gas supply to main burner

When reset button is released pressure exerted by spring returns it to starting position

Figure 9.11 Thermoelectric valve in the open position

When appliances are not going to be used for any length of time, such as leaving the house vacant when on holiday, then the valve can be manually reset by the user by turning the reset button to the off position. After a short period of time a small noise will be emitted from the valve – this indicates that the valve has fully re-seated and should happen within 30 seconds of turning off the pilot, but can be different from appliance to appliance (check with manufacturer's instructions).

Flame conduction and rectification systems

These electronic flame-protection methods are now being used much more on domestic appliances. The chemical reaction in a flame produces ions, which are electrically charged particles. Together with an electrode, these ions rectify the AC and produce a small DC output, which in turn operates a relay activating the gas valve.

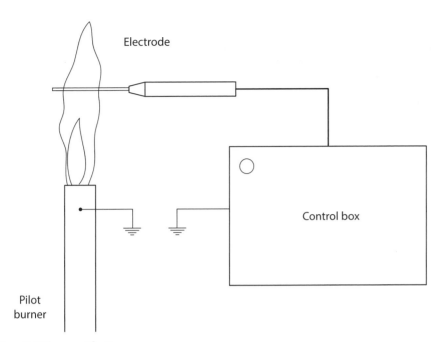

Electrode

Control box

Pilot
burner

Figure 9.12 Flame rectification

Figure 9.13 Flame rectification on a boiler

The shut-down of the gas valve is instantaneous in the event of flame failure. Faults in the system can be identified using a suitable test instrument (usually a multimeter). Readings should be in line with manufacturer's requirements.

In terms of the electrical connections, the polarity of the AC supply has to be correct – phase (live) cable to phase connection on the boiler, neutral connection to neutral, and the presence of a positive earth – as failure to do this will result in the boiler not functioning.

Interrupter devices

Interrupter devices are incorporated into a thermocouple and connected to an extra safety device, such as one of these types of overheat device:

- overheat thermostat
- fusible links
- spillage monitoring device (e.g. TTB – see the section on 'Spillage detection devices' page 299).

The interrupters work by cutting (interrupting) the electrical current which has been generated in the thermocouple. When this has been disrupted it causes the pilot flame to shut down.

Interrupters can be in one of two positions: position 1 is shown in Figure 9.14 and is the equivalent of a fusible link which acts as an overheat thermostat where the appliance has no electrical controls. An example of this would be a flueless water heater. This allows the electrical charge that is being generated by the thermocouple to operate the gas valve and when the heater overheats the links breaks and the circuit is broken so the gas valve closes.

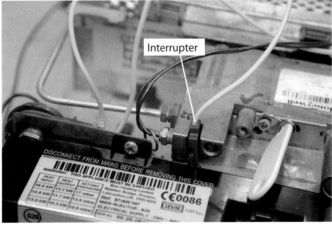

Interrupter

Figure 9.14 Interrupter position 1

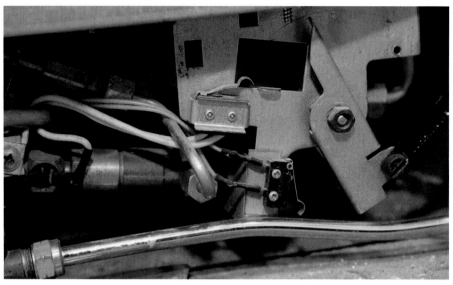

Figure 9.15 Interrupter position 2 – alternative connection

Position 2 as shown in Figure 9.16, represents a spillage monitoring device such as a TTB. The TTB detects heat that would occur when an open flue from a boiler is in a down-draught condition (see Chapter 7). As before this interrupts the circuit but this time the interrupter is connected to the valve itself.

Atmosphere sensing devices (ASD)

ASDs are also known as vitiation-sensing devices or oxygen-depletion devices (ODD). They operate by a flame from the sensing port, which heats the thermocouple (Figure 9.16) which then operates a thermo-electric FSD. They are fitted to appliances such as flueless space or water heaters or open-flued appliances (e.g. gas fires and back-boiler units) that can be a danger in terms of products of combustion affecting the air in a room.

When domestic appliances are designed, they allow for any excess of air to be funnelled into the appliance and into the atmosphere through the flue. Where combustion products spill into the room in which the appliance is installed, complete combustion of products will still occur for a period of time, even though oxygen levels within the room are being depleted. The carbon dioxide (CO_2) levels increase during this time and oxygen levels decrease so the appliance is now operating on incomplete combustion, which produces CO. When the atmosphere/combustion air is contaminated like this, with oxygen levels falling, the specially designed pilot lifts away, as in Figure 9.18 on page 299, cooling the thermocouple and eventually closing the thermo-electric valve.

ASDs have an intervention level of 200 ppm (parts per million) which is 0.02 per cent of CO concentration in the room in which the appliance is installed.

Under the GSIUR, the installation and servicing of appliances with ADS devices must be done by a competent operative. Regulation 26(9) states that the operative has to ensure that there is sufficient combustion air and that the device is working effectively.

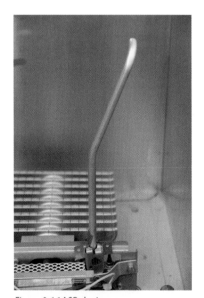

Figure 9.16 ASD device

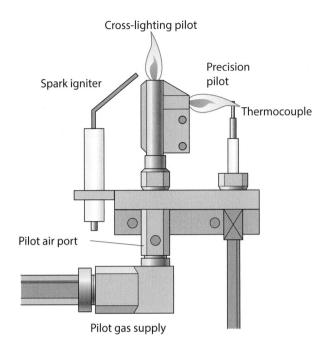

Figure 9.17 ASD working correctly

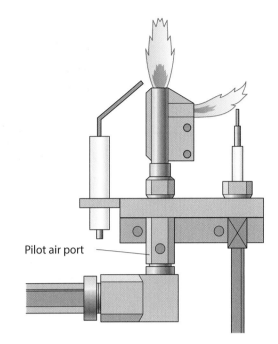

Figure 9.18 ASD when oxygen is depleted causing flame to lift off thermocouple

During the annual servicing of an appliance, you should check the device according to the manufacturer's instructions. These pilots have no serviceable parts and must be replaced rather than repaired to make them work.

If a customer complains that the ASD keeps 'going out' then there is a good chance that the device is doing what it is meant to do, indicating a fault with the appliance. Further investigation needs to be carried out to find the reason why.

Spillage detection devices

These are thermal backflow detection devices, commonly known as TTBs, (standing for the Dutch term 'themishe terugslag beveiliging'). They are located on or near the edge of a draught diverter and are normally fixed to a flange or bracket. The devices are not adjustable and are pre-set by the manufacturer.

Spillage detection devices are similar in appearance to heat sensors on central heating pipes; they are connected by two cables and are linked into a thermocouple which uses the interrupter principle described in 'Interrupter devices' (page 297).

These devices work by detecting adverse flue conditions, as in open flues when it is in a downdraught condition (see Figure 9.19 on page 300). Heat is generated by the downdraught and the sensor detects this and will interrupt the thermocouple, shutting down the appliance by causing the pilot to extinguish.

You should check the position of the TTB during installation and servicing of an appliance. Also ensure that the leads are connected to the TTB correctly, according to the manufacturer's instructions.

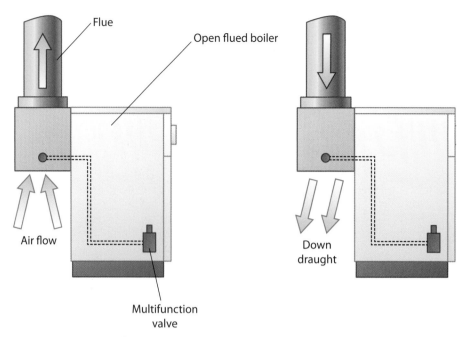

Figure 9.19 Flue in normal flow conditions and downdraught conditions

Multifunctional control valves

A multifunctional control is a combination of several control devices all contained in one unit and fitted to an appliance, the most common of which is a boiler. The multifunctional control often includes the:

- main control gas cock
- pressure regulators
- solenoid valve
- flame supervision device (thermocouple)
- thermostat
- ignitor.

When installing an appliance such as a boiler, these devices allow you to adjust the burner pressure and the pilot. Some models allow parts of the valve to be replaced, such as solenoid valves and FSDs where present.

Another type of multifunction valve is the modulating gas valve. These valves adjust the pressure to the burner depending upon the load that is required. The most common type of appliances that have these fitted is the high-efficiency condensing boilers. Where they are installed on combination boilers the valve will give a different rate to the burners when the boiler is only heating water for hot water, than it will when feeding the central heating too.

Not all of these valves can be adjusted; where they can, they must be adjusted only in accordance with manufacturer's instructions, this is to ensure that the appliance is working correctly and is rated to its optimum output. These adjustments are made using a combustion analyser rather than a manometer. A small change in setting may produce a large change in combustion products including CO levels. Where there are no instructions for adjustments then the manufacturer must be contacted for technical advice.

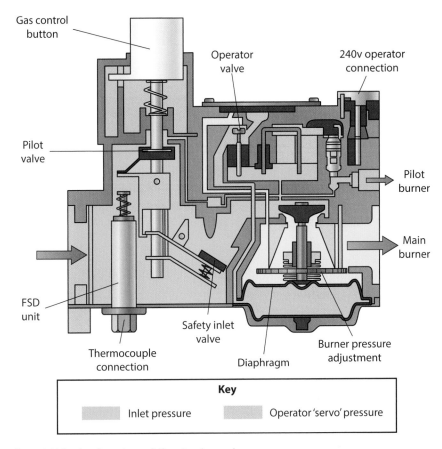

Figure 9.20 Section through a multifunctional gas valve

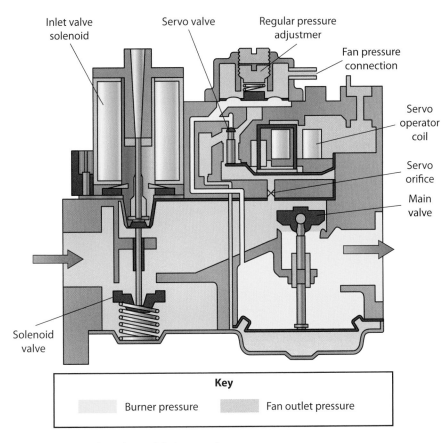

Figure 9.21 Section through a modulating gas valve

Progress check

1 Which of the Gas Safety (Installation and Use) Regulations deals with controls?

2 As a gas operative you are required to never leave an appliance which may cause injury or death to the user. What else must you ensure before leaving?

3 Why is a low-pressure cut-off valve installed where LPG bottles are supplying the gas?

TYPES AND OPERATING PRINCIPLES OF THERMOSTAT/THERMISTOR

Thermostats are electrical devices that control the level of heat emitted from appliances and can be of either a bi-metallic or a liquid vapour type. Thermostats work by breaking the contact on an electrical circuit.

The bi-metallic type is normally attached/surface-mounted in such a way as to sense temperature within pipework. Whereas the liquid vapour phial would be located into a pocket within, for example, the waterways of a boiler.

Bi-metallic

There are two types of bi-metallic thermostats: rod and strip. They work on the principle that two dissimilar metals have different expansion rates.

Strip

Strip thermostats are made up of two dissimilar metals attached together so that the outer metal has a higher rate of expansion than the inner metal. When heated, because they are joined, the metals bend together (see Figure 9.23).

These are normally in the shape of a coil as in Figures 9.22 and 9.23. Figure 9.22 shows the strip before heat is applied.

Figure 9.22 Bi-metal strip

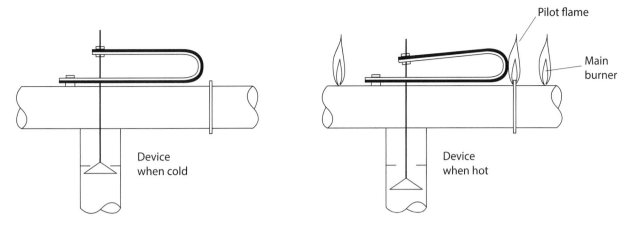

Figure 9.23 Section through a bi-metallic strip with no pilot lit and with pilot lit

Rod

Rod thermostats will be found on older cooker ovens and work on the same principle as a bi-metallic strip in that they have two dissimilar metals with differing expansion rates. These thermostats work in conjunction with vapour pressure devices (see 'Vapour pressure devices' page 293).

When the oven is cool, the gas to the oven is turned on and the FSD is sensing a flame at the pilot, the valve will open the seating and gas will flow to the burners.

When the thermostat gets up to temperature, as set by the control knob, the sensing rod detects the temperature rising and closes the valve onto the seating. The valve then goes into a **by-pass rate**, through the by-pass screw. A common cause of oven faults not going into by-pass rate is the by-pass screw holes being blocked by grease.

When the temperature decreases sufficiently then the rod senses this and the oven burners again go into full flame until it is up to temperature again.

Key term

By-pass rate
A low rate of gas going to the burners to sustain temperature in the oven.

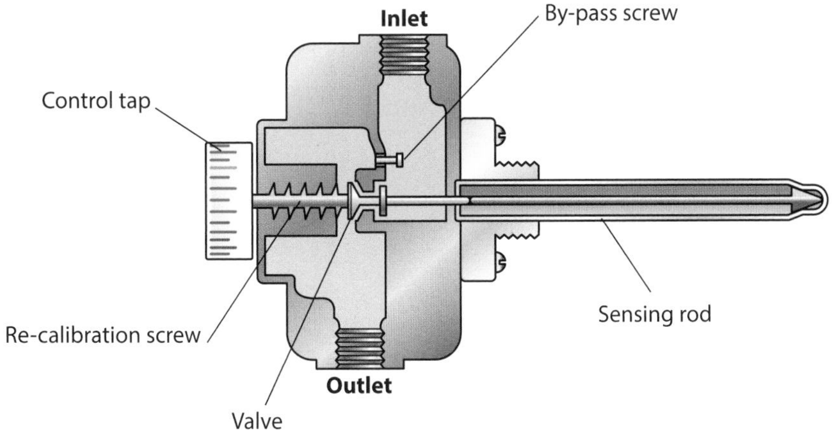

Figure 9.24 Section through an oven rod thermostat

Case study

Angus has gone to a job and the customer explains that the oven lights ok but, after a while, the flame goes out.

1 Which control would Angus need to check is working?

2 How would Angus check that the control is working?

3 What could be causing the fault to occur?

Check your answers with your tutor.

Liquid expansion valve

Liquid expansion valves are mostly found on gas cookers. The temperature is sensed by the remote phial, which contains liquid. It can be seen from Figure 9.25 that, on heating, the bellows expand and close the gas valve to the burner. Take note of the bypass too, as this maintains a small amount of gas to keep the oven burner lit even if the main valve is fully closed. When the oven cools, the thermostat phial detects this and the bellows contract, opening the gas valve again.

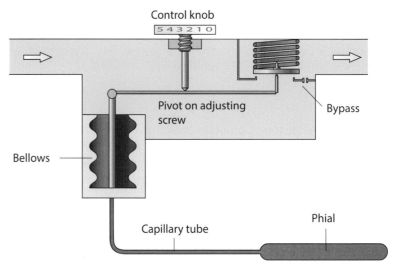

Figure 9.25 Section through a liquid expansion valve

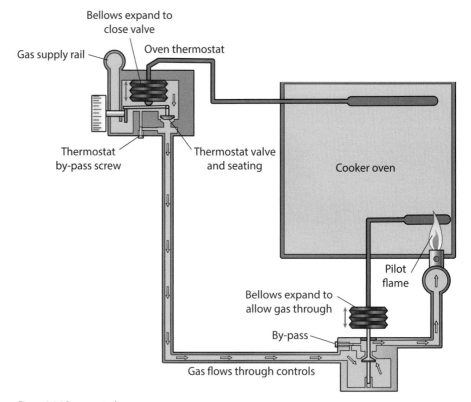

Figure 9.26 Oven controls

Vapour pressure thermostat

The vapour pressure thermostat is used mainly on refrigerators. The sensing phial is fitted to the evaporator and when the cabinet is cooled to the correct temperature, the gas rate is reduced to a minimum bypass rate.

How a vapour pressure thermostat works

The phial of the thermostat is filled with a gas that condenses at approximately 0°C e.g. sulphur dioxide. When the temperature of the cabinet is above the set temperature, the gas pressure in the phial increases. This in turn expands the bellows, pushing the gas valve off the seating.

As the cabinet cools, the gas in the phial condenses, reducing the pressure and allowing the bellows to contract and the gas valve to close. A minimum gas rate is maintained through the fixed bypass.

Electrical overheat/limit thermistors

Thermistors use internal electrodes that sense surrounding heat and measure it through electrical impulses. They are made with semi-conductor materials using temperature-sensitive resistance. There are two kinds of thermistors: negative thermal coefficient (NTC) and positive thermal coefficient (PTC). With the former, the resistance of the thermistor decreases as the temperature increases, and with the latter, the resistance increases as the temperature increases. There are several layers of tiny electrodes within the body of the thermistor, all of which are connected by semi-conducting materials. Thermistors show a large resistance for a small temperature change, and most of them are made from manganese, cobalt, nickel and copper oxides. Silicon and germanium are also used in the production of thermistors.

The thermistor is used to detect heat on pipework, being attached to specially designed clips. These can also sense overheating of the pipework, such as when there is an air-lock in the system and water is not moving through a boiler. They will then shut off the supply to the appliance and in some cases cause an overheat button to lock out the appliance. The appliance would then need to be manually reset after investigation of the fault.

Knowledge check

1 Which electromechanical device automatically opens a valve when a circuit is made and then closes itself when the circuit is broken?

a Solenoid valve

b Ventilation sensing device

c Vitiation sensing device

d Flame suppression device

2 Fill in the missing word: 'A mechanical thermostat works by moving a valve using the expansion and contraction of … to control the supply of gas to an appliance to limit specific temperature levels.'

a Solids

b Liquids

c Vapours

d All of the above

3 When a lack of oxygen is present a device which shuts off the pilot and the main gas valve is called a(n):

a Auto solenoid valve

b Gas relay valve

c Ventilation sensing device

d Vitiation sensing device

4 A thermoelectric flame supervision device performs which of the following?

a It generates electricity via a thermocouple to hold open a valve

b It shuts down an appliance's gas supply if the flame is extinguished

c Both A and B

d It prevents electricity holding open a multifunctional control via a thermocouple

5 A device designed to cut off the gas supply to an appliance by the use of a flame acting by expanding or vaporising a liquid is a:

a Gas pressure flame supervision device

b Control supervision device

c Vapour pressure flame supervision device

d Piezoelectric flame supervision device

6 A device which uses an AC current passed through a flame to produce a rectified DC current to maintain a flame at the burner is called a:

a Flame rectification supervision device

b Flame conduction supervision device

c Flame detection device

d Thermoelectric supervision device

7 SSOV stands for:

a Secondary shut-off valve

b Safety shut-off valve

c Solenoid shut-off valve

d Safety supply overpressure valve

8 What fault on a flame rectification supervision device might cause the burner to lock out?

a A damaged or wrongly positioned probe

b Poor earthing

c Poor lead connections

d All of the above

9 What device would an interrupter be incorporated into?

a A cooker lid

b A thermocouple

c Flame rectification device

d Atmosphere sensing device

10 Where on an open-flued boiler is a TTB positioned?

a Near the burner

b Next to the pilot

c In the draught diverter

d On the gas inlet supply

Chapter 10

Checking and setting burner pressures and gas rates

THIS UNIT COVERS

- methods of determining gas rates at appliances
- how to check the gas pressure in installations
- methods of determining and/or setting working pressures
- calculating gas rates
- identifying and rectifying faults discovered during testing

INTRODUCTION

Gas supplies to appliance burners, whether they are burning natural gas or liquefied petroleum gas (LPG), must be supplied at the pressures which have been specified by the manufacturer, so as to ensure that that the expected heat output and the efficiency of the appliance is achieved. This chapter will have an emphasis on natural gas systems; for more information on LPG systems, see Chapter 13.

METHODS OF DETERMINING GAS RATES AT APPLIANCES

Gas rating of appliances is a test to determine the volume of gas which is being used by the appliance. This test converts the time it takes to pass a known quantity of gas through a meter to an appliance into a heat input.

When you carry out checks on appliances you must be capable of verifying that the correct gas pressure and gas rates are available at the appliance. Where this is not possible you need to know how to make the necessary adjustments to the burner pressure to achieve this. This is a requirement of Regulation 26(9) of the Gas Safety (Installation and Use) Regulations.

When commissioning or fault finding in appliances, it is an essential part of the process to undertake gas burner pressure tests and the gas meter volume measurements, otherwise known as 'rating'.

Bear in mind that domestic natural gas installations are designed to ensure that there is no more than 1 mbar drop in pressure between the meter outlet and the appliance. Also remember that natural gas is supplied into the dwelling at 20 mbar; where an appliance requires a lower pressure than this, it would be regulated down at the appliance.

When carrying out a gas rating test on an appliance, it is important that you know if the rate is correct to the manufacturer's specifications data or to the data plate on the appliance. To check the rate, it may be necessary to convert from a gross rate to a net rate. To do this you divide the gross rate by 1.11 for natural gas.

Example

If the rate of an appliance is 24 kW (gross) then the net rate for natural gas would be calculated as:

24 ÷ 1.11 = 21.62 kW (net)

For propane you would divide by 1.09 and for butane by 1.08. See the section on 'Calorific value' (page 88) for more detail.

HOW TO CHECK THE GAS PRESSURE IN INSTALLATIONS

Most domestic gas appliances use atmospheric burners, like a Bunsen burner. Gas is delivered from an injector, mixed with air in the mixing tube, ignited and burned at the burner head. To make sure we have satisfactory combustion (a good flame) it is essential for the gas pressure to be kept constant and for there to be sufficient air.

The pressure at the inlet of any appliance is required to be 20 mbar in order to draw (entrain) the right amount of primary air for correct combustion. Some appliances, however, are designed to operate at different pressures: boilers, for example, may be able to operate at a range of pressures. These appliances are fitted with an individual appliance regulator which is sometimes built into the **multifunctional control valve**.

When you put an appliance into use you will have to check and adjust the appliance regulator. It is your responsibility to ensure that the meter regulator is set correctly, but only the gas supplier is allowed to adjust it if it is wrong. So how do you check it? You need to connect a 'U gauge' to the meter test point and put any appliance on full, or a cooker with three burners on full. Only *appliance needs to be on* to check the **working pressure**.

The pressure recorded should be **21 mbar ±1 mbar**. That is to say, if the working pressure is 20 mbar to 22 mbar, then the meter regulator is working satisfactorily. If the recorded pressure is outside these limits, switch off the supply and contact the Gas Supplier on 0800 002 001 (England and Wales).

As an additional check it is now worth trying the **standing pressure**. All you need to do is turn all appliances off and record the pressure. It is likely to be approximately 21 mbar to 25 mbar – if it is vastly different to these figures, then you are likely to need to call out the gas supplier. Never break the regulator seal yourself.

Regulation 14(6) of the Gas Safety (Installation and Use) Regulations generally restricts the breaking of regulator seals to the gas transporter or supplier unless you have been authorised to do so by the gas transporter/ supplier. On completion you must reseal using an appropriate seal to prevent setting of the regulator.

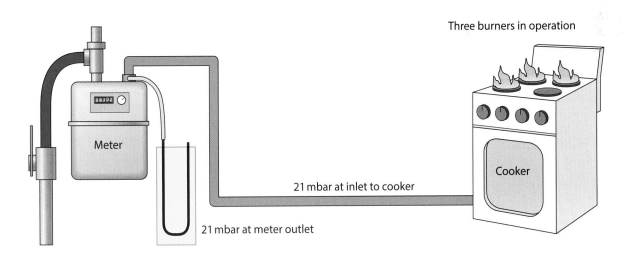

Three burners in operation

21 mbar at inlet to cooker

21 mbar at meter outlet

Meter

Cooker

Figure 10.01 Checking working pressures

METHODS OF DETERMINING AND/OR SETTING WORKING PRESSURES

Now that you have confirmed that the meter regulator is working correctly, you need to check and adjust appliance pressures.

Use of manufacturer data (appliance burner pressure)

The correct burner pressure for each appliance is given:

- on the data plate (fitted to all appliances)
- in the manufacturer's instructions (always left with the householder).

If this information is not available, then it is not possible to carry out the work and replacement instructions are required.

Methods of checking burner pressures

It is important that before taking any measurements with the manometer that you set it to zero (see Figure 10.02).

When using a water-filled manometer, 'U' gauge, for measuring, take a reading from both legs of the gauge, from the bottom of the meniscus, and divide by two. This will give you a more accurate reading.

Check manufacturer's instructions before using electronic manometers and make sure they have been calibrated. You need to look at the appliance instructions to determine where to test for burner pressures. This could be before and after the regulator or on the multifunctional valve.

The pressures to any appliance (before the regulator) should never be more than 1 mbar below the correct working pressure at the meter regulator. Remember that this will usually be between 20 mbar and 22 mbar. If it is less than this, the pipework may be undersized, partially blocked or kinked, and the installation does not therefore meet with the standards laid out in the regulations. You must rectify the fault or shut down the installation and make it safe.

Fixed-rate appliances (non-modulating)

In order to check and alter the pressure to a fixed-rate appliance, follow the steps on page 311.

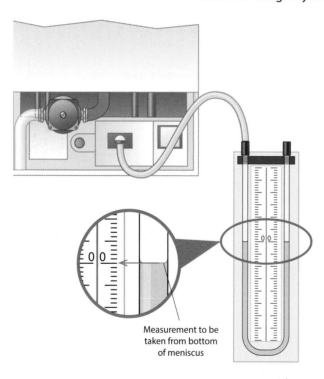

Measurement to be taken from bottom of meniscus

Figure 10.02 Taking burner pressure of non-modulating boiler

How to check the burner pressure

Checklist

PPE	Tools and equipment	Source information
• Overalls • Protective footwear • Barrier cream	• Manometer • Flat head screwdriver • Leak detection equipment • Electrical isolation testing equipment	• Manufacturer's instructions

1 Ensure before removing the boiler cover that you have safely isolated the appliance from the electrics – test using electrical test equipment. You will however need the electricity later to check the burner pressure and CO/CO_2 ratio.

2 Identify the location of the burner pressure test point and governor.

3 Check the manufacturer's instructions for the correct burner pressure reading.

4 Connect the manometer to the test point. Ensure the manometer is set to zero and turn on the appliance. Make sure when you take the reading that your eyes are level with the meniscus to ensure an accurate reading.

5 If the burner pressure is correct, carry on with the rest of the boiler checks.

6 If the burner pressure requires adjustment, using a screwdriver turn the governor screw until the required pressure is read on the manometer.

7 Turn off the burner and remove the manometer from the test nipple and spray the test nipple with leak detection liquid to ensure it does not leak from the grub screw.

8 Check the CO/CO_2 ratio to ensure that the adjustment is causing the burner to work

9 Confirm with the customer that everything is satisfactory or, if not, what further repair work may be necessary.

How to take a working pressure

Checklist

PPE	Tools and equipment	Source information
• Overalls • Protective footwear • Barrier cream	• Flue gas analyser • Spanners • Screwdriver • Leak detection equipment • Manometer • Brush set • Vacuum cleaner	• Manufacturer's instructions

1 After checking the appliance works correctly, isolate the electrical supply before removing the boiler cover.

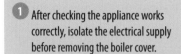

2 Check that your manometer is set to zero and identify the inlet pressure test point on the boiler (air/gas ratio valves do not normally have a burner test point).

3 Identify the test point from the boiler manufacturer's instructions and loosen the test point grub screw – do not remove it or let it fall as they are very small and can be easily damaged.

4 Only use the correct sized screwdriver otherwise the grub screw can be easily damaged when opening and closing – do not apply undue pressure.

5 Connect the manometer tube to the test point ensuring it is secure and then turn the electrics back on and fire the boiler.

6 The reading on the pressure test point should be 20 mbar or within 1 mbar of that reading to ensure the appliance is getting sufficient gas.

7 Turn off the boiler and electrics, remove the manometer tube from the test point shut-down grub screw and test that it is gas tight using leak detection fluid.

8 Continue with the rest of the boiler checks.

would divide your final figure (13.56) by a constant figure of 1.1 to give a net figure of 12.32 kW (net).

2 min gas flow m³	Heat input kW (gross)	Heat input kW (net)	2 min gas flow m³	Heat input kW (gross)	Heat input kW (net)	2 min gas flow m³	Heat input kW (gross)	Heat input kW (net)	2 min gas flow m³	Heat input kW (gross)	Heat input kW (net)
0.002	0.65	0.59	0.052	16.8	15.14	0.102	32.95	29.68	0.152	49.10	44.23
0.004	1.29	1.16	0.054	17.44	15.71	0.104	33.59	30.26	0.154	49.74	44.81
0.006	1.94	1.75	0.056	18.09	16.30	0.106	34.24	30.85	0.156	50.39	45.40
0.008	2.58	2.32	0.058	18.73	16.87	0.108	34.88	31.42	0.158	51.03	45.97
0.010	3.23	2.91	0.060	19.38	17.46	0.110	35.53	32.01	0.160	51.68	46.56
0.012	3.88	3.50	0.062	20.03	18.05	0.112	36.18	32.59	0.162	52.33	47.14
0.014	4.52	4.07	0.064	20.67	18.62	0.114	36.82	33.17	0.164	52.97	47.72
0.016	5.17	4.66	0.066	21.32	19.21	0.116	37.47	33.76	0.166	53.62	48.31
0.018	5.81	5.23	0.068	21.96	19.78	0.118	38.11	34.33	0.168	54.26	48.88
0.020	6.46	5.82	0.070	22.61	20.37	0.120	38.76	34.92	0.170	54.91	49.47
0.022	7.11	6.41	0.072	23.26	20.95	0.122	39.41	35.50	0.172	55.56	50.05
0.024	7.75	6.98	0.074	23.9	21.53	0.124	40.05	36.08	0.174	56.20	50.63
0.026	8.4	7.57	0.076	24.55	22.12	0.126	40.70	36.67	0.176	56.85	51.22
0.028	9.04	8.14	0.078	25.19	22.69	0.128	41.34	37.24	0.178	57.49	51.79
0.030	9.69	8.73	0.080	25.84	23.28	0.130	41.99	37.83	0.180	58.14	52.38
0.032	10.34	9.32	0.082	26.49	23.86	0.132	42.64	38.41	0.182	58.79	52.96
0.034	10.98	9.89	0.084	27.13	24.44	0.134	43.28	38.99	0.184	59.43	53.54
0.036	11.63	10.48	0.086	27.78	25.03	0.136	43.93	39.58	0.186	60.08	54.13
0.038	12.27	11.05	0.088	28.42	25.60	0.138	44.57	40.15	0.188	60.72	54.70
0.040	12.92	11.64	0.090	29.07	26.19	0.140	45.22	40.74	0.190	61.37	55.29
0.042	13.57	12.23	0.092	29.72	26.77	0.142	45.87	41.32	0.192	62.02	55.87
0.044	14.21	12.80	0.094	30.36	27.35	0.144	46.51	41.90	0.194	62.66	56.45
0.046	14.86	13.39	0.096	31.01	27.94	0.146	47.16	42.49	0.196	63.31	57.04
0.048	15.5	13.96	0.098	31.65	28.51	0.148	47.80	43.06	0.198	63.95	57.61
0.050	16.15	14.55	0.100	32.3	29.10	0.150	48.45	43.65	0.200	64.60	58.20

Table 10.02 Natural gas rating chart for metric meters up to 6 m³/hr

Case study

Rob has installed a combination boiler and is commissioning the boiler.

1 List the procedure that he should follow to ensure the boiler is working efficiently.

2 What is the formula for calculating the gas rate when gas is supplied through a metric meter?

3 How many revolutions should an imperial meter do when Rob is gas rating the appliance?

Check your results with your tutor.

IDENTIFYING AND RECTIFYING FAULTS DISCOVERED DURING TESTING

During routine servicing or a call-out to a customer's appliance, you may come across the faults while testing and would need to be able to determine the possible cause of the fault. Such faults include excessive pressure loss at the appliance or incorrect gas rate at the appliance. During tests for these faults it may reveal that the appliance is working correctly but may not be giving the correct heat output as specified in the manufacturer's instructions data tables.

Excessive pressure loss at the appliance

If during your test you cannot get sufficient pressure at the burner look for the following:

Undersized gas supply pipework from the meter to the appliance – this will require the pipework being re-sized and re-piped from meter to the appliance. The installation is classed as 'at risk' (AR) situation 5.10 under the GSIUR (see Table 11.02, situation 2.8, page 346).

Meter regulator incorrectly set – you need to contact the Emergency Service Providers as the installation is classed as 'not to current standards' (NCS) situation 2.8 (see Table 11.02).

Blockage in supply to appliance – for this you will need to trace which section of pipe the blockage is at. You can do so by checking the pressure at individual appliances that are connected to the gas installation.

Incorrect gas rates at the appliance

When commissioning or servicing, you may come across incorrect gas rate faults. Here are some examples of what to look for and how to deal with your findings:

- A blocked or partially blocked burner injector – for this you should clean the burner injector
- Incorrect injector size – check with manufacturer's data, ask the responsible person for permission then change the injector. Re-check burner pressures and heat inputs.
- Incorrect appliance or burner inlet pressure – you'll need to adjust the pressure with procedure depending on type of valve fitted.

Pressure absorption

When gas within pipes is static, the pressure is the same throughout its entire length. As soon as the gas begins to flow, the pressure falls progressively from the entry point to the exit. This is called 'pressure absorption' or the more common term that operatives use is 'pressure loss'. This is a result of the friction as the gas flows against the pipe wall. The rate at which the pressure drops is constant, provided that the pipe diameter and material are the same throughout the length.

Effects of meter pressure absorption under full load conditions

The gas supplier must provide a minimum of 19 mbar at the outlet of the emergency control valve (ECV) during peak flow conditions. The pressure absorption across the primary meter must not exceed 4 mbar at maximum flow rate, which provides a minimum of 15 mbar. If you also take into consideration the 1 mbar drop across the system, this would then give a pressure at the appliance of 14 mbar – this is a pressure at which manufacturers permit the appliance to operate safely. On, for example, a boiler where the working pressure is taken at the gas valve, there is usually another 1.5 mbar allowance for pressure drop through the appliance pipework and valve, this may then give a reading of 12.5 mbar at the gas valve.

Ideally there should be 21 mbar ± 2 mbar at the meter. Should there be a shortfall in pressure, you must report it to the gas transporter. Where the pressure is affecting the safe operation of an appliance, you must:

- disconnect and cap off the appliance
- contact the gas transporter

Never leave an appliance in a condition that may cause harm to the users of that appliance.

Knowledge check

1 One of the differences in calculating the gas rating on a U6 meter and an E6 meter is:.

   a On a U6 you use 2 mins of time and on the E6 you use 2 revolutions of the test dial

   b On a U6 you use two revolutions of the test dial and on the E6 you use 2 mins of time

   c On a U6 you use one revolution of the test dial and on the E6 you use 2 minutes of time

   d On a U6 you use 2 minutes of time and on the E6 you use 1 revolution of the test dial

2 When using a water-filled manometer the reading for the gas pressure in relation to each leg of the manometer tube should be taken:

   a One from the bottom and one from the top of the meniscus

   b Both from the top of the meniscus

   c Both from the bottom of the meniscus

   d Both from half way down the meniscus

3 The gas rate of an appliance is calculated using which of the following information?

   a Pipe length

   b Calorific value

   c Number of seconds for one revolution of the test dial

   d All of the above

4 $Q = \dfrac{\text{Appliance rating} \times 3600}{\text{Calorific value of the gas}}$

What does Q stand for in this gas formula?

   a Exact gas rate

   b Appliance gas pressure

   c Supply pipe sizing

   d Maximum number of appliances

5 To convert a gross heat input to a net heat input you would:

   a Divide the heat input by 1.11

   b Multiply the heat input by 1.11

   c Add 1.11 to the heat input

   d Use any of the above

6 Where can the technical data for the commissioning of a gas appliance or component normally be found?

   a BS 6871

   b Manufacturer's instructions

   c Gas Safety Regulations

   d Building Regulations

7 What is the maximum pressure absorption across a meter during maximum flow rate?

   a 1 mbar

   b 2 mbar

   c 4 mbar

   d 8 mbar

8 If you were attempting to calculate the working pressure of a gas cooker installation this would best be achieved when:

   a Three rings of the hob are on full

   b One ring of the hob are on full

   c Two rings of the hob are on full

   d All rings of the hob are on full

9 A boiler is connected to an E6 meter, the first reading is 566785.020 m^3, the second meter reading after 2 mins is 566785.090 m^3. Which of the following is the net heat input to the boiler:

   a 21.47 kW

   b 20.37 kW

   c 22.34 kW

   d 24.34 kW

10 To convert kW to Btus, the constant you should calculate it with is by:

   a Dividing by 3,412

   b Adding 3,412

   c Multiplying by 3,412

   d Subtracting 3,412

Respond to unsafe situations

INTRODUCTION

This chapter deals with the Gas Industry Unsafe Situations Procedure (GIUSP) and the regulations and procedures that enable you to carry out these duties. Examples of the types of forms and warning notices/labels that you would need to fill in or refer to when in a particular working situation are given. Example unsafe situations you may encounter when servicing or maintaining systems and appliances are also given, along with an explanation of the action you should take and the regulation and class of situation that is relevant to that particular scenario.

LEGISLATION FOR UNSAFE GAS SUPPLIES OR APPLIANCES

The GIUSP is a guidance document available to all gas engineers. At the time of writing the latest edition of the GIUSP is Edition 6, issued 5 March 2012. It details how to deal with unsafe gas installations based on the level of risk that is posed to occupiers of the premises/dwellings, including you, the responsible person and members of the public. As a gas engineer the GIUSP helps you to comply with the main gas safety laws which apply to all installations immaterial of the type of gas and whether it is in a domestic dwelling or business premises. Always check to make sure you are using the most recent edition of GIUSP.

All gas businesses and operatives must be familiar with their obligations under Part 1 of the GIUSP. You must also be aware of any additional safety and technical guidance that you may need to use in addition to the information found in the regulations.

There are a number of statutory regulations which all companies or individual registered engineers must be aware of:

- the Gas Safety (Installation and Use) Regulations 1998
- the Gas Safety (Management) Regulations
- the Gas Safety (Rights Of Entry) Regulations
- Reporting of Injuries, Diseases and Dangerous Occurrences Regulations.

The Gas Safety (Installation and Use) Regulations 1998 (GSIUR)

The general points covered in the GSIUR are discussed in Chapter 2 and the regulations are available in 'Approved Code of Practice and Guidance – Safety in the installation and use of gas systems and appliances' (L56) – downloadable from the HSE website www.hse.gov.uk/pubns/priced/l56.pdf

Here the focus is on how the regulations deal with responding to unsafe situations.

The GSIUR are for the installation of gas fittings in all domestic premises, commercial premises (e.g. hospitals, educational establishments, offices, hotels, restaurants) mobile catering units, leisure accommodation vehicles (including caravan holiday homes and hired touring caravans), inland waterway craft hired out to the public and sleeping accommodation, wherever it is located.

Regulation 3(1) of the GSIUR states that:

No person shall carry out any work in relation to a gas fitting or gas storage vessel unless he is competent to do so.

This means that any gas business carrying out gas works in any of the premises mentioned above must be registered with the Gas Safe Register and their operatives must hold a valid certificate of competence for each work activity they wish to undertake e.g. CENWAT for boiler and water heaters.

Under the GSIUR you are required to make judgements on the level of risk of an installation, this relates to regulations 26(9), 34(3) and 34(4).

Regulation 26(9) states that:

Where a person performs work on a gas appliance he shall immediately thereafter examine:

a) The effectiveness of any flue

b) The supply of combustion air

c) Its operating pressure or heat input or, where necessary, both

d) Its operation so as to ensure its safe functioning, examples are: safety controls, flueing, ventilation, gas rate of the appliances and burners pressures are correct.

In addition to this, you have a duty of care to take all reasonably practicable steps to notify any defect and unsafe situation to the responsible person and, where that is different, the owner of the premises in which the appliance is situated. This would be the case where a user is living in rented accommodation, in which case you would inform the landlord. Where neither of these cases are reasonably practicable, in the case where liquefied petroleum gas (LPG) is supplied to an appliance, then inform the supplier of gas to the appliance or the transporter.

Regulation 34(3) and 34(4) places a duty on the gas operative who identifies any defects to an appliance, gas fitting, gas supply, service pipework, gas meters and storage vessels (unsafe or dangerous) to report the defects to the responsible person. Again where these are not available the gas supplier or transporter must be contacted.

When an unsafe situation has been reported to the responsible person, you must record this on the appropriate paperwork and ensure the appliance or installation is correctly labelled according to the unsafe situation – whether it be deemed 'at risk' (see page 333), 'immediately dangerous' (see page 332) or 'not to current standards' (see page 334).

The importance of documenting unsafe situations cannot be stressed enough. These documents provide an accurate system of recording the actions which have been taken by the operative in the event of an unsafe appliance or installation situation. This information may be needed later; an example is where a user has reinstated an appliance which has been deemed unsafe without undertaking the necessary remedial works, causing an incident. The regulations have taken this into account by making it an offence for the user/owner/responsible person or any other person to use a gas appliance or installation where they have been advised that it is unsafe.

The Gas Safety (Management) Regulations

These regulations deal with gas operatives who work for the emergency service providers (ESP). These operatives carry out work on behalf of the Public Gas Transporter (PGT). Currently the regulations require British Gas Plc to set up and operate this emergency service. The ESP may be a different company, depending on the area of the country you live in.

Transco and other ESPs respond to calls regarding:

- reported or suspected gas escapes
- reports of 'fumes' indicating spillage or incomplete combustion.

When the ESP are called to a suspected spillage or combustion problem their operatives adopt either a 'Concern for Safety' policy or identify it as an 'immediately dangerous' or 'at risk' situation.

In each of these cases the appliance is labelled accordingly; the label states that the appliance should not be used until it has been examined by a Gas Safe registered installer.

Where other situations occur as in the case of unsatisfactory workmanship then in compliance with the GIUSP these defects must be reported (see the section on 'Reporting of Injuries, Diseases and Dangerous Occurrences Regulations (RIDDOR)').

Gas Safety (Rights of Entry) Regulations 1996

These regulations apply to authorised officials working for the PGT who are empowered to enter premises to make an unsafe installation safe. As already mentioned, Transco is the responsible PGT and ESP.

Authorised officials base their rights of entry on whether there is a danger to life or to the property. Where the appliance has been disconnected following a rights of entry, it is an offence under the regulations for an appliance to be reconnected without the permission of the PGT.

When an immediately dangerous situation has been identified and the responsible person refuses to allow the appliance or installation to be made safe, then you or the authorised official should contact the National Emergency Service Call Centre. This number should be located on a label at the meter.

The National Emergency Service Call Centre will need all of the details recorded on your documentation about the unsafe situation. They should give you a reference number which you should record on your documentation.

Attending a gas-related incident site

Where operatives are called to or encounter a gas-related incident, it is extremely important that the incident scene is *not* disturbed.

You should immediately contact the appropriate ESP for natural gas or the supplier for LPG and inform them of the incident. With the least possible disturbances to the incident scene and where possible and safe to do so, the installation should be made safe, for example by turning off the gas supply at the appropriate ECV and, if necessary, ventilating the premises. In cases of fire and explosion, or where an immediately dangerous situation is evident, the gas installation must be disconnected and sealed.

In non-domestic premises, the responsible person should take the decision whether or not to shut down the installation or process. This is essential where issues of process safety are involved, for example the cooling down of a furnace.

It is important to record all actions undertaken, as they will assist those parties involved in any subsequent incident investigation.

Attending the site after a gas-related incident

When attending the site, operatives should question the gas user or responsible person and check the installation for any gas safety warning label(s) to determine whether a RIDDOR reportable incident has occurred. If working at a site where it is known that there has been a gas-related incident, do not carry out any work other than making the installation safe, without first liaising with the HSE and the gas supplier to ensure that investigations are complete.

It is extremely important that the incident scene is *not* disturbed.

Reporting of Injuries, Diseases and Dangerous Occurrences Regulations (RIDDOR)

Under RIDDOR as a gas installer you have a legal obligation to report any of the events which are covered by these regulations to the appropriate authority, the Health and Safety Executive (HSE). The following gives a guide to the reporting of a dangerous gas fitting. For information on the reporting of injuries or dangerous occurrences, see Chapter 1.

What you must report

If you are a gas engineer registered with the Gas Safe Register, you must provide details of any gas appliances or fittings that you consider to be dangerous, to such an extent that people could die or suffer a major injury because the design, construction, installation, modification or servicing could result in:

- an accidental leakage of gas
- inadequate combustion of gas
- inadequate removal of products of the combustion of gas.

Examples of these in dangerous gas installations include:

- serious gas escapes caused by poor workmanship or materials
- a defective flue, where the products of combustion (POC) are not completely cleared
- a flued gas appliance installed without a flue
- inadequate ventilation
- safety devices being made inoperative, for example by a flame supervision device (FSD) being wedged open

- the wrong type of gas being used with an appliance, for example an LPG boiler being connected to natural gas
- an appliance installed with an incorrect flexible connector, for example. connecting a gas hob with a flexible tap connector
- an open-flued appliance installed in a bathroom or shower room after 24 November 1984
- any open-flued appliance installed legally before 24 November 1984 which is proved to have serious flueing or ventilation defects.

Examples of reportable flammable gas incidents include where a person is killed or suffers a major injury as a result of:

- a gas explosion
- a fire
- carbon monoxide poisoning

HSE

Health and Safety Executive

Health and Safety at Work etc Act 1974
The Reporting of Injuries, Diseases and Dangerous Occurrence Regulations 1995

Zoom 100% ⬍ KS i ?

F2508G1 - Report of a flammable gas incident

A report of a flammable gas incident will ask for the number of injured persons and their names. You will need this information to complete and submit the form.

About you and your organisation

*Title	*Forename	*Family Name
MR	John	Pipe

*Job Title: Gas installations   *Your Phone No: 01923 123890

*Organisation Name: J.P. Gas services

Address Line 1: 27 Governor Road   (eg building name)

Address Line 2:   (eg street)

Address Line 3:   (eg district)

*Town: London

County: Middlesex

*Post Code: LO1 10TD   Fax Number:

*E-Mail: JPgas@yahoo.co.uk

☐ Remember me   ?

[Next]   [Form]

Page 1 of 5

Figure 11.01 Gas incident report form

- gas asphyxiation
- gas-related injuries.

This includes employees of the company who have installed the installation and members of the public.

When reporting incidents you must use the correct form. These forms can now be completed online at www.hse.gov.uk/riddor/report.htm#online and include:

- Form F2508G1 – Report of a flammable gas incident (Figure 11.01)
- Form F2508G2 – Report of a dangerous gas fitting (Figure 11.02).

A telephone service is available, but only for reporting fatal and major injuries: Incident Contact Centre 0845 300 9923, Monday to Friday 8.30am to 5.00pm. You must ring this number immediately to report a death or major injury.

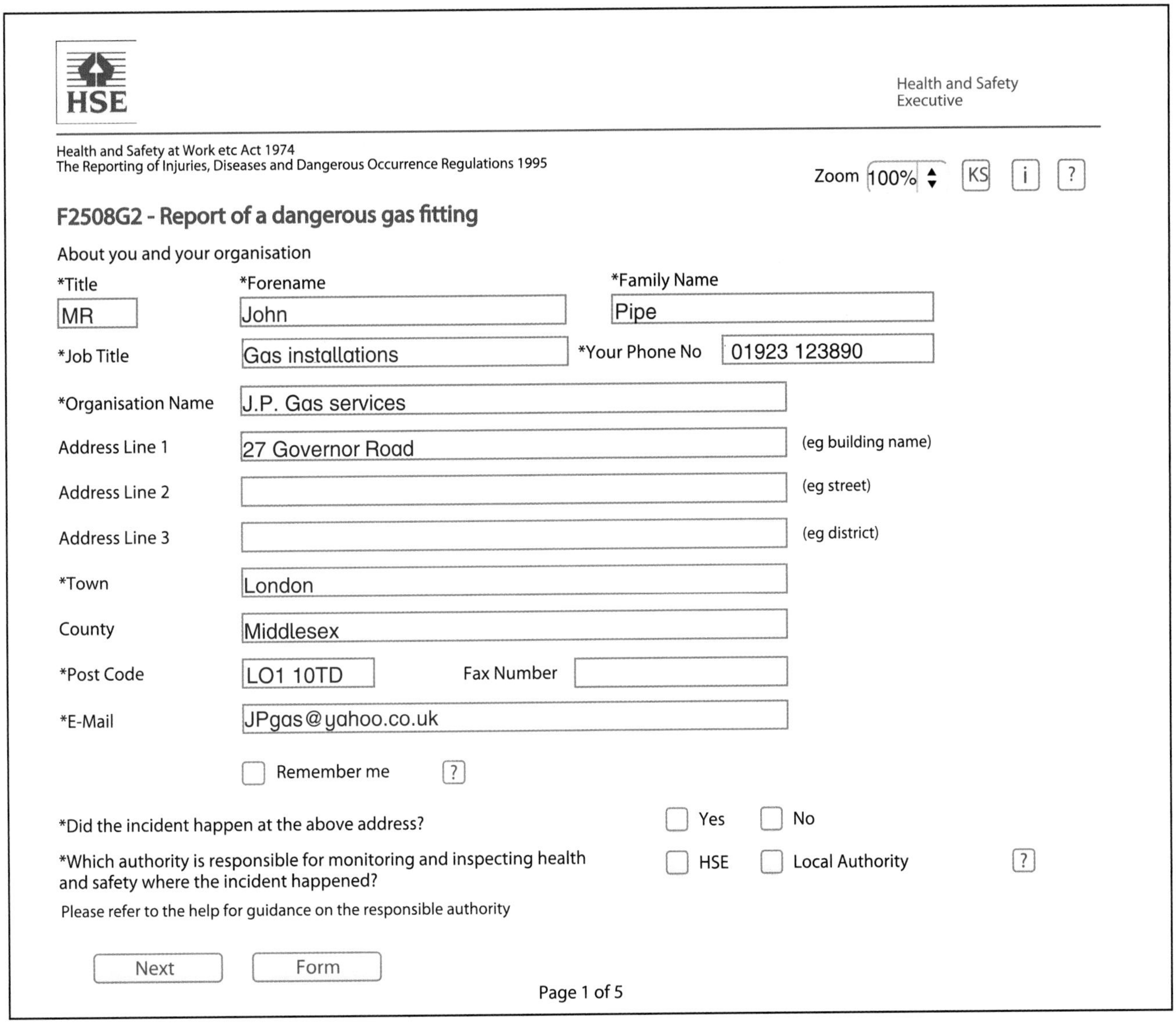

Figure 11.02 RIDDOR reporting form for a dangerous gas fitting

Where the incident is out of normal working hours, follow the guidance on the HSE website: www.hse.gov.uk/contact/outofhours.htm

Copies of reports will be sent to you after completion by the HSE. Your company is advised to keep these reports for six years to cater for any future civil litigation.

DEALING WITH GAS EMERGENCIES – RESPONSIBILITIES AND ADVICE

Whether you are on site when the unsafe situation occurs or are away from the scene, you have certain responsibilities as a gas operative. You may need to offer a gas user advice over the telephone if they report a gas escape or you may need to contact the gas supplier and be aware of their responsibilities in such situations.

Advice for the gas user when you are not on site

If a gas user reports a gas escape while you are not on site, you should give them or, if different, the responsible person the following instructions and guidance.

Turn off emergency controls

For **natural gas** you should advise the user to turn off the gas supply at the meter at the emergency control valve (ECV) unless the meter is located in a basement or cellar. If there is a smell of gas in a basement or cellar, then advise them to evacuate the building.

For **LPG in bulk storage supply**, advise them to shut off the ECV outside the building and the gas isolation valve on top of the above-ground storage vessel(s), or underground storage vessel(s).

For **LPG meter installations**, instruct the user to shut off the ECV at the meter installation outside the building.

For **LPG cylinder-fed installations**, advise them to shut all cylinder valves.

Eliminate ignition sources

Advise the customer not to smoke or light matches under any circumstances. You must also instruct them not to operate any electrical lights or switches, as these can ignite gas escapes and cause an explosion.

Reduce gas concentrations via ventilation

For natural gas escapes, tell the user to open windows and doors to get rid of gas by ventilating the property. As LPG is heavier than air it sinks to ground level, so where this gas is escaping, advise the user to vent at low levels also.

If the person reporting the gas escape can smell, hear, see or feel gas outside the property, you should ask them to *close* the doors and windows of their house.

Your responsibilities in unsafe gas situations

When you are informed of a smell of gas or a gas escape is detected on a site you are visiting, you should take the following actions:

- turn off the ECV
- extinguish all naked flames and remove sources of ignition
- inform everyone not to operate switches (on or off)
- ventilate the property by opening windows and doors
- test the installation for gas tightness.

When conducting a gas tightness test under these circumstances, then no pressure drop is allowed.

Responsibilities of the gas user in unsafe gas situations

The gas user or responsible person for the premises has a responsibility to inform the ESP if they know of any gas escape or fumes into the property and must immediately take reasonable steps to turn the gas supply off at the meter or ECV to prevent any further escape of fumes or gas into the property. They also have a duty to ventilate the property of fumes or gas escapes.

If there is still an escape or smell of gas after the user/responsible person has isolated the gas supply, then they must immediately contact the relevant emergency contact centre (see Table 11.01 on page 330).

Responsibilities of the gas transporter in an unsafe gas situation

When the Gas Transporter (GT) receives notice regarding an unsafe situation, the GT will send an authorised officer to the site address to make safe. Should the gas user still not give authority for the installation/appliance to be made safe, the GT will then take the appropriate action which could lead to the termination of the gas supply.

Report gas escapes

Where there is a suspected gas escape, contact the Gas Emergency Service Provider (ESP) for a natural gas escape, or in the case of LPG, contact the gas supplier. Table 11.01 gives details of ESP contact information.

Region	Gas type		Contact details	Telephone details
England, Wales and Scotland	Natural gas		Contact the gas emergency contact centre	0800 111 999
	LPG	Bulk and metered supplies		See telephone number on the bulk storage vessel or at the meter
		Cylinder supplies	For cylinder supplies on caravan parks and hire boats, the site owner and/or boat operator may also have responsibilities. Advice may be obtained form the gas emergency company identified on the cylinder through their emergency contact details	See gas supplier emergency contact details in the local telephone directory
Northern Ireland	Natural gas		Contact the gas emergency contact centre	0800 002 999
	LPG	Bulk and metered supplies		See telephone number on the bulk storage vessel or at the meter
		Cylinder supplies	For cylinder supplies on caravan parks and hire boats, the site owner and/or boat operator may also have responsibilities. Advice may be obtained form the gas emergency company identified on the cylinder through their emergency contact details	See gas supplier emergency contact details in the local telephone directory
Isle of Man	Natural gas and LPG		Manx Gas Ltd	0800 1624 444
Channel Islands – Guernsey	Mains gas and LPG*		Contact Guernsey Gas Ltd	01481 749000
Channel Islands – Jersey	Mains gas and LPG*		Contact Jersey Gas Company Ltd	01534 755555
*Mains gas in the Channel Islands is an LPG and air mixture				
Gas emergency contact details of the four main suppliers of LPG in the British Isles are shown below				
Calor: 08457 444 999			**BP:** 0845 607 6118	
Flogas: 0845 7200100			**Shell:** 0870 7359999	

Table 11.01 ESP contact information

Rights of entry to properties

In accordance with GIUSP 4.3, as a gas operative, you work in premises by invitation from the gas user or responsible person. As a result, any actions you may wish to take must be with the gas user or responsible person's permission.

Rights of entry to properties are dealt with in the Gas Safety (Rights of Entry) Regulations. These regulations apply to natural gas and make provision for an officer who is authorised by the gas transporter to take immediate action if they have reasonable cause to suspect that gas conveyed by the gas transporter is escaping, or may escape. This applies to *any* premises (domestic or non-domestic and includes factories) that gas is conveyed to, where gas has escaped and may enter premises, including escapes of POC, within their gas transporter's network.

Where the above criteria apply then the regulations allow for the officer authorised by the gas transporter, to enter the premises to carry out any work necessary to prevent the escape and take any other steps to avert any danger to life. The authorised officer must be able to produce an authenticated document to prove their authority on request.

Action of the LPG supplier

The Gas Safety (Rights of Entry) Regulations do not generally apply to LPG installations. You should contact the LPG supplier who may have a contractual right of entry agreement between the customer and the supplier. The gas supplier will have a duty under the GSIUR to respond to situations where gas is escaping. The gas supplier for LPG installations is normally the company that refills the storage vessel(s) or cylinder(s). On caravan parks and sites this may be the park owner or landlord.

Although LPG gas suppliers have duties to attend gas escapes, where these involve suspected emissions of carbon monoxide from a gas appliance, their duty is limited to giving advice on how to prevent the escape or emissions, and the need for examination and, where necessary, repair by a competent person.

Case study

Chris has received a telephone call from a customer who complains that they can smell gas. What information should Chris give the customer for a) a natural gas system, b) an LPG in bulk storage system, c) an LPG meter installation and d) an LPG cylinder-fed installation?

Check your answer with your tutor.

PRIORITIES IN GAS EMERGENCIES

Under the GSIUR you have a duty to take appropriate action regarding unsafe situations. Remember, you would be in contravention of the regulations if you failed to act on an unsafe situation. The priority for gas operatives when encountering an unsafe situation is to safeguard life and property.

Safe working

Only enter a dwelling if it is deemed safe to do so. Do not put yourself or others at risk by entering a dwelling where carbon monoxide is suspected at dangerous levels without the appropriate equipment.

Secure the escape

Where an escape of gas has occurred, you have a duty to ensure that the leak is sealed off. In the case of fumes escaping, you have a duty to ensure that the appliance that is the cause of the fumes is made safe and isolated.

Leave the site safe

Where possible and with the gas user/responsible person's agreement, you should make every effort to rectify the situation(s) and make the appliance(s) and/or installation safe to use at the time of the visit. Where this is not possible, you must take the following actions:

- Explain to the gas user/responsible person that the appliance and/or installation is, in your opinion, 'immediately dangerous' and must be

Progress check

1 Where a gas operative is on site and there is a smell of gas, list the actions that should be taken.

2 List the types of incident that must be reported to RIDDOR.

3 When a gas operative performs work on a gas appliance, what must they also examine? List the actions they should undertake.

disconnected from the gas supply until the situation has been rectified and that further use would contravene the GSIUR.

- Attach a suitably worded 'DO NOT USE' warning label to the appliance and/or installation.
- Complete a 'warning notice' and ask the gas user or responsible person to sign it as a record of receipt. Give a copy to the gas user or responsible person.
- With permission, immediately disconnect and seal the appliance and/or installation with an appropriate fitting.
- If the gas user or responsible person does not allow disconnection, try to turn off the appliance and/or installation. In the case of natural gas, make immediate contact with the Gas Emergency Contact Centre. Obtain a job reference number from the operator and the time of the contact for your records.

DEALING WITH UNSAFE SITUATIONS WITH APPLIANCES AND INSTALLATIONS

Immediately dangerous (ID) situations

Where you find an installation or appliance that would cause an immediate danger, which cannot be rectified, you must make safe and isolate installation and/or appliance and attach the relevant warning labels and complete a 'notice of immediate danger'.

Notices and warning labels

Before disconnection of the appliance or installation from the gas you must seek permission from the customer or owner. After permission is granted a suitably worded 'DO NOT USE' warning label must also be securely attached to the appliance or installation in a prominent position.

In some cases the customer or owner may not give you permission to disconnect from the gas supply. In these circumstances, you should turn off the appliance or installation at the isolation device and attach a warning label. Immediately contact the Gas Emergency Contact Call Centre or, for LPG, contact the gas supplier.

On completion of disconnection or operation to make safe the appliance, you must immediately complete a 'notice of immediate danger'. Only the operative to have carried out the disconnection or make safe should complete the notice. The notice should clearly indicate that an 'immediately dangerous' (ID) situation exists, with a note of the fault and the action taken. A copy of the notice should be given to the owner, landlord or responsible person.

The notice of ID must contain the following information, which the Gas Emergency Contact Centre would ask for:

- confirmation that it is an ID and/or rights of entry disconnection request
- name of the person reporting, the Gas Safe Register registration number of the business and the operative's individual number
- name of the responsible person for the property

Figure 11.03 Immediately dangerous warning label

■ the address at which the ID situation exists

■ details of the ID situation

■ type of appliance and/or installation

■ location of appliance/installation within the property.

At risk (AR) situations

An appliance and/or installation 'at risk' (AR) is one where one or more faults exist and which, as a result, if operated, may in the future constitute a danger to life or property.

Notices and warning labels

Where possible, and with the gas user/ responsible person's agreement, you should try to rectify the appliance/installation to make it safe to use at the time of the visit. If this is not possible, you should take the following action:

■ Explain to the gas user/responsible person that the appliance and/or installation is, in your opinion, AR and that it should not be used. Where it is continued to be used in these circumstances it would be the gas user/ responsible person's own responsibility and may be a breach of the law.

Figure 11.04 Concern for safety notices and labels

- Attach a suitably worded 'DO NOT USE' warning label in a prominent position on the appliance/installation.
- Complete a 'warning notice' and ask the gas user/responsible person to sign it as a record of receipt. Give a copy to the gas user/responsible person and keep a copy in your file.
- With the gas user/responsible person's permission, turn off the appliance and/or installation. Where permission is refused, you should draw the gas user/responsible person's attention to the fact that it may be an offence to use a gas appliance/installation that has the potential to be dangerous.
- Where the gas user/responsible person refuses to sign the warning notice, it is recommended that this is recorded on the warning notice.
- If the gas user is not the owner of the appliance and/or installation, a copy of the warning notice should also be provided to the owner, landlord or managing agent.

A number of companies, in particular meter installers, are now adopting a 'concern for safety policy'. Meter installers have a limited scope and are not ACS qualified to work on appliances. When they install a meter they may encounter the following:

- new appliances
- older appliances
- appliances which need re-commissioning following a repair because of an ID or AR situation
- appliances where, because of their limited scope, the installer is prevented from making an adequate assessment of its safety.

Because these appliances must be commissioned in accordance with the GSIUR, then the meter should be connected to a sealed meter outlet and a capped meter warning notice should be fitted to meter saying 'WARNING The outlet of this meter has been capped off with a sealing disc'.

The sealing disc can only be removed by a Gas Safe registered installer, once all appliances have been checked and commissioned by the registered installer who has the competencies for those appliances e.g CKR1 for cookers.

Not to current standards (NCS) situations

Gas safety standards are constantly reviewed and improved, as a result of new research findings, incident experience or changes in technology. It is a requirement that gas installations meet these standards and comply with the legislation that is applicable at the time of installation. When changes to standards and legislation have occurred, those installations are considered safe for continued use; with few exceptions, there are no requirements for the gas user or responsible person to upgrade them. However, you should assess existing installations against current standards and/or requirements and, provided the installation is operating safely, make a judgement about the advice to give the gas user or responsible person.

You may deem an installation or appliance 'not to current standards' (NCS). Where multiple NCSs have been identified this may escalate to an AR situation. This does not include those situations that fall outside the requirements of the GSIUR.

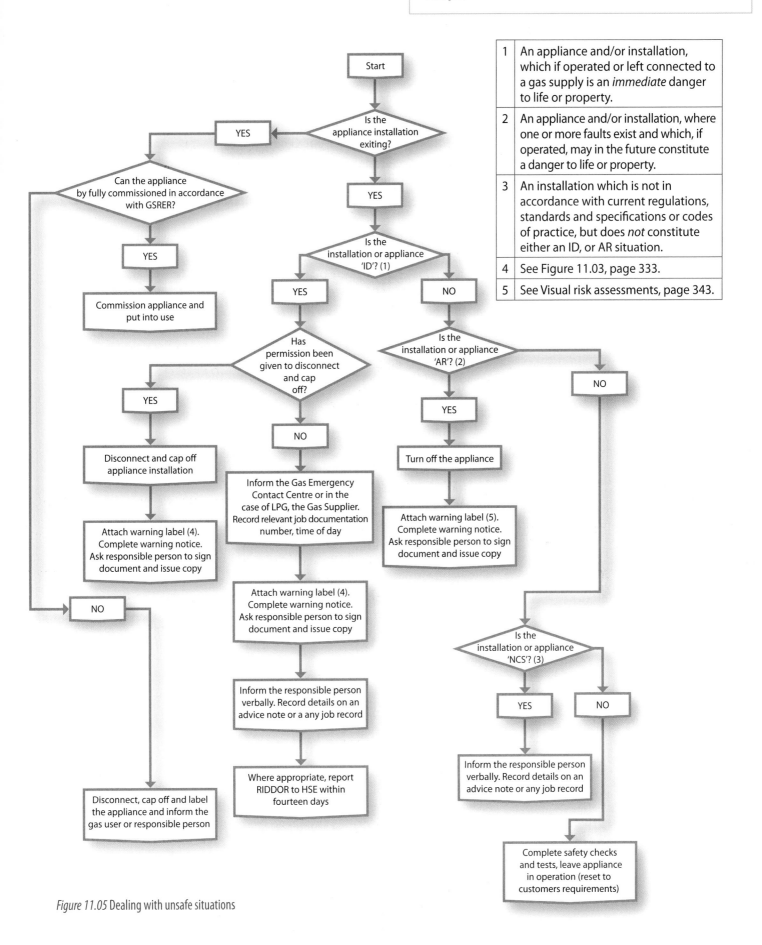

1	An appliance and/or installation, which if operated or left connected to a gas supply is an *immediate* danger to life or property.
2	An appliance and/or installation, where one or more faults exist and which, if operated, may in the future constitute a danger to life or property.
3	An installation which is not in accordance with current regulations, standards and specifications or codes of practice, but does *not* constitute either an ID, or AR situation.
4	See Figure 11.03, page 333.
5	See Visual risk assessments, page 343.

Figure 11.05 Dealing with unsafe situations

Notices and warning labels

The NCS category covers many different circumstances, broadly categorised as:

1 One or more flueing and/or ventilation NCS situations.
2 NCS situations that contravene GSIUR.
3 Other NCS situations where industry standards have changed since the original installation, but the appliance is operating safely.

You must notify the gas user or responsible person of any category 1 and 2 NCSs, where they relate to a gas appliance. For category 3 NCSs, you should use your judgement when deciding whether to report, depending upon the usefulness of the information.

When notifying, particularly for categories 1 and 2, it is recommended that you write an 'advice notice' and keep a copy of the notice for future reference. Failure to notify may be construed at a later stage as negligence where criminal proceedings are in place.

Examples of NCS installations

- Pipework which is not sleeved as it passes through a wall into a property, where there is no sign of corrosion on the pipework.
- Where the pipework installation has been undersized.
- Insufficient supports (clips) for pipework installations.
- A flue where the first bend is less than a 600 mm rise vertical from the draught diverter and where manufacturer's instructions are not available.
- Flues that have been installed with 90° bends.
- Where compartment ventilation is inadequate or incorrectly installed.
- Polyethylene (PE) pipework which has been installed in buildings without adequate protection against ultra-violet (UV) radiation.
- Appliances which have been installed too close to combustible surfaces without signs of damage (scorching).
- Terminals that are incorrectly sited (see Figure 7.19 page 212).
- An undersized flue, where the appliance is still operating correctly.

GENERAL NOTICES AND WARNING LABELS TO AVOID UNSAFE SITUATIONS

General notices and labels are provided as guidance for the gas user. The information provided on the notice is important for the user as they provide details of whom to contact in the event of gas escapes or emergencies.

Such labels and notices include meter and compartment labelling; appliance commissioning, appliance servicing and landlords' safety certificates.

Meter labelling

British Standard (BS) 6400 sets out the requirements for meter notices. Warning notices should be prominently displayed and of a durable form to protect against damage (e.g. weather-resistant). Labels which conform to BS 4871 are of a suitably durable type.

Emergency notices

Where a primary meter is fitted adjacent to the ECV, a permanent emergency notice (see Figure 11.06) must be fitted by the installer on or near the meter to inform the consumer:

- to shut off the supply of gas if there is a gas escape at the premises
- to immediately notify the Gas Emergency Service on 0800 111 999, if gas continues to escape
- not to reinstate the supply until remedial action has been taken by a competent person to prevent gas escaping again
- of details of the emergency service contact, including the emergency telephone number
- of the date the notice was first displayed.

This is a requirement of Regulation 15(1) of the GSIUR.

When the primary meter is not adjacent to the ECV then an emergency notice should be fitted on or near the ECV with the words 'GAS EMERGENCY CONTROL'. The notice must also display the same information as shown in Figure 11.06.

Regulation 15(2) of the GSIUR requires that where a meter is installed more than 2 m from the ECV, the installer should fit a notice at or near the meter indicating where the ECV is situated, as in Figure 11.07.

Regulation 16(2) requires that the GT fits a prominent notice on each primary meter indicating that they are supplied through a service pipe which supplies gas to more than one primary meter.

Where installations incorporate a secondary meter, a label should be fitted by the installer at the primary meter indicating the number and location of any secondary meters which are supplied through the primary meter. Where the secondary meter(s) are fitted then at each meter should be attached a label stating 'secondary meter' (Figure 11.08).

Regulation 17(1) requires that a line diagram is posted in a prominent position which is permanent on or near the primary meter or gas storage vessel and near to all emergency controls which are connected to the primary meter. The drawing must include the configuration of all meters, installation pipework and emergency controls.

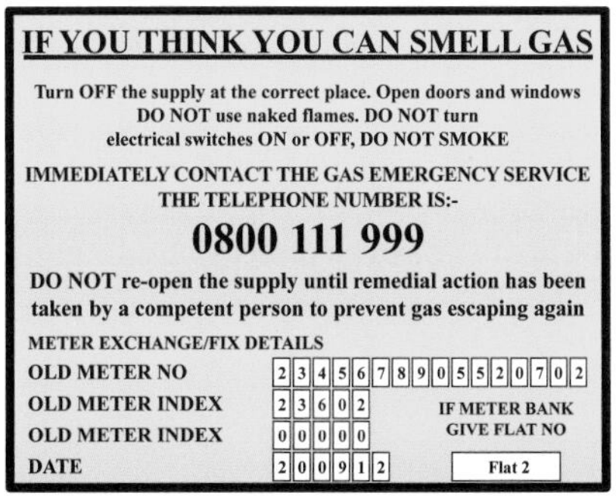

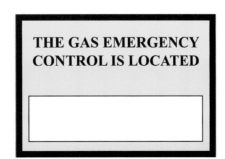

Figure 11.06 A permanent emergency notice

Figure 11.07 Gas ECV location notice

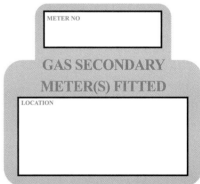

Figure 11.08 Secondary meter notice

Regulation 17(2) requires that if the configuration of the pipework or installation is changed then the diagram must be altered to show those alterations.

Meter housing notice for medium pressure systems

The installer should ensure that the meter housing has a notice to indicate:

- how the outlet pipe and vent pipe should not exit through the rear of the housing (see Figure 11.09)
- the housing details
- the maximum operating pressure and gas family which the housing is intended to serve (e.g. '2nd family gas with maximum operating pressure not exceeding 2 bar)

Figure 11.09 Medium pressure housing notices

Compartment labelling

In multi-occupancy dwellings meters are generally situated in a common compartment. The compartment is built to house all meters that serve all of the dwellings. When the installer fits the meter it is important that they ensure meter numbers and readings are taken accurately and they correspond to the dwelling they are serving. Each meter must have the label shown in Figure 11.10.

Appliance commissioning certificates

During the commissioning of an appliance – whether it is for a landlord or a household user – you will need to complete an appliance commissioning certificate. A copy of this should be left with the user or responsible person and with your company for its records.

Case study

Derek is undertaking a landlord's check of appliances. The tenant has installed a second-hand gas fire, which was not on your list of appliances from a previous check. After inspection of the appliance you have found that the fire has been incorrectly fitted and that the FSD has been disabled.

a What category is the situation classed as?

b What action should Derek take?

Check your answers with your tutor.

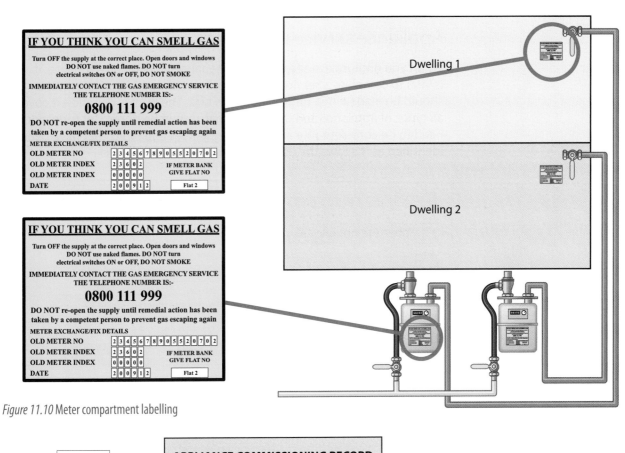

Figure 11.10 Meter compartment labelling

APPLIANCE COMMISSIONING RECORD

Job no.

Invoice no:

Registered business details GasSafe Reg no.

Gas operative: _____ (Print Name)

Gas Safe ID: _____

Company: _____

Address: _____

Postcode: _____ Tel no.: _____

Job address

Name (Mr, Mrs, Miss, Ms) _____

Address: _____

_____

_____

Postcode: _____ Tel no : _____

Client/landlords details (if different)

Name (Mr/ Mrs/Miss/Ms) _____

Address: _____

_____

Postcode: _____ Tel no. : _____

Rented accommodation Yes/No No. of appliances tested

	Details of the work carried out

				Appliance details				Inspection details						
	Location	Appliance type	Make	Model	Landlord's appliance Y/N/NA	Appliance inspected Y/N	Flue type OF/ RS/FL	Operating pressure in mbars or heat input kW/h or Btu/h	Combustion analyser reading (if applicable)	Safety device(s) correct operation Y/N/NA	Ventilation provision satisfactory Y/N	Visual condition of flue and termination satisfactory Y/N/NA	Flue performance checks Pass/ Fail/NA	Appliance serviced Y/N	Appliance safe to use Y/N
1															
2															
3															
4															

Defects Identified	Remedial action taken	Warning advice notice issued Y/N (If Y state serial no.)

Gas installation pipework satisfactory visual inspection Yes/No

Emergency control accessible Yes/No

Satisfactory gas tightness test Yes/No

Equipotential bonding satisfactory Yes/No

This safety record is issued by : Signed _____ Print name _____

Received by: Signed _____ Tennant/Landlord/Agent/Home owner

Date : _____

Figure 11.11 Appliance commissioning record

Appliance service certificates

When you undertake a service or maintenance work on an appliance, it is vital to complete a service form, identifying the variety of checks that should be made when undertaking this task. The form is generic – it covers all types of appliance and the checks that should be done. This form would need to be completed for individual appliances and faults that have been identified as it allows the user to see the results of the check.

Job No	

SERVICE/MAINTENANCE CHECK LIST

This form should not be used as a landlord's safety check

Registered business details		**Job address**	
Gas Safe Reg No			
Company: _____		Name:_____	
Address:_____		Address:_____	
_____		_____	
Postcode: _____ Tel no. _____		Post code:_____ Tel. No:_____	
Gas tightness test carried out state Yes or No		Rented accommodation state Yes or No	
Gas tightness test indicate Pass or Fail		Type of work state service or maintenance	

Appliance details			**Comments**
Type			
Make			
Model			
Location			

Appliance checks	YES	NO	N/A	Remedial action and/or defect
Burners/injectors				
Heat exchanger				
Ignition				
Electrics				
Controls				
Gas/water leaks				
Gas connections				
Appliance seals (case etc.)				
Gas pipework				
Fan(s)				
Fireplace opening/void				
Closure plate				
Flame picture				
Location				
Stability of appliance				
Return air/plenum				

Safety checks				Remedial action/type of defect
Ventilation				
Flue termination				
Flue flow test				
Spillage test				
Working pressure				
Safety devices				
Combustion analysis reading				Record reading here:

Findings			YES	NO
Is the appliance/installation safe for use?				
If NO, issue a Warning/Advice Notice (record serial no. of form and attach warning label)	Serial no.:			
The following remedial work is required:				

Gas user signature:_____ Gas operative signature:_____

Print name: _____   Print name: _____
Date: _____

Date:_____   Gas Safe card serial no.: _____

Figure 11.12 Appliance service and maintenance checklist

The user or responsible person should be made aware that their appliances should be serviced at least annually. A copy of this form should be given to the user/responsible person and a copy retained by the company.

LANDLORD'S GAS SAFETY RECORD

TENANT'S DETAILS	LANDLORD'S (OR LETTING AGENT'S) DETAILS
Name: _____	Name: _____
Address: _____	Address: _____
_____	_____
_____	_____
Telephone no.: _____	Telephone no.: _____

METER, PIPEWORK AND TIGHTNESS TESTING DETAILS

Gas supplies and pipework satisfactory		Meter/Emergency control accessible		Installation passes tightness test	

Answer YES or NO

	Location	Type	Make	Model	Flue type OF/RS/FLs	Landlord owned	Inspected	Safety controls working	Ventilation satisfactory	Flue visually satisfactory	Flue termination satisfactory	Appliance serviced	Flue flow test	Flue spillage test	Burner pressure or heat input	Safe to use Y/N
1																
2																
3																
4																
5																
6																

Answer YES or NO

Answer Either P (Pass) or F (Fail)

Numer of appliances tested	

Please remember the next safety check is due in 12 months' time		Warning label attached, notice issued
	Details of any faults found	Action taken
1		
2		
3		
4		
5		
6		
		Answer YES or NO

Date issued: _____ Gas Safe no: _____

Company name: _____	Signed for by:
Address: _____	Signature: _____
_____	
Telephone no.: _____	Tennant [ ] Landlord [ ]
Operative's name: _____	Managing agent [ ]
Operative's signature: _____	Please tick as appropriate

Figure 11.13 Landlord's safety record

Landlords' safety certificates

Landlords have a responsibility to ensure that gas appliances in their rented accommodation have been checked yearly by a registered Gas Safe installer. Landlords also have a responsibility to ensure that all gas appliances are in a safe condition and the installation is gas tight prior to re-letting the property.

Regulation 36 of the GSIUR imposes two main duties on landlords, concerning:

- annual safety checks on gas appliances/flues
- ongoing maintenance.

The regulations require that the landlord keeps a record in respect of the appliances to which these regulations apply. The records must include the following:

- the date of the inspection
- any defects that have been detected
- any remedial action taken on those defects.

Tenants should have access to the safety records. The safety checks that are carried out should ensure:

- the flue is working correctly – terminated correctly, of sound construction and free from leaks
- the burner pressure or heat input is correct (in accordance with the manufacturer's instructions)
- there is an adequate provision of ventilation
- all gas safety devices fitted to appliances are working correctly and are up to current standards (e.g. FSDs are working correctly)
- the installation with all appliances connected passes a tightness test.

When all of these safety checks have been completed by the gas operative, the operative is to complete the necessary paperwork. Three copies of the paperwork are required: one for the landlord, one for the tenant and one must be kept by the installer for their future reference.

The landlord has a responsibility to ensure that the operative who carries out the inspection is Gas Safe registered and has the competencies to carry out the necessary checks.

The tenant must be able to have access to the report within 28 days of the safety checks taking place. Where a tenant does not have any of the appliances within their dwelling then the landlord must display a copy in a prominent position for the tenants to see. This display copy need not contain the installer's signature, but must indicate that the original has been signed by the installer.

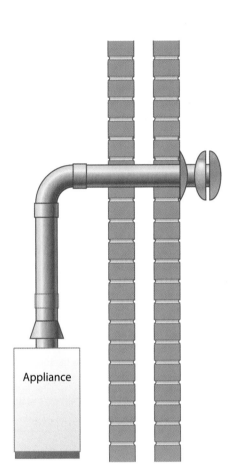

Figure 11.16 Wall-faced flue termination which is not allowed

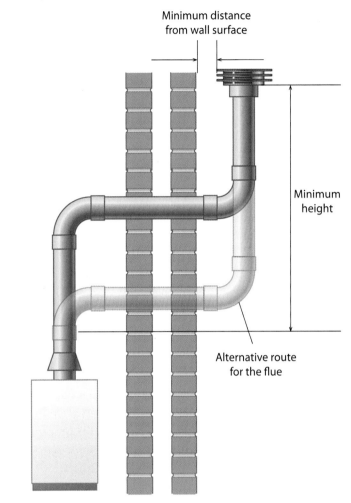

Figure 11.17 Wall adjacent flue terminations

Knowledge check

1 How might you classify the following situation? A room-sealed appliance terminal is not sited in accordance with manufacturer's instructions:

a Not to current standards
b At risk
c Immediately dangerous
d Immediately dangerous and RIDDOR reportable

2 A gas operative will normally complete a warning notice:

a During a landlord's safety inspection
b When an appliance or equipment is deemed to be not to current standards
c When an appliance is found to be at risk or immediately dangerous
d Both a and b

3 If there are signs of scorching where a gas fire has been fitted on a carpet, this type of risk is categorised as:

a ID – immediately dangerous
b AR – at risk
c AIR – at immediate risk
d NCS – not to current standards

4 A primary gas meter fitted without a regulator would be classified under the Gas Industry Unsafe Situations Procedure as:

a At risk
b Immediately dangerous
c Safe to use
d Not to current standards

5 Unsleeved and undamaged gas pipework passing through a cavity wall will be classed under the Gas Industry Unsafe Situations Procedure as:

a At risk
b Safe to use
c Immediately dangerous
d Not to current standards

6 What would be the classification (under the Gas Industry Unsafe Situations Procedure) for a flued gas appliance terminating into a closed space e.g. conservatory?

a AR – at risk
b AIR – at immediate risk
c ID – immediately dangerous
d NCS – not to current standards

7 Under which one of the regulations would authorised officials working for the public gas transporter have the power to enter a property in the event of a suspected gas escape?

a Gas Industry Unsafe Situations Procedures
b Gas Safety (Rights of Entry) Regulations
c Gas Safety (Installation and Use) Regulations
d All of the above

8 Which one of the following instances would be RIDDOR reportable under the Gas Industry Unsafe Situations Procedure?

a 'Let-by' of an ECV
b Flue termination located where products of combustion can enter premises
c Inadequately supported pipework
d 90° bends or horizontal runs fitted to open-flued appliances

9 If a gas operative is not on site and a user is reporting a smell of gas (gas escape), what advice should the gas operative give to the user?

a Report it to the HSE
b Open the windows
c Tell them it's nothing to worry about
d Turn off the ECV

10 The action to be followed when an installer identifies an Immediately Dangerous situation is:

a Fix a warning label the appliance/installation
b Disconnect the appliance/installation with the responsible person's permission
c Issue a warning notice
d All of the above

Chapter 12

Appliances

THIS UNIT COVERS

- **the different types of appliance – boilers, water heaters, gas fires, gas cookers, warm air heaters and gas-fired tumble dryers**
- **installing appliances**
- **commissioning appliances**
- **servicing and maintaining appliances**

INTRODUCTION

The correct installation, commissioning and maintenance of gas appliances is essential to ensure the safety of the public both in their homes and in the work place. As an engineer working on appliances, you therefore have to thoroughly understand the principles of installation and commissioning, and be able to recognise the symptoms of faulty equipment. The manufacturer's instructions must always be followed – the information provided in this chapter is for general guidance only.

BOILERS

The boiler does not in fact make water boil because bringing water to boiling point in the domestic setting would be quite dangerous. But to control the temperature of water for heating purposes and the provision of hot water is

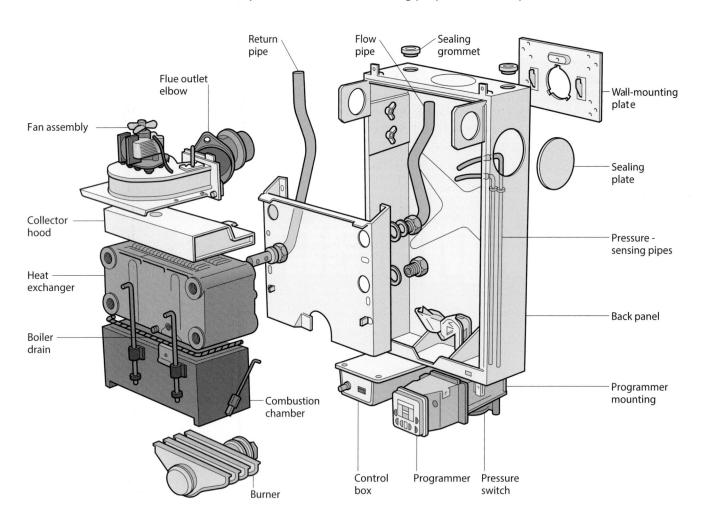

Figure 12.01 Typical wall-mounted boiler with fanned flue

a boiler's main function. They have evolved over the years into many types and designs. The most common models found in domestic premises today are covered in this section.

Types of boiler

Domestic boilers have an output of between 9 kW and 70 kW, and are designed to meet the maximum heating load in winter and provide hot water at the same time.

Many boilers have 'range-rated' burners, which means they can be adjusted to suit the size of the system. This is done by altering the burner pressure in accordance with the manufacturer's instructions or data plate.

Floor-standing boilers

Typical floor-standing boilers were fitted in kitchens with a cast-iron heat exchanger and an open flue. The floor-standing boiler is very heavy and takes up a lot of floor space on the ground. The trend now, however, is towards wall-mounted fan-flued boilers with low water content, integral pumps and control systems, and is a much lighter design.

Wall-mounted boilers

These are often fitted with light, compact heat exchangers that may contain as little as one litre of water. They are known as 'low water-content' boilers and require a positive flow of water at all times. If the water circulation slows down too much, the water can overheat and boil. This produces a noise like the 'singing' of a kettle, which is known in the trade as 'kettling'. A bypass is often fitted to these types of boiler to ensure that the flow of water is always above the minimum, therefore preventing overheating.

The bypass is usually a 15 mm pipe with a regulating valve or automatic bypass valve normally fitted between the main flow and return, as shown in Figure 12.02.

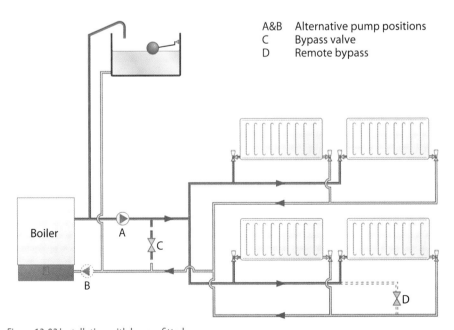

A&B   Alternative pump positions
C   Bypass valve
D   Remote bypass

Figure 12.02 Installation with bypass fitted

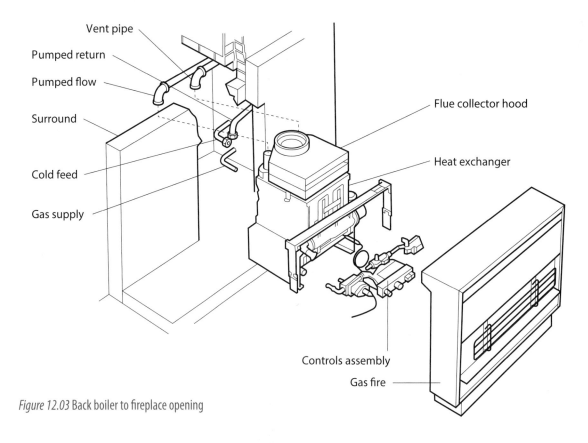

Vent pipe

Pumped return

Pumped flow

Surround

Cold feed

Gas supply

Flue collector hood

Heat exchanger

Controls assembly

Gas fire

Figure 12.03 Back boiler to fireplace opening

A further problem with low water-content boilers can be found when a boiler shuts down but the residual heat (in the cast-iron heat exchanger) continues to heat the water and boil. In this case a pump overrun thermostat is fitted to allow the pump to run after the boiler has shut down to remove the residual heat from the heat exchanger.

Back boilers

Back boilers are fitted into the fireplace opening behind a fireplace surround to which it is normal to have a gas fire fitted on the face (see Figure 12.03 above). Although they are concealed and save wall- or floor-space, they have a limited heat output of approximately 15 kW. They were very common in the 1970s and 1980s but due to changes in technology are not a common installation today.

Figure 12.04 Back boiler with combustion cover removed

Combined heating and hot-water units

Combined heating and hot-water units have a boiler and hot-water storage cylinder all contained within one unit; some may also contain the feed-and-expansion cistern or a sealed expansion vessel. All controls and connections are built into the unit, as shown in Figure 12.05.

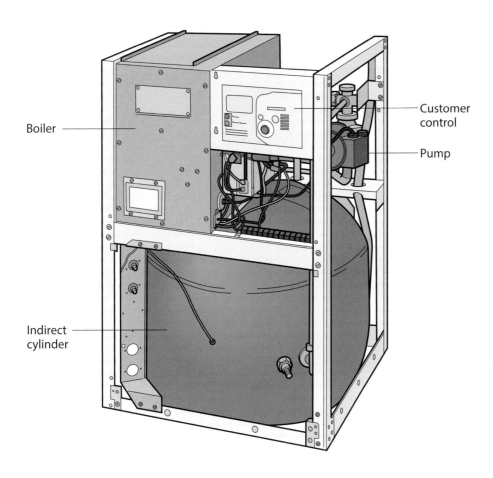

Figure 12.05 Combined heating and hot-water unit

Combination (combi) boilers

Combination boilers supply central heating via a radiator system and domestic hot water on an instantaneous water-heater basis.

The central heating normally uses a sealed system, but the important feature is that any hot-water demand takes priority over the central-heating demand. A big advantage of these appliances is that there is no space needed for a hot-water cylinder. All hot water is heated instantly; this does, however, have limitations on the hot-water delivery rate to the taps, and of course another disadvantage is that the heating system does not get any heat while hot water is being drawn off. (See detailed drawings of a combination boiler on page 376 and 377.)

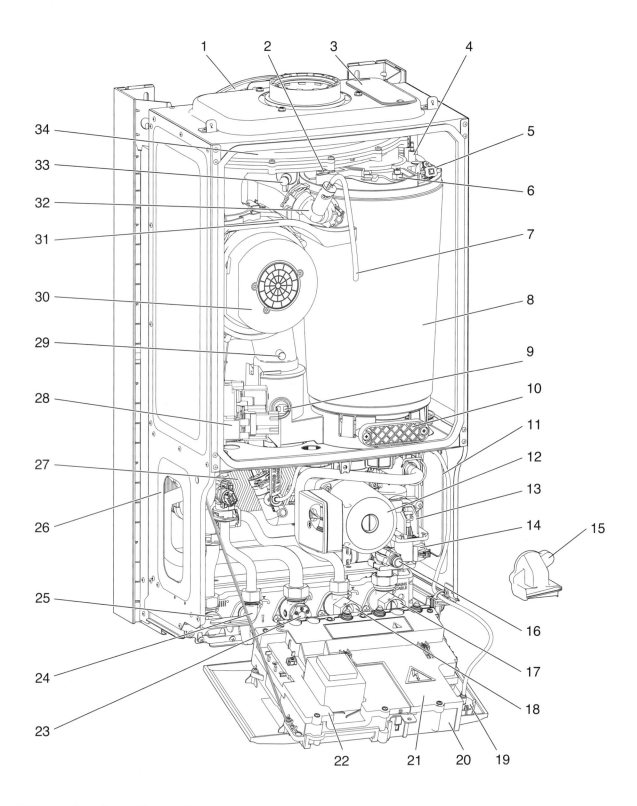

Figure 12.06 Layout of a modern combination boiler

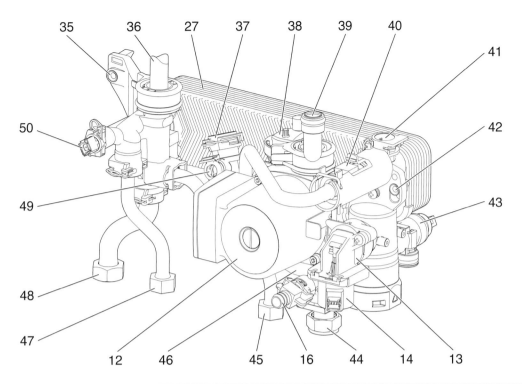

1	Expansion vessel	26	Left side hand-hold for lifting boiler
2	Flow temperature sensor (NTC)	27	Plate to plate heat exchanger
3	Removable panel – for servicing	28	Gas valve
4	Electrode assembly	29	Flue air pressure switch connection (not used)
5	Overheat thermostat	30	Fan
6	Securing nut – air/gas manifold clamp	31	Fan guard
7	Silicon tube – heat exchanger air vent	32	Manual vent point
8	Heat exchanger	33	Fan pressure test point
9	Flue over-heat thermostat	34	Air/gas manifold
10	Access panel – heat exchanger/sump cleaning	35	Compact hydraulic – left mounting point
11	Right side hand-hold for lifting boiler	36	Flow connector from heat exchanger
12	Pump	37	Expansion vessel hose connection point
13	Diverter valve assembly (body)	38	Auto air vent
14	Diverter valve actuator (stepper motor)	39	Return connection to heat exchanger
15	Diverter valve protective cover	40	Flow turbine
16	Drain point	41	Unused port
17	Central heating (CH) return isolator	42	Compact hydraulic – right mounting point
18	Domestic cold water mains (DCW) isolator	43	Pressure relief valve
19	System pressure gauge	44	CH return connection to service valve
20	Control panel (in service position)	45	DCW in connection to service valve
21	Cover – external wiring connections	46	Internal by-pass
22	Cover – transformer and PCB	47	DHW out connection
23	Gas inlet connection 22 mm	48	CH flow connection to service valve
24	Domestic hot water (DHW) connection	49	Pressure gauge connection point
25	CH flow isolator	50	DHW temperature sensor (NTC)

Figure 12.06 Layout of a modern combination boiler sharing hydraulic block components (continued)

The combination boiler is really only a central-heating boiler that becomes an instantaneous water heater on demand of any hot water. Figures 12.07 and 12.08 help show the two different circuits that are controlled by a diverter valve activated by opening a hot tap.

There can be sizeable differences in the controls used in combination boilers to separate and provide heat to the two circuits – here we can only provide an overview. It can be beneficial to attend manufacturer training courses to cover specific combination boilers, as their service and maintenance can often be quite complex.

When there is no demand for central heating, the boiler only fires when the domestic hot water is drawn off.

If there is a demand for central heating, the boiler supplies the heating system. When hot water is drawn off, the full output from the boiler is directed via the diverter valve to the calorifier or plate heat exchanger to give maximum supply to hot water.

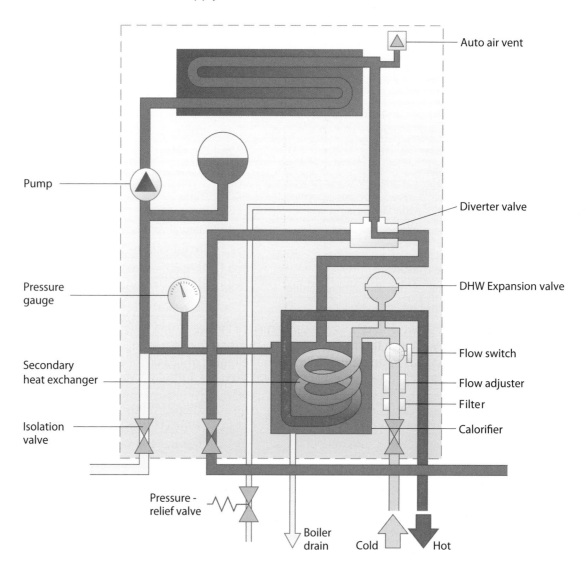

Figure 12.07 Combination boiler – domestic hot water circuit

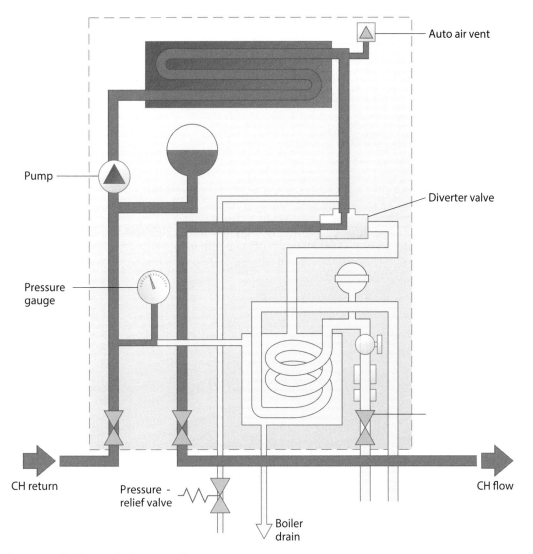

Auto air vent

Pump

Diverter valve

Pressure gauge

CH return

Pressure - relief valve

Boiler drain

CH flow

Figure 12.08 Combination boiler – central heating circuit

Condensing boilers

Condensing boilers are approximately 94 per cent efficient in their condensing mode. They achieve this higher efficiency by using the heat from the normally wasted flue gas temperature, which is in the region of 175 to 250 °C. Extra tubes/heat exchangers take this extra heat from the flue and use it to heat the water entering the boiler. Since the flue gases are now as low as 75 °C, the majority of water vapour in the flue products becomes liquid giving up its latent heat, hence the term 'condensing'. See the section on 'Installing condensing boilers' (page 386) for information on ways of dealing with the condensate, which is a weak acid that must be discharged safely to waste.

Due to the acid that is produced, condensing boilers need to be made from different materials to their traditional counterparts – typically stainless steel is used. The flue system must always fall back to the boiler to permit the condensate to fall back and be discharged through the condensate pipe. A forced-draught burner enables the fan combustion speed to be made directly proportional to the gas rate while it modulates over the range. This allows the efficiency of the boiler to be maintained and reduces carbon dioxide emissions compared to non-condensing boilers.

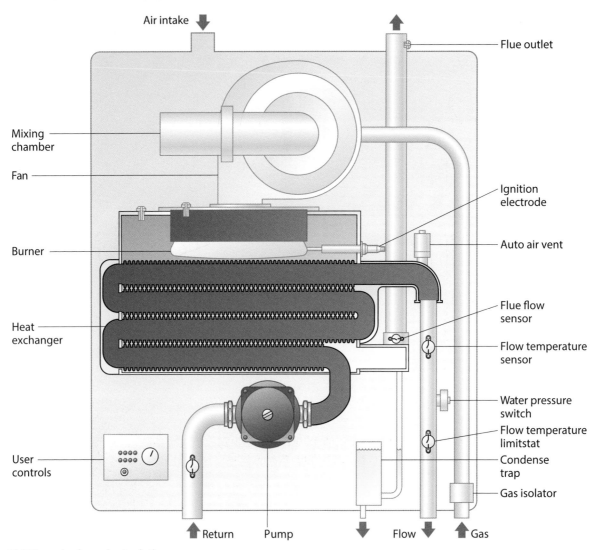

Air intake

Flue outlet

Mixing chamber

Fan

Ignition electrode

Burner

Auto air vent

Flue flow sensor

Heat exchanger

Flow temperature sensor

Water pressure switch

Flow temperature limitstat

User controls

Condense trap

Gas isolator

Return   Pump   Flow   Gas

Figure 12.09 Example of a condensing boiler

Burners and controls

All boilers use a combination of mechanical and electrical control devices. The most modern boilers now use electronic ignition, and many light use an intermittent pilot, although some existing ones still use a permanent pilot.

Permanent pilot

A permanent pilot works as follows:

- gas is supplied through a service cock to the inlet of the multifunctional valve
- a piezo igniter is used to light the pilot
- the pilot heats a thermocouple which connects to the multifunctional valve (gas valve)
- a thermostat (Figure 12.10) calling for heat then gives power to the gas valve, which opens
- the main burner receives gas, which is lit by the pilot

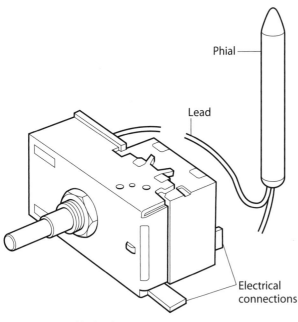

Phial

Lead

Electrical connections

Figure 12.10 Typical boiler thermostat

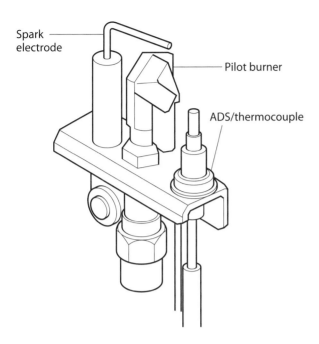

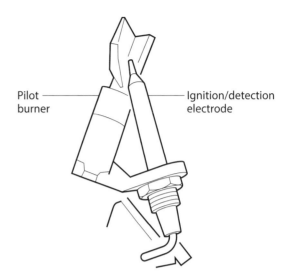

Figure 12.11 Pilot assembly with atmospheric sensor device (ASD)

Figure 12.12 Intermittent pilot assembly

- if the pilot fails then the thermocouple cools, the gas valve is de-energised, closes and fails safe
- the boiler thermostat controls the temperature of the water by opening and closing the gas valve according to the temperature reached in the boiler.

Intermittent pilot

An intermittent pilot works as follows:

- the boiler thermostat calls for heat
- the control board/box activates spark ignition
- the spark occurs at the pilot assembly
- the pilot solenoid valve opens to let gas through to pilot
- the pilot ignites and the ignition/detection electrode detector senses flame
- the spark stops and the main solenoid opens
- gas flows to the main burner and is lit by the pilot.

Air-pressure switch

Fan-flued boilers use an air-pressure switch as a safety cut-off device should the fan fail at any time. Because the fan is used to assist with the removal of products of combustion (POC) and the supply of combustion air, it is critical to the safe operation that the fan is working satisfactorily at all times. The main feature of the air-pressure switch is that it prevents the boiler lighting sequence in the event of the fan not operating and causing sufficient air pressure.

Figure 12.14 on page 382 shows how all the components link to the PCB within a typical boiler.

Figure 12.13 Boiler with combustion cover removed, showing air-pressure switch

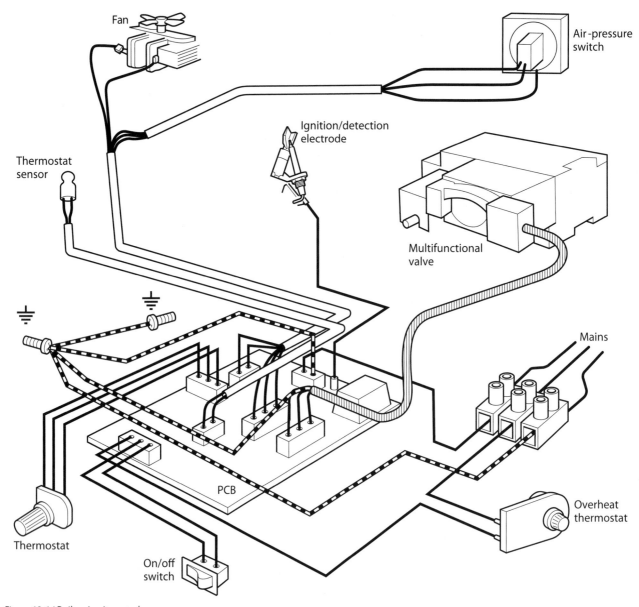

Figure 12.14 Boiler circuit control

Overheat thermostat

Figure 12.14 shows the boiler controls and an overheat thermostat; this is fitted as standard on most low water-content boilers. The overheat thermostat is wired in series with the normal temperature control thermostat to provide added appliance protection in the event of a control-stat failure. In addition, with combination boilers and many low water-content boilers there are other thermostats installed to sense temperature at various points in the appliance.

There can also be water-pressure switches installed, which control the operation of components such as the pump. Where multiple controls such as this are included in boilers, they are usually controlled by a number of PCBs.

When you attend college your tutor will provide you with a practical demonstration of the range of boilers, including an overview of key boiler controls.

Installing boilers

It is now common practice to install boilers with wall frames that allow the pipework to be installed onto a jig before the boiler needs to be fitted. This is to prevent boiler theft or damage prior to the customer moving into a new house. The plumber/gas engineer can fit the boiler and commission the day before the customer moves in to the property. It also enables the boiler to be replaced without major re-piping of the system.

The components of a modern boiler are expensive and need to be fitted into a building when it is least likely to be exposed to debris and dust from building work.

A typical layout of the pipework is shown in Figure 12.17 on page 384, however it will differ from manufacturer to manufacturer. The gas pipe size should be that as defined by the boiler manufacturer which overides any gas pipe sizing carried out by the engineer and should never be less than that specified by the manufacturer.

The space around the boiler shall be at least the minimum specified in the manufacturer's installation instructions. The floor or wall on which the boiler is to be mounted should be protected as detailed in the manufacturer's instructions.

The space around the boiler should be adequate:

- to ensure sufficient air circulation for any draught diverter to operate
- to ensure sufficient air for combustion and cooling
- for maintenance and servicing.

If the boiler manufacturer's installation instructions do not give any specific advice, then any internal surface of the boiler compartment which is combustible should be at least 75 mm from any part of the boiler or be lined with non-combustible material. Methods of determining whether a material can be described as combustible or non-combustible are given in BS 476-22.

You should ensure the compartment allows sufficient access for inspection and servicing of the boiler and any ancillary equipment. A notice should be fixed in a prominent position within the compartment to warn against its use as a storage cupboard. The compartment should be fitted with a door that allows the boiler to be removed along with any ancillary equipment. If the boiler compartment houses an open-flued boiler, neither the door nor the air vents should communicate with a room containing a bath or shower.

Open-flued boilers installed in compartments communicating directly with a room used or intended to be used for sleeping accommodation should have a heat input of less than 14 kW gross (12.6 kW net) and shall include a safety shut-down control.

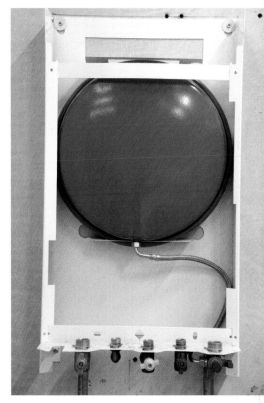

Figure 12.15 Boiler installation jig

Figure 12.16 Inside a boiler (spur has cover removed)

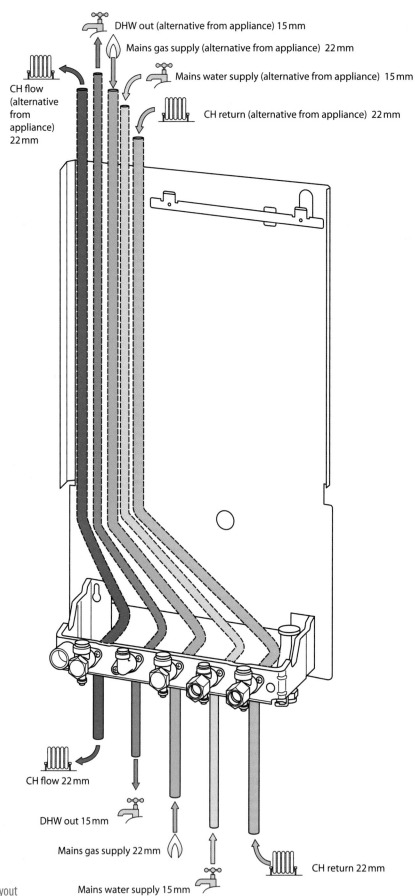

DHW out (alternative from appliance) 15 mm

Mains gas supply (alternative from appliance) 22 mm

Mains water supply (alternative from appliance) 15 mm

CH flow
(alternative
from
appliance)
22 mm

CH return (alternative from appliance) 22 mm

CH flow 22 mm

DHW out 15 mm

Mains gas supply 22 mm

Mains water supply 15 mm

CH return 22 mm

Figure 12.17 Installation pipe layout

How to install a boiler

Checklist

PPE	Tools and equipment	Source information
• Overalls • Protective footwear • Barrier cream	• Electric drill • Level • Spanners • Screwdriver • Leak detection equipment • Manometer	• Manufacturer's instructions

1 Using the manufacturer's installation instructions, identify a good location to install the boiler. Mark out the fixing positions, level the fixing bracket and fix to the wall using the recommended fixings by the manufacturer.

2 Pipe up the water and gas connections on the boiler jig and pressure test before fitting the boiler into position.

3 Most boilers require two people to lift the boiler to ensure secure location onto the boiler brackets. Ensure the boiler is properly engaged and secure.

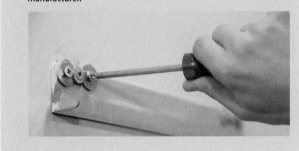

4 You can now connect the pipework to the boiler fittings ensuring that you do not cross thread and all the appropriate washers are in place.

5 Fill the system and boiler with water, checking for any leaks. Only when fully satisfied, fit the flue sections to the boiler ensuring flue integrity is maintained throughout its length.

6 Fit the boiler casing and electrical controls then get ready to start the commissioning process.

Installing condensing boilers

Condensate pipes from condensing boilers must be installed as per manufacturer's instructions. There are special requirements, as shown in Figures 12.18 to 12.22 from BS 6798.

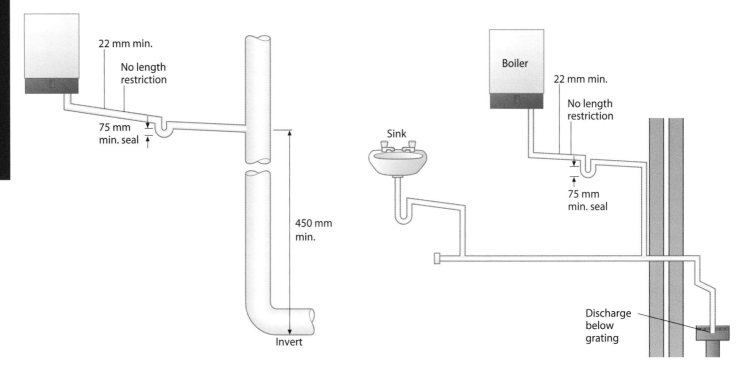

Figure 12.18 Condensate pipe to a soil/vent stack

Figure 12.19 Condensate pipe to a gully via a sink discharge pipe

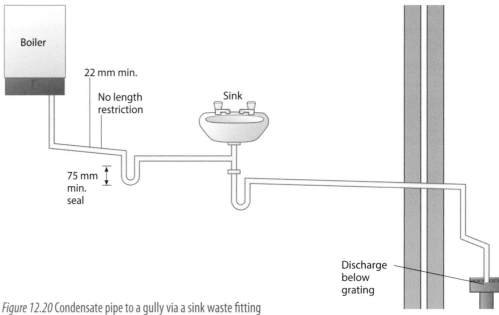

Figure 12.20 Condensate pipe to a gully via a sink waste fitting

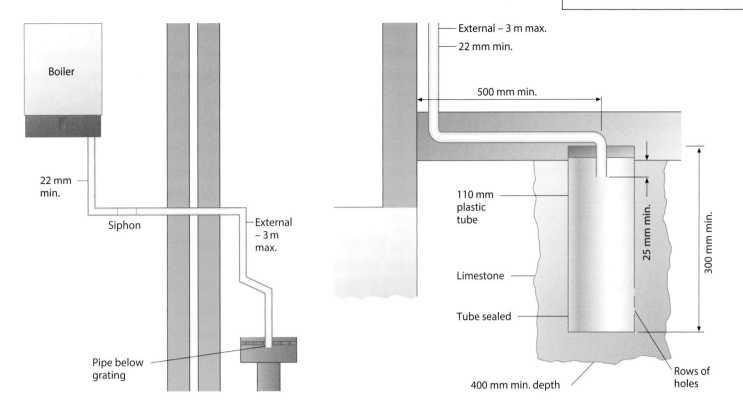

Figure 12.21 Condensate pipe to a gully via condensate siphon

Figure 12.22 Condensate pipe to a purpose-made condensate absorption point

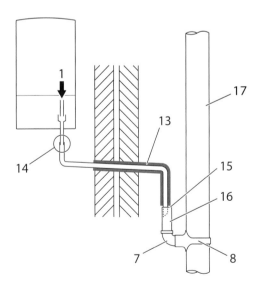

Figure 12.23 Disposal into a rainwater down pipe

8	PVCu strap on fitting
13	Insulate and increase pipe size
14	Pipe work transition
15	External air break
16	Air gap
17	External rain water pipe into foul water
18	43 mm 90° male/female bend

Condensate drainage pipe

The condensate drainage pipe can be run in a standard drainpipe material, such as polyvinylchloride (PVC), unplasticised PVC (PVC-U), chlorinated PVC (PVC-C), acrylonitrilebutadiene-styrene (ABS) or polypropylene (PP).

Condensate removal by gravity

The condensate drainage pipe should have a nominal outside diameter of 22 mm, or as recommended by the appliance manufacturer. Any trap in the condensate drainage pipe should be fitted within the dwelling to prevent freezing and ensure access for cleaning.

Some boilers have a siphon fitted as part of the condensate trap arrangement. This will reduce the risk of the condensate freezing if part of the condensate drainage pipe is run externally. If you can't find an alternative route, the external condensate pipework should be kept and insulated with external grade insulation, which is vermin- and waterproof. Alternatively, increasing the diameter to 32 mm will offset the possible effect of freezing. Pipe runs in a loft or garage space should also be classed as an external environment and treated the same.

If a boiler is sited below drain level, a condensate pump can be fitted to pump the condensate to the drain. The pipe should be fixed at intervals of 0.5 m horizontally and 1.0 m vertically, and run to a fall of 50 mm per 1 m length.

Commissioning boilers

Before it is commissioned, inspect the boiler installation to ensure that the work has been carried out to the boiler manufacturer's instructions, the relevant sections of the current Gas Safety (Installation and Use) Regulations and BS 7671.

Pay particular attention to the Gas Safety (Installation and Use) Regulations requirements that:

- the provision of ventilation air and combustion air is adequate
- the chimney is correctly constructed
- the general condition of the boiler and the installation is adequate
- the gas fittings and other works for the supply of gas are adequate.

Pre-commissioning checks

Once you have isolated the electricity supply to the boiler:

- check that the service and water pipes are correctly connected to the manifold of the boiler rig to ensure a tight seal
- check the boiler data plate to make sure the gas type specified matches that of the gas supply
- check the pressure relief drain pipe is correctly and securely fitted.

If you do not carry out each of the above steps, it could result in:

- electrocution if you don't turn off the power to the boiler
- a water or gas escape if you don't connect the supplies properly
- an explosion or damage to the boiler controls if the wrong type of gas is connected
- hot water pouring into the room if the pressure relief pipe isn't secure.

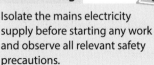

Safe working

Isolate the mains electricity supply before starting any work and observe all relevant safety precautions.

The commissioning procedure

The commissioning of any boiler has to be done to the manufacturer's instructions. The following text is therefore for general guidance only.

The combustion settings on gas-fired boilers have been **factory set** and therefore do not require adjustment. All you should do in this regard when commissioning a new boiler is check that these setting have not altered.

You also don't need to measure the combustion values of the boiler as long as there is a meter installed to allow the gas rate to be checked at the meter.

It is not your responsibility to adjust the air/gas ratio valve on the boiler. If it is not deemed to be set up correctly, then the boiler manufacturer will send one of their engineers to adjust or replace it under warranty.

First of all, check that the boiler has been installed according to the installation instructions.

Check the integrity of the flue system and flue seals as described in the boiler manual. Damage to flues, such as deflected flue pipe or damaged seals, can normally be seen from a visual inspection.

Next check the integrity of the boiler combustion circuit and its relevant seals. The combustion chamber seals should have no sign of damage and be properly attached – they may come loose in transit.

You should then check the inlet pressure. The following procedure can be used in most cases:

- ▪ Turn off the boiler gas isolation valve.
- ▪ Undo the screw on the inlet pressure test point but don't remove it.
- ▪ Connect a manometer onto the test nipple, making sure it is secure.
- ▪ Slowly open the boiler gas isolation valve.
- ▪ Measure the inlet working pressure with the boiler on maximum. For combination boilers this will often have to be set to the hot water mode while running a tap to get maximum gas use.

With the temperature set to maximum, check the working pressure at the gas valve inlet point is no less than 20 mbar.

The gas pressure at the boiler inlet test point must not be less than the pressure reading at the meter, minus 1 mbar for natural gas and 2.5 mbar for LPG.

If the pressure drops are greater than shown in Figure 12.25, then there is a problem with the pipework sizing or partial blockage in the system.

Checking the gas rate

Take a reading of the gas rate at the meter after 10 mins operating at maximum and check it against that given in the manufacturer's manual. Where a gas meter is unavailable (e.g. for LPG appliances), the CO/CO_2 levels

Key term

Factory setting
A particular setting on a piece of equipment that should not be adjusted as it can compromise the guarantee.

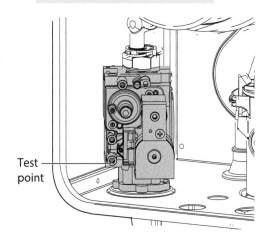

Figure 12.24 Test point on a boiler

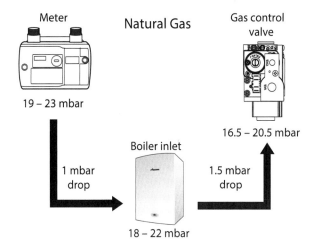

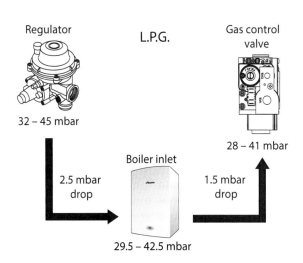

Figure 12.25 Lowest pressure drops for natural gas and LPG

must be checked against the units shown in the setting of the air/gas ratio supplied by the manufacturer. If the gas rate and pressure are all correct, return the boiler to normal operation.

Once the gas rate reading has been taken:

- slowly close the gas isolation valve
- carefully remove the manometer from the test point
- re-screw the gas inlet pressure test point
- open the gas isolation valve slowly
- ensure that the test point does not leak by using leak detection fluid
- replace the outer cover of the boiler.

If left in the service mode, the majority of boilers will automatically return to normal settings after around 15 mins.

Case study

During a routine check a gas engineer failed to secure a manometer. It fell to the ground, pulling the tubing from the end of the test point, allowing gas to escape into the room. The gas caught fire and flames shot out of the test point melting cables and part of the PCB board before the engineer was able to turn off the gas valve. Thankfully on this occasion, only the appliance was damaged – nobody was hurt. This illustrates the importance of securing the manometer in position and ensuring all tubing is securely fixed.

Safe working

Turn off the gas supply and isolate the mains electricity supplies before starting any work and observe all relevant safety precautions.

Safe working

After changing of any components, always check for gas tightness to ensure no gas leaks or disturbance has occurred. Where relevant, carry out functional checks as described in the manufacturer's commissioning procedures.

Servicing and maintaining boilers

Service and maintenance work must only be carried out by a competent engineer who is Gas Safe registered. You should never service a boiler without a CO/CO_2 analyser to hand.

Filling a central heating system

1. Check that the expansion vessel is pre-charged to the manufacturer's instructions.
2. Using a pressure gauge, insert into the Schroeder valve fitted to the expansion vessel.
3. If the reading is okay, open the cold water supply and then open the filling valve/or loop.
4. Let the pipework come up to pressure and then open the radiator valves, one radiator at a time, bleeding each radiator as you go.
5. Ensure that the heat exchanger is bled of air.
6. Check that the automatic air release valve has vented out the air as sometimes they can jam. This can be very dangerous because if the boiler were to fire it would cause an explosion or melt the heat exchanger, as there will be no water to remove the heat.
7. Check that the system is fully charged and has no leaks by carrying out visual checks.
8. Check for any movement on the boiler pressure gauge over the course of an hour.
9. If system remains at set pressure you should remove the filler loop or filler key to ensure you are complying with water regulations.

Flushing a central heating system

1. Isolate the boiler from the gas and electrical supplies.

2. Open all drain valves and empty the system while the water is hot.

3. Close the drain valves and introduce a suitable chemical flushing agent in accordance with the manufacturer's instructions.

4. Turn on the boiler as stated by the flushing agent manufacturer instructions to the normal operating temperature of the system and leave it running for the time stated.

5. The system is now ready to drain down and flush out any debris and remaining chemicals.

An alternative to the above procedure is using a power flush.

Adding an inhibitor

1. Check all drain valves are closed and that all radiator valves are in the open position.

2. Add to the heating system water a suitable inhibitor for the type of heat exchanger or an inhibitor/anti-freeze mixture if the system may become exposed to freezing conditions, in accordance with the manufacturer's instructions.

3. The **pH** value of any heating system water must be less than 8 or the appliance warranty will become invalid for most types of boilers.

4. Refill the system and vent radiator.

5. Check the heat exchanger is clear of air.

6. Check for leaks and re-set the pressure to the operating pressure.

7. Turn on the system.

8. Record on the warranty card the date, the type of inhibitor used and the name of the engineer who carried out the procedure.

Key term

pH
The pH scale measures how acidic or alkaline a substance is. Pure water has a neutral reading of 7 on the pH scale. Anything above this level of pH is classed as acidic and therefore likely to increase corrosion within a boiler. The concentration of the inhibitor should be checked every 12 months or if the system water has been lost through leaks or a change to system pipework.

The concentration of inhibitor in the system should be checked every 12 months or sooner if system content is lost. Adding sealing agents to system water is not recommended, as it can cause problems with deposits in the heat exchanger.

Inspection

Before servicing or carrying out maintenance work on a boiler, conduct a general inspection of the boiler. Much of the inspection procedure is similar to the commissioning procedure because you are basically checking to see that the appliance is still operating as it was set up to do at the commissioning stage.

The first thing to check is that the terminal and terminal guard, if fitted, are clear of any obstruction and damage.

Next, if the appliance is in a compartment or cupboard, check that the manufacturer's required servicing space around the appliance is correct. Otherwise you may find it impossible to carry out any work on the appliance.

Then check all the joints in the system for leaks. If leaks are found, drain down and repair the leak before re-commissioning the system. Finally, turn on the appliance and check its performance against the manufacturer's specification.

When checking the gas inlet pressure, do so with all other gas appliances working in the system. Other appliances can sometimes interfere with the safe working of the boiler by reducing the gas pressure reaching the boiler. Do not continue with the other checks if the correct gas inlet pressure cannot be achieved.

Checking flue integrity

The flue pipe system and the flue gases need to be checked to ensure that the boiler is working correctly. The flue system and performance of the boiler can be checked via the **flue turret** sample points, as seen in Figure 12.26.

Turn the boiler on to maximum and insert the probe of the flue gas analyser into the air intake test point. Allow the reading stabilise first, as you will have disturbed the flue when you removed the sampling point cap and this unsettles dust which can then give a false reading.

After the reading has stabilised, check that:

- oxygen levels are equal to or greater than 20.6 per cent
- CO_2 levels are less than 0.2 per cent.

Fan pressure test

The purpose of a fan pressure test is to check the fan is still working efficiently: the more worn the motor, the slower the fan turns, causing the pressure to drop.

Set the boiler to maximum so that you have a set point to compare your readings. Remove the boiler cover and connect the digital manometer to the fan pressure test point as described in the manufacturer's instructions (see Figure 12.27). After taking the pressure reading, put the test point back together.

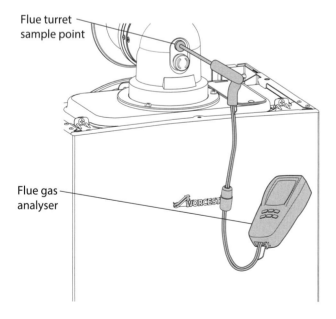

Flue turret sample point

Flue gas analyser

Figure 12.26 Flue integrity test

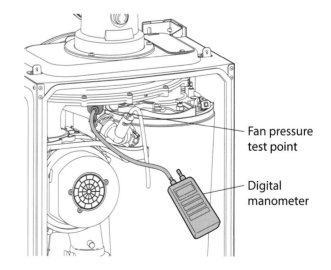

Fan pressure test point

Digital manometer

Figure 12.27 Fan pressure test

A typical pressure reading will be negative but it depends on type of boiler. Refer to the chart in Figure 12.28 for an example of fan pressure readings for a Worcester Bosch boiler.

If the boiler reading is wrong, carry out the following checks:

- Check the syphon is not blocked on the condensing line.
- Clean the sump and heat exchanger with a suitable brush.
- Clear the flue path of restrictions such as leaves.
- Check the fan pressure readings again. If no improvement, adjust the air/gas ratio according to the manufacturer's instructions.

Flue gas analysis

Analysing the flue gas requires you to carry out a combustion test, as follows.

Connect the flue gas analyser to the flue gas sampling point as shown in Figure 12.26. Next, run the boiler at maximum output for a minimum of 10 mins to ensure everything in the boiler is up to temperature, which in turn ensures as complete combustion as possible. Take a sample reading, allowing the reader to stabilize. Print off or record the reading and check it against the manufacturer's data.

If the boiler fails the combustion test, take a further two sample readings to confirm the results.

Further checks include:

- the air intake of the boiler for any restrictions from lint etc
- the operation of the diaphragm in the fan, to see if it is working correctly
- the condition of the heat cell/heat exchanger – look for any blockages in the condensate disposal line.

When cleaning and replacing any boiler parts, always refer to the manufacturer's instructions.

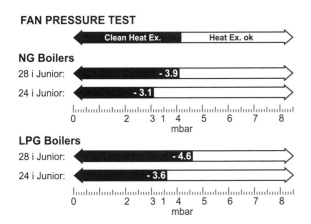

Figure 12.28 Fan pressure readings

Fault finding and diagnosis

The modern boiler will have a built-in diagnostic programme that will use a screen with numbers or flashing lights to help you identify the fault. Do not, however, rely on this programme alone, as the diagnostic programme can itself be faulty. You therefore need to be able to diagnose problems yourself.

Unless the fault is immediately obvious, there are some standard preliminary checks that you can carry out. First, check the electric circuit – is the boiler getting power? If not, the fault may well be in the heating controls circuit. All boilers are wired differently and use different voltage and current. As you check and/or replace the circuits and components, you are going to disturb the wiring, so on completion be sure to check:

- earth continuity
- for a short circuit
- polarity
- resistance to earth.

General fault-finding checks:

- Always discuss the problem with the customer first as they are likely to be able to help you identify possible causes. This at least gives you a starting point. This also introduces customer care and helps build up a relationship with the customer and works towards getting repeat business.
- You need to identify what type of system and boiler you are working on so that you have as much information as possible to identify the fault.
- Look at the boiler to see any visual signs of faults, such as loose wiring, signs of spillage, low system pressure and fault report on boiler panel.
- If it is safe to do so, turn on the boiler and watch for the reported fault to occur. Then turn off the system. Make it safe – don't forget to isolate the electrical supply.
- With the help of a manufacture's fault-finding chart or Table 12.01, identify the probable cause.

Table 12.01 on page 395 is not an exhaustive list.

Re-commissioning boilers

- Re-light the boiler as per the manufacturer's instructions.
- Check the burner pressure as stated on data plate.
- Check the gas rate.
- Check the operation of the flue (remember, open flues require a spillage test).
- Check the boiler controls and safety devices are functioning correctly.
- Report back to the user and return the manufacturer's instructions.
- Advise the customer of the on-going need for regular maintenance.

In the event that a parts failure occurs on the appliance, the manufacturer's instructions will provide a detailed fault diagnosis and testing procedure for key system components – so refer to the instructions. A complete range of parts is usually available as replacements in the event of a failure. Again, a key document to refer to is the installation instructions.

Fault	Possible causes	Solutions/checks
No power at control board	Permanent mains supply to boiler Fuse replacement Transformer coil	Otherwise replace control board
Boiler not operating during central heating demand (HW okay)	Live demand (from external room thermostat/timer) Fascia-mounted timer (if fitted) CH knob in winter position Diverter valve Control board	
Boiler operating without live demand (from external room thermostat timer)	Some older thermostats (containing capacitors) may give a low voltage return when the thermostat contacts are open	Check that there is no permanent live from another source
Boiler not operating during hot water demand (CH okay)	Ensure cold inlet and DHW outlet are correctly piped i.e. are not crossed piped Check diverter valve, motor and control board	Re-pipe installation correctly Replace diverter valve/motor
Boiler not operating during any demand.	Fan Control board	Replace fan control board
Ignition lockout	Gas present and at correct pressure? Combustion CO_2 level Blocked flue? Blocked condensate pipe or frozen condensate? Gas valve adjustment? Ignition? Electrodes/harness/connections? Check for condition and resistance of leads	Gas valve Check that there is voltage to each solenoid Check the resistance of any solenoid Unblock flue Unblock frozen condensate pipe Re-adjust gas valve Clean electrodes Replace electrodes/harness/connections Otherwise replace control board
Flue overheat	Heat exchanger baffles removed and not refitted	Fit baffles
Heat exchanger Overheat Dry fire	Heat exchanger blocked Water pressure All air vented Pump/harness/connections Water leaks/blockage Safety thermostats/low voltage wiring harness/connections	Unblock exchanger Recharge system pressure Fix leak or blockage Replace safety thermostat/wiring harness Otherwise replace control board
Volatile lockout – fan does not run	Temperature sensors Code plug missing or not inserted properly?	Check condition and continuity of leads Fan – 230 V AC across the live (purple) and neutral (brown) Fan lead Check continuity Replace or insert code plug
Internal fault		Replace control board
Kettling	Pump Blockage of pipework Blockage of heat exchanger	Replace pump Unblock pipework Unblock heat exchanger

Table 12.01 Fault finding, possible causes and solutions – boilers

De-commissioning boilers

Case study

A plumber attended a housing association rented flat to de-commission an iron boiler. The plumber wasn't properly qualified and failed to de-commission the appliance correctly. The water left in the back boiler led to the boiler exploding and the resident of the flat, who was in another room at the time of the explosion, was hospitalised with severe burns. The plumber and housing association were sentenced for health and safety breaches.

What should the plumber have done which would have prevented this accident?

For a boiler to be properly de-commissioned you will need to safely and thoroughly remove all water and fuel from the system. You will also need to safely isolate the electrical controls. A reverse of the commissioning procedure is often used in de-commissioning and the boiler can be left in position to be re-used at a future point. A full de-commission requires the full removal of the boiler so that it cannot be re-used or be at risk of being re-commissioned by an unqualified person. If you are in any doubt about the safety of the procedure, ask the manufacturer's for advice.

WATER HEATERS

Types of water heater

There are two categories of water heaters: storage and instantaneous. The most common being the instantaneous type.

The storage water heater comes in two forms: integral hot water heater and a circulator usually found in warm air units (see the section on 'Warm air heaters', page 435). The integral hot water storage unit is usually used in commercial premises and consists of a storage vessel with a gas burner underneath and an open flue above. Because of the size they require a large floor area and are therefore not often used in domestic situations.

Instantaneous water heaters

An instantaneous water heater provides instant hot water on opening a tap; there is no storage of hot water. The water is heated to the required temperature as it passes through the heat exchanger. The temperature is determined by the speed that it passes through so that on full water flow the temperature will be lower. This because, in winter, as the water is colder coming in to the house, even if the water heater is at full gas rate, the water will be slowed down to allow it to be heated up to the correct temperature. This increase in heating time is sometimes reported as a fault by the customer.

There are two main types of instantaneous water heater:

- single point – as the name suggests, it supplies only one point and is normally flueless
- multi-point – these are large heaters that can supply several outlets and require a flue, which is normally room-sealed.

Progress check

1 Why do low water-content boilers require a positive flow of water at all times?

2 Under what circumstances does the air-pressure switch prevent the boiler lighting sequence from taking place?

3 What is the condensate drainage pipe nominal diameter?

4 Complete the following statement: 'You must not service a boiler if a … is not available.'

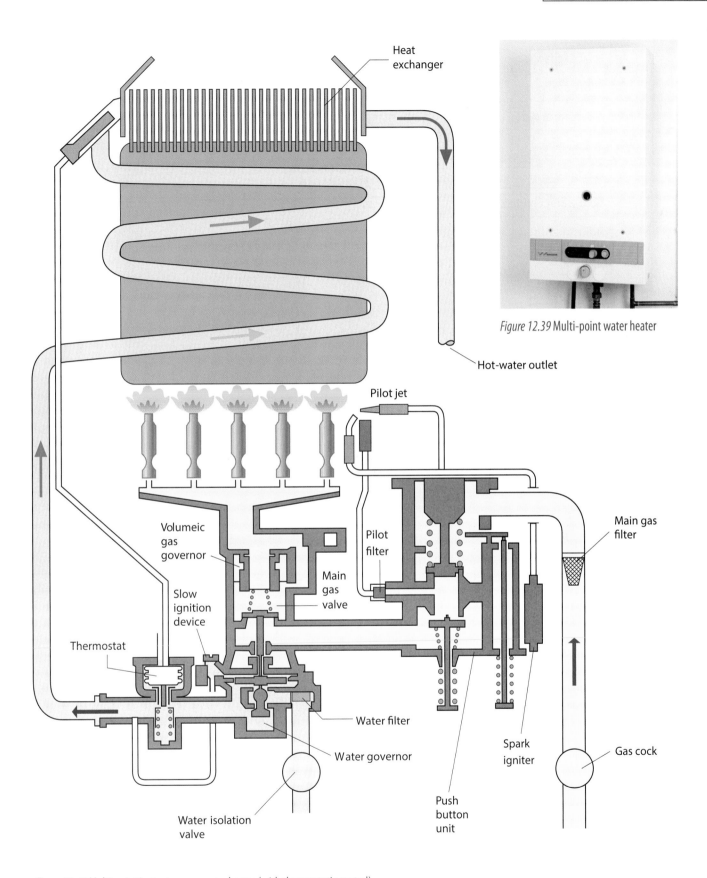

Heat exchanger

Hot-water outlet

Figure 12.39 Multi-point water heater

Pilot jet

Volumeic gas governor

Slow ignition device

Thermostat

Main gas valve

Pilot filter

Main gas filter

Water filter

Water governor

Water isolation valve

Push button unit

Spark igniter

Gas cock

Figure 12.40 Multi-point instantaneous water heater (with thermostatic control)

1	Heat exchanger
2	Pilot gas pipe
3	Gas control slide
4	Piezo igniter
5	Gas inlet
6	Cold water connecting point
7	Adjustment screw for min. water flow
8	Flue hood
9	Combustion chamber seal
10	Main gas burner
11	Pilot assembly

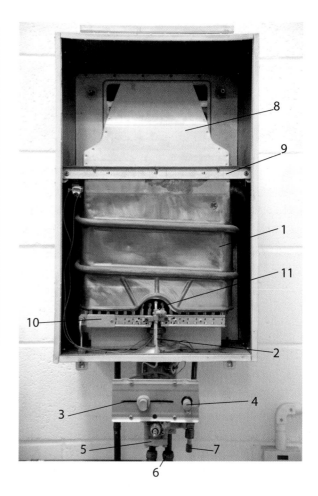

Figure 12.41 Water heater

Commissioning water heaters

Inspection

Before it can be commissioned the water-heating installation should be inspected to ensure that the work has been carried out as stated in the manufacturer's instructions and the Gas Safety Regulations.

Key requirements are:

- adequate ventilation air and combustion air – check vents are the correct size and type
- correctly constructed and terminated flue – as per manufacturer's instructions
- satisfactory general condition of the appliance and the installation – look for signs of damage to the case seals and missing screws
- adequate gas fittings/installation – carry out a tightness test and check the burner assembly with leak detection fluid
- electrical fittings/work are in accordance with BS 7671 – conduct checks for:
  — earth continuity
  — short circuits
  — polarity
  — resistance to earth.

The commissioning procedure

- Ensure all components and appliances are complete by conducting a visual inspection and checking fixings are secure.
- Light the pilot and check the size and position of the flame with that specified in the manufacturer's instructions.
- Fit a manometer gauge to burner test point, ensuring it is secure and will not fall.
- Turn on the heater/tap to light the main burner. This is best carried out by asking a customer or colleague to operate the tap while you observe the burner lighting up in order to witness any faults.
- Check that all the gas connections/controls are sound.
- Check/adjust the burner pressure by checking the manometer reading against that in the manufacturer's instructions. Adjust the gas regulator with a screwdriver until optimum burner pressure is reached.
- Compare the gas rate with that on the data plate. Do not allow anyone to adjust the flow or turn on any other appliances that use water while the test is being carried out.
- Adjust the slow ignition if required. Allow time for the device to settle before making any re-adjustments.
- Carry out a spillage test (open flues only).
- Check the flame supervision device (FSD).
- Check the ASD, if fitted. For single point water heaters you need to place a non-combustible plate over the exhaust outlet to simulate faulty conditions. Always perform this operation in a well-ventilated room to prevent breathing in CO.
- Determine temperature rise and flow rate of water using flow-measuring gauge and temperature sensor.
- Demonstrate the workings of the appliance to the customer and leave them with the relevant documents (e.g. the guarantee and operating/manufacturer's instructions).

Servicing water heaters

When carrying out a service, always refer to the manufacture's instruction. The following list is for general guidance only.

- Conduct a tightness test.
- Isolate the gas/electricity/water supplies to the appliance.
- Remove the outer casing/panels to access burner.
- Remove the burner and control assembly.
- Remove the burner injector and blow clean. You can use compressed gas bottles for this but remember to protect your eyes when doing so.
- Take out the pilot/thermocouple/ignition assembly, place them on a tray to avoid losing any components then clean.
- Clean the main burner and lint arrestor. This can be done with a brush or may require a thorough clean in water.

- Take the flue cover off and clean the flue ways. Use a lint-free rag and clean all surfaces. Don't allow debris to fall onto the components – cover them with sheeting or newspaper.
- Clean the heat exchanger with a soft brush. Protect components from falling debris with newspaper or sheeting.
- Remove all debris from the combustion chamber using, where available, a small portable vacuum cleaner.
- Check the condition of all components and replace if necessary.
- Reassemble the components, checking any gas connections.
- Ensure all appliance seals are sound by conducting a visual inspection or, for positive pressured appliances, using a smoke pen.
- Conduct a further tightness test.

Servicing a water heater

Checklist

PPE	Tools and equipment	Source information
• Overalls • Protective footwear • Barrier cream	• Manometer • Spanners • Screwdriver • Flat non-combustible plate	• Manufacturer's instructions

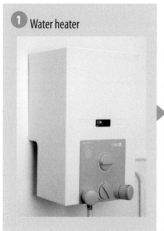

1 Water heater

2 Cover removed

3 Removing heat exchanger

4 Burner removed

5 Checking pilot assembly

6 Setting burner pressure

7 Checking ASD (atmospheric sensing device)

Fault finding and diagnosis

Problem	Possible cause	Solution/checks
Incorrect water temperature	Incorrect gas rate	Check the gas supply to appliance gas isolation valve Open the gas isolation valve Check the burner pressure Check the inlet pressure
	Water section sticking or faulty	Clean or replace
Incorrect water flow rate	Incoming supply valve closed	Open supply valve
	Low water pressure	Check the water pressure is above 0.2 bar
	Water section sticking or faulty	Clean or replace
Noise	Scale in heat exchanger	Descale and service
	High gas rate	Check the burner pressure
	Low water flow rate	Check the water flow rate
Pilot flame will not light or stay lit	No gas supply	Connect gas supply
	Gas isolation valve closed	Open gas isolation valve
	Air in gas line	Purge the line
	Pilot injector blocked	Clean/replace the injector
	No spark	Clean or replace the electrode, piezo unit or cable
Main burner will not light	Gas pressure low	Check the inlet and burner pressure
	Gas isolation valve partially closed	Open the gas isolation valve
	Low water rate	Check the water rate
	Water section diaphragm faulty	Replace the diaphragm
	Gas valve faulty	Replace the gas valve
Explosive ignition	Reduced pilot flame	Clean or replace the pilot injector assembly
	Faulty slow ignition valve	Replace the slow ignition valve

Table 12.02 Fault finding, possible causes and solutions – water heaters (from main water heater manufacturer instructions)

Re-commissioning water heaters
- Re-light the water heater as per manufacturer's instructions.
- Check the burner pressure as stated on the data plate.
- Check the gas rate.
- Check the operation of the flue (open flues require a spillage test).
- Check the air pressure.
- Check that the water heater controls and safety devices are functioning correctly.
- Check the ASD, where fitted.
- Report back to the user and return the manufacturer's instructions.
- Advise the customer of the need for regular maintenance (fan flued).

Progress check

1 What are the two main types of instantaneous water heater?

2 On what principle do most water heater control valves work?

3 What type of safety device safely shuts down an instantaneous water heater before there is a dangerous build-up of products in a room?

4 What is explosive ignition caused by?

GAS SPACE HEATERS

British Standards classification of gas space heaters

Gas space heaters can be classified according to the three types detailed in the British Standards.

Convector heaters, fire/back boilers and heating stoves

The first category according to BS 7581 Part 1 is gas fires, convector heaters, fire/back boilers and heating stoves (2nd and 3rd family gases). These should have the following characteristics:

- flue size: minimum of 125 mm across the axis of the flue is normally required
- location: normally in front of the closure plate, which is fitted to the fireplace opening to prevent too strong a draught from going up the flue
- ventilation: purpose-provided ventilation not normally required up to 7 kW input.

For this type of appliance, the radiating surface can be either radiant(s) or imitation fuel giving a live-fuel effect.

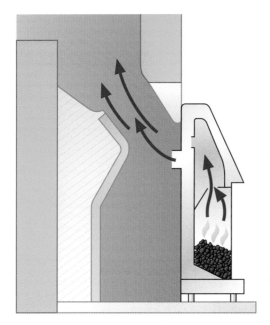

Figure 12.42 Radiant/convector gas fire

Figure 12.43 Radiant convector fire

Inset live fuel-effect fires

Inset live fuel-effect (ILFE) gas fires belong to BS 5871 Part 2 and have the following characteristics:

- flue size: a minimum of 125 mm across the axis of the flue is normally required
- location: either fully or partially inset into builder's opening or fireplace recess – for a recess, the **chair brick** at the back of the opening might have to be removed, depending on the appliance design
- ventilation: purpose-provided ventilation is not normally required up to 7 kW input.

Key term

Chair brick
A chair-shaped brick fitted in coal or wood-burning openings to direct the fumes through the main chimney and remove excess heat.

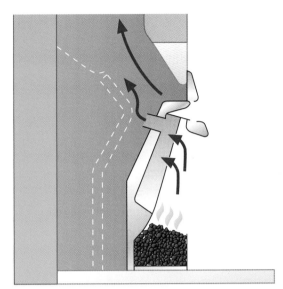

Figure 12.44 Inset live fuel-effect

Figure 12.45 Inset live fuel-effect

Decorative fuel-effect fires

Decorative fuel-effect (DFE) gas appliances belong to BS 5871 Part 3 and should have the following characteristics:

- location: within a builder's opening, fireplace recess or flue box, or under an independent canopy – for a recess, the chair brick may have to be removed, depending on appliance design, as it may be blocking the flue spigot
- ventilation: purpose-provided ventilation of at least 100 cm^2 is normally required up to 20 kW input
- flue size: a minimum of 175 mm across the axis of the flue is normally required, with a flue similar to those for solid-fuel appliances, including:
  - an existing masonry chimney provided it is in good condition (can be relined if necessary)

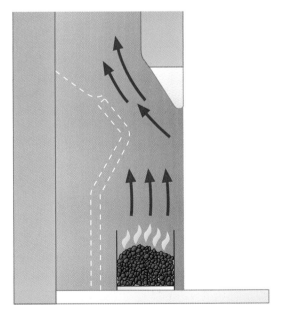

Figure 12.46 Decorative fuel-effect fire

Figure 12.47 Decorative fuel-effect fire

— a single- or double-walled metal flue pipe (manufacturer-approved)

— an existing masonry chimney suitably lined for a solid-fuel appliance

— a lined masonry chimney provided it is of a suitable material for the appliance

— a pre-cast flue block suitable for solid fuel

— a factory-made system to BS 4543.

Make sure terminals do not restrict the exit and safe dispersal of flue products. Any chimney pot should not be less than 175 mm across its axis to allow flue gas to escape while not being so large as to allow birds to nest in it. Some examples of unsuitable terminals are shown in Figure 12.48.

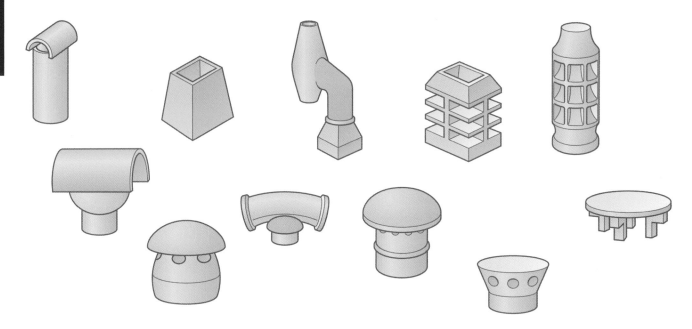

Figure 12.48 Unsuitable terminals

Independent gas fired flueless fires, convector heaters and heating stoves (less than 6 kW)

Independent gas-fired flueless fires, convector heaters and heating stoves of nominal input not exceeding 6 kW belong to BS 5871 Part 4.

For installations other than in a greenhouse, consider the following:

■ potential for problems of condensation

■ whether the appliance input is such that the appliance may be installed in its intended location in conformity with the input and ventilation requirements

■ presence of any flues which might have an adverse effect on the appliance performance unless closed off

■ presence of other heating appliances in the room space

■ position of heating appliances in relation to probable position of fixtures, furniture and curtains

- availability of gas supplies
- availability of electrical supply (where applicable)
- positioning of air vent
- installation of an independent electrical CO alarm conforming to BS EN 50291 in accordance with BS EN 50292 and the appliance and/or alarm manufacturer's instructions.

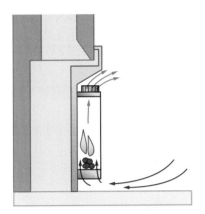

a) 'Inset' type flueless appliance Normally installed into recess or disused flue.
NOTE Disused flues should be suitably de-commissioned

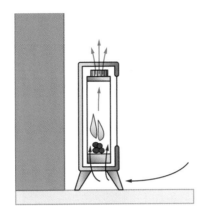

b) Freestanding flueless appliance

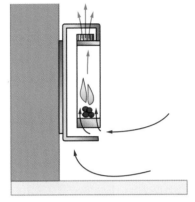

c) Wall-mounted flueless appliance

Figure 12.49 Examples of flueless appliances

Types of gas space heater

Gas space heaters can provide heat to rooms according to the needs of individuals. The types available are:

- Gas fires – these may be radiant or radiant/convector, natural or fanned flue.
- Convectors – normally wall-mounted and room-sealed but may be open-flue. Flueless convectors are less common now as they caused problems with condensation.
- Cabinet heaters (mobile) – these are flueless radiant heaters using LPG, so you need to hold specific qualifications to deal with them.

Radiant fires

A typical radiant gas fire has an outer decorative casing. The inner firebox contains the radiants and a firebrick with a canopy that leads the products to the flue spigot. Radiants are available in many shapes and sizes, and are designed to aid complete combustion and radiate maximum heat output to the room. The radiants are often surrounded by chrome-plated reflector panels, which give maximum heat to the room and help to keep the fire casing cool. Although efficient, these types of fire are not often installed as new, but there are still many installed in the housing stock.

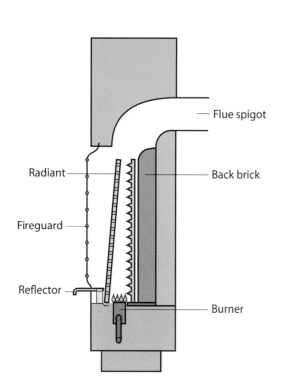

Figure 12.50 Typical radiant fire

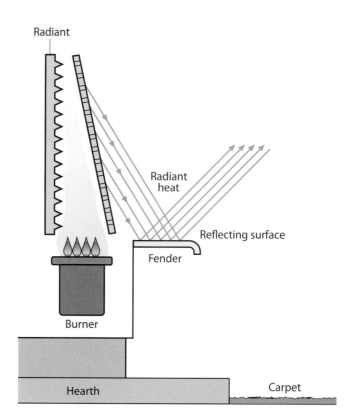

Figure 12.51 Section through gas fire showing the radiant, reflector and burner

The burner may be of two main types: simplex or duplex. A simplex burner has all the gas supplied by a single injector that is controlled by a simple tap. Duplex burners allow radiants or pairs of radiants to be controlled separately. These give low or high heat settings, as required by the user so offer the user more flexibility and are energy saving. For more information on burners, see the section on 'Types of gas burner' (page 106).

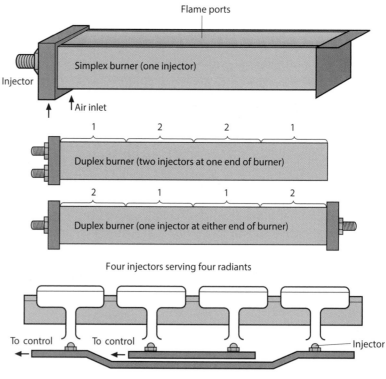

Figure 12.52 Typical layout of burners to a gas fire

Radiant convector fires

This is probably the most popular type of domestic fire, with the open-flued model being the most common. Convected heat gives a higher heat efficiency by allowing air from the room to circulate around the firebox, become heated and then pass into the room.

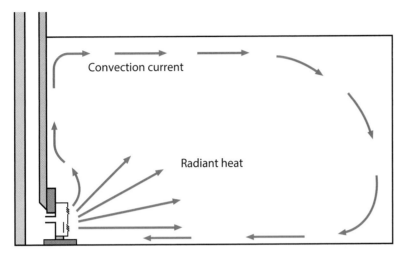

Figure 12.53 Radiant and convected heat from a radiant convector fire

Some radiant convector fires are enclosed (built-in) with a glass front. These slightly reduce the radiant heat but improve the overall efficiency. Note the integral draught diverter, as can be seen in Figure 12.55.

Safe working

Glass fronts on gas fires can become stained and brittle. Over use of detergents and bleach can discolour and/or damage the glass, at which point it must be replaced to avoid incident.

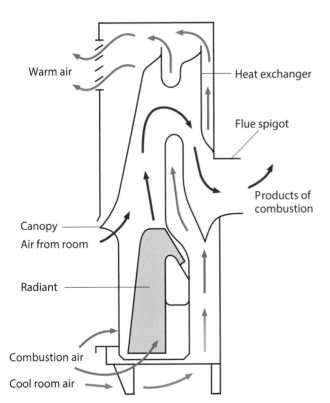

Figure 12.54 Convected heat from a gas fire

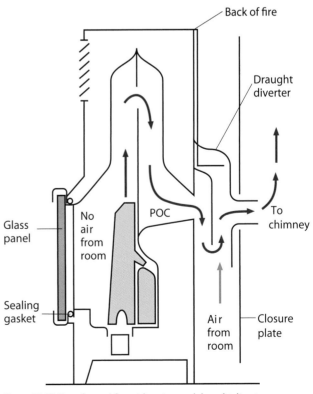

Figure 12.55 Glass-fronted fire with an integral draught diverter

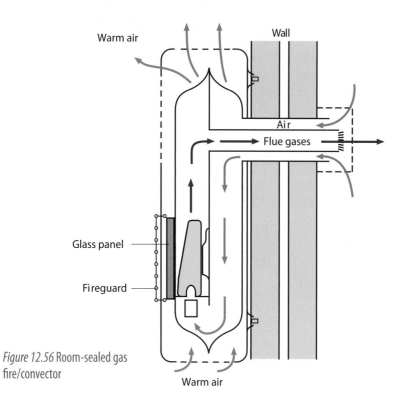

Figure 12.56 Room-sealed gas fire/convector

Room-sealed wall heaters

Room-sealed wall heaters are a safer option than open-flued conventional fires as there is no risk of any downdraught from a chimney causing POC to spill into the room. The principle of room-sealed flues is simply applied to a radiant and/or convector fire; all other aspects of the heater are similar.

Heating stoves

Heating stoves are available in two basic types: free-standing with a top flue outlet and those fitted with a rear flue spigot to a closure plate similar to a gas fire. Heating stoves must not be confused with wood-burning stoves, which, because of their higher temperatures, require different flue options. Some examples of installations are shown in Figures 12.57, 12.58 and 12.59.

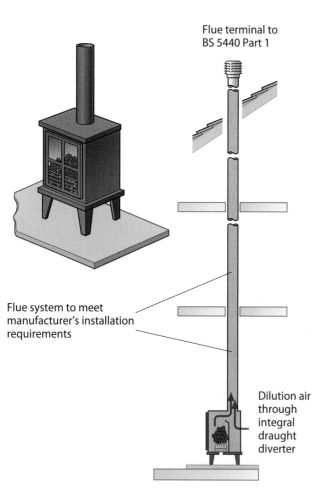

Figure 12.57 Installation of heating stove connected to a factory-made chimney or flue system

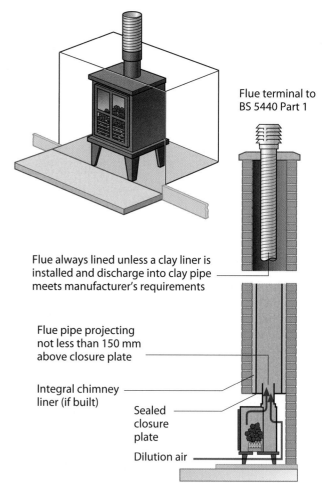

Figure 12.58 Installation of a free-standing heating stove fitted in a builder's opening

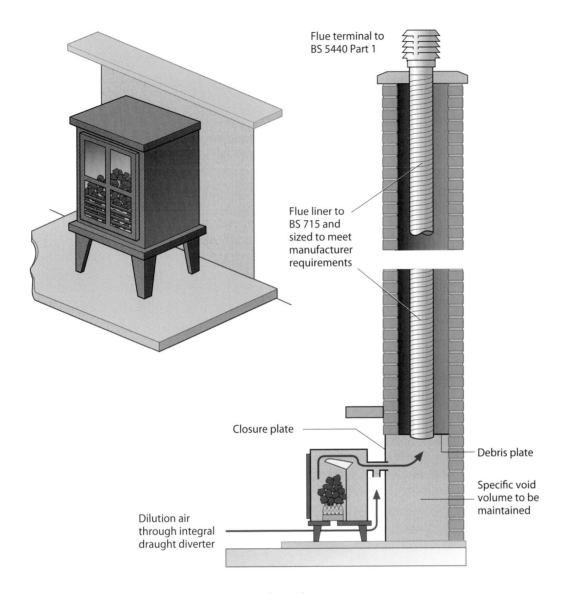

Flue terminal to
BS 5440 Part 1

Flue liner to
BS 715 and
sized to meet
manufacturer
requirements

Closure plate

Debris plate

Specific void
volume to be
maintained

Dilution air
through integral
draught diverter

Figure 12.59 Installation of a free-standing heater using a closure plate

Live fuel-effect (LFE) fires

There are several types of gas fire that use a flame effect, and these are becoming increasingly popular. The radiants are replaced by realistic coals or logs; these give a real-fire effect. The coals/logs must be placed in position exactly as per the manufacturer's instructions in order to give complete combustion.

Inset live fuel-effect (ILFE) fires

ILFEs are usually inset into the fireplace opening, so the flame effect is open and efficiency is only about 40 to 50 per cent. However, there is a convection chamber that helps to recover some of the heat. Note the path of the POC, the convected air and the flow of room air in Figure 12.60 on page 416. A great deal of heat escapes to the atmosphere with these fires, so they should never be installed as the main heating source.

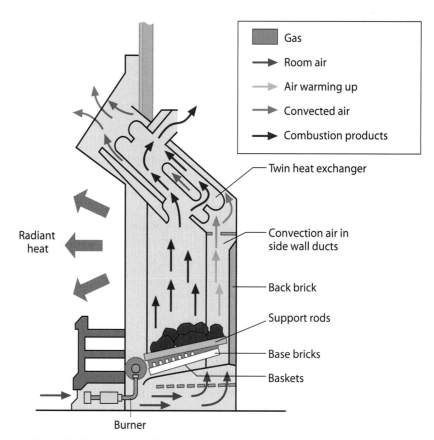

Figure 12.60 Inset live fuel-effect fire

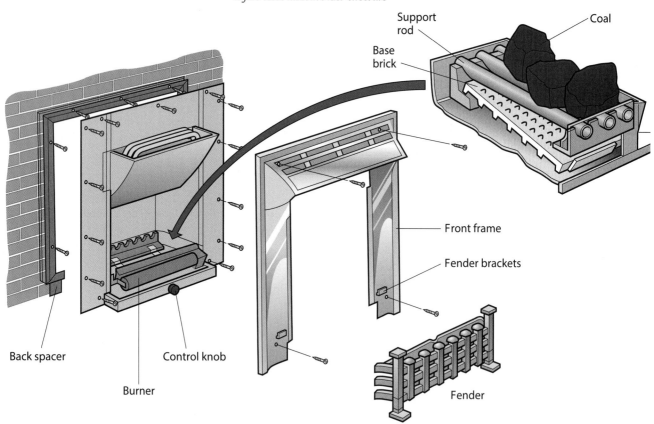

Figure 12.61 Typical components of an ILFE gas fire

Decorative fuel-effect (DFE) fires

DFE appliances produce only radiant heat and have a very low efficiency. They used to be extremely popular, though they are generally used as a decorative focal point in a room that already has another form of heating. The flue should have a cross-sectional dimension of not less than 175 mm unless certified in the manufacturer's instructions. Typical fireplaces into which appliances can be installed are shown in the case types in Figure 12.63, but special installation considerations apply. You should also note that open flue ways draw the air from the room, creating a cold draught, so the position of vents needs careful consideration. See the section on 'Installing gas space heaters' (page 420).

It is also becoming increasingly popular for DFE fires to be fitted under a **fire canopy**, as shown in example 4 in

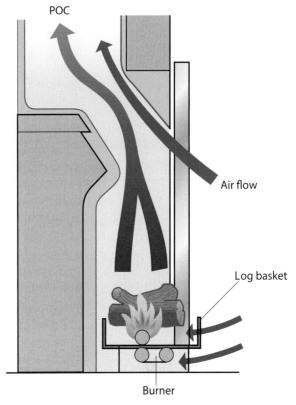

Figure 12.62 Decorative fuel-effect fire

> **Key term**
>
> **Fire canopy**
> A purpose-designed opening with sides that gather the POC from a gas fire into the main flue. Often used as a decorative feature but must be made from non-combustible material.

1 Builder's opening

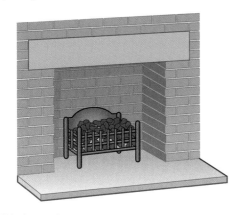

2 Fireplace recess

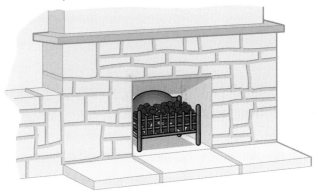

3 Raised builder's opening

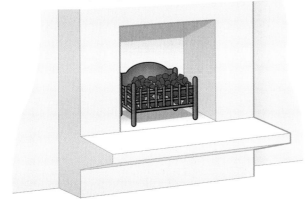

4 Builder's opening with independent canopy

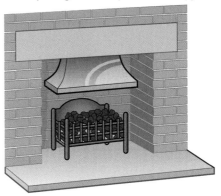

Figure 12.63 Typical fireplaces

Figure 12.63 and examples 5 and 6 in Figure 12.64. Particular care should be taken with the constraints of these, as shown in Figure 12.65.

Refer to BS 5871 Part 3 and take note of how to size the canopy for example 6. Ask your tutor if anything is unclear.

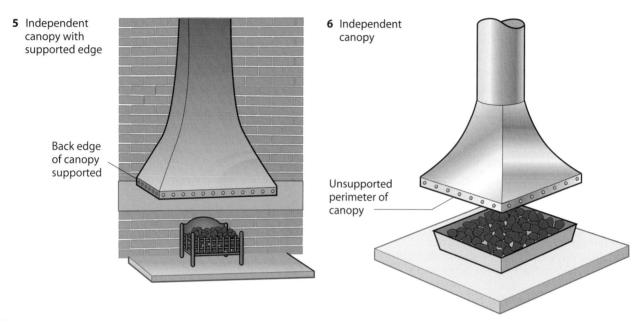

5 Independent canopy with supported edge

Back edge of canopy supported

6 Independent canopy

Unsupported perimeter of canopy

Figure 12.64 Canopy details

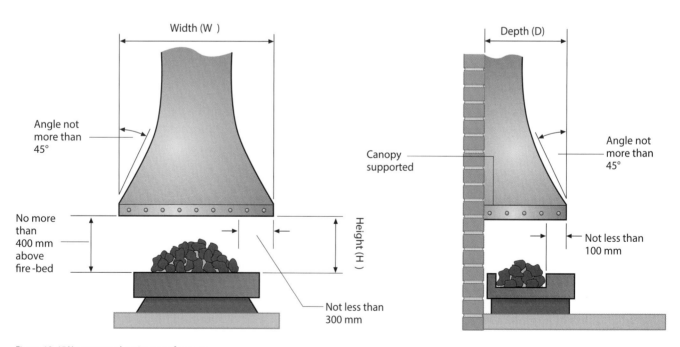

Width (W)

Angle not more than 45°

No more than 400 mm above fire-bed

Height (H)

Not less than 300 mm

Depth (D)

Angle not more than 45°

Canopy supported

Not less than 100 mm

Figure 12.65 Unsupported perimeter of canopy

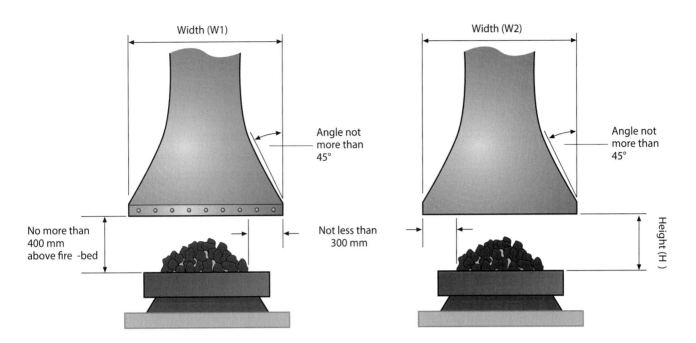

Width (W1)

Angle not
more than
45°

No more than
400 mm
above fire -bed

Not less than
300 mm

Width (W2)

Angle not
more than
45°

Height (H)

Figure 12.66 Unsupported perimeter of canopy

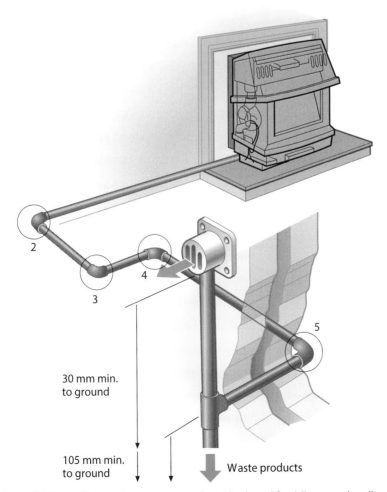

2

4

3

5

30 mm min.
to ground

105 mm min.
to ground

Waste products

Figure 12.67 General layout of a condensing gas fire with a fanned flue (elbows numbered)

Condensing gas fires

Condensing gas fires rely on a fan flue to entrain combustion air from the room, through the heat exchanger and out through a small diameter (28 mm) PVC-C flue pipe. As with all fan flue appliances, there must be a safety device, such as a pressure switch, fitted to shut down the burner in the event of fan failure.

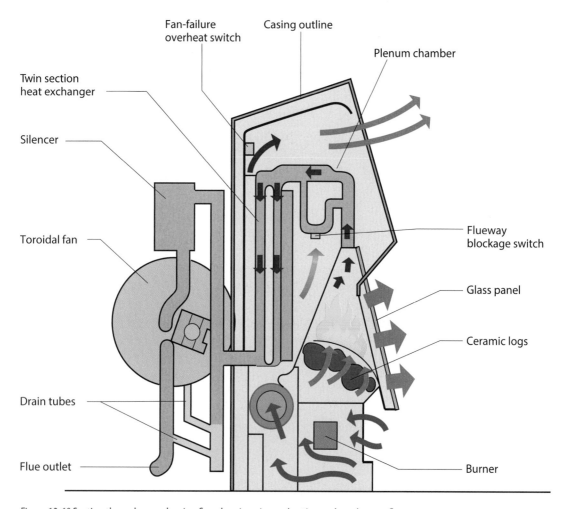

Figure 12.68 Section through a condensing fire, showing air, combustion and condensate flows

Installing gas space heaters

When fitting gas space heaters it is essential to follow the manufacturer's instructions for the appliance. Reference to the British Standards can be made as follows:

- BS 5781 Part 1: Gas fires and convector heaters
- BS 5781 Part 2: Inset live fuel-effect fires
- BS 5781 Part 3: Decorative fuel-effect gas fires.
- BS 5781 Part 4: Independent gas fired flueless fires, convector heaters and heating stoves

In addition, it is necessary to satisfy the requirements of the Gas Safety (Installation and Use) Regulations. It is important to note that:

- Open-flued or flueless appliances must *never* be installed in bathrooms or shower rooms.
- Appliances over 12.7 kW net (this includes combined gas fire/back-boiler units) must not be fitted in a room used for sleeping, unless they're fitted with an ASD.

When installing gas fires it is also crucial that you ensure the flue to which the appliance is fitted is suitably constructed and is in a proper, safe condition for the operation of the gas fire. This can be done by carrying out visual checks and smoke tests. You must also ensure that there is a gas tap for shutting off the appliance.

Gas fire installation checklist

The following checklist may help when installing a new gas fire, but always refer to the manufacturer's instructions.

- Visually inspect the flue and determine suitability by checking it with a torch for blockages, suitable materials and loose mortar joints.
- Carry out a flue-flow test. Be sure to warm the flue prior to introducing the smoke pellet.
- Check ventilation is suitable (take into account other appliances in the room). Measure the vent size carefully as often appliances are added to the room after the vent was installed.
- Test the existing installation for tightness. The gas installation may already have a leak, so it is important to test before carrying out any work.
- Fit the fire and pipework. Follow the manufacturer's instructions to the letter – failure to do so could lead to criminal proceedings at a later date.
- Test the completed installation for tightness as work on existing pipes can disturb the set-up and cause a leak.
- Purge each appliance as the whole system has been subjected to air infiltration. Relight and check the operation of each appliance.
- Light the fire and check/adjust the burner working pressure. It is important that the gas burner pressure is set to the manufacturer's instructions – some appliances can be adjusted, others cannot.
- Check for spillage. Light the appliance and after giving the flue chance to warm up, carry out the spillage test with a smoke match. Gas fires can be subject to atmospheric pressures and this should be considered in your findings. Always re-test on another day if weather conditions are adverse.
- Advise the customer as to the fire's correct use and need for regular maintenance.
- Leave the manufacturer's manual with the customer so that they can pass it to the next engineer to work on the appliance.

Closure plate

A closure plate is designed for specific fires (square outlet, or round or rectangular flue outlet), and it is sealed to the fireplace with heat-proof tape to resist temperatures up to approximately 100 °C. The plate often includes a **relief opening**, which varies for each particular appliance.

Key term

Relief opening
Stops excess draughts from lifting the flame from the burner port, especially in windy conditions.

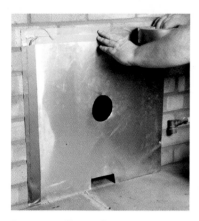

Figure 12.69 Closure plate

Before fitting the closure plate, examine the flue in detail to make sure that:

- ■ it is not blocked
- ■ it has been swept if solid fuel has been used previously
- ■ no dampers or restrictions exist (dampers are only fitted to coal fires and should be removed or fixed open)
- ■ the catchment or void space is adequate so that falling debris will not block the flue spigot (refer to Table 12.03)
- ■ all openings in the fireplace recess for pipework etc. are sealed
- ■ the flue serves only one room – if the flue is combined with another fireplace, there is potential for downdraught to occur and fumes can spill into another room.

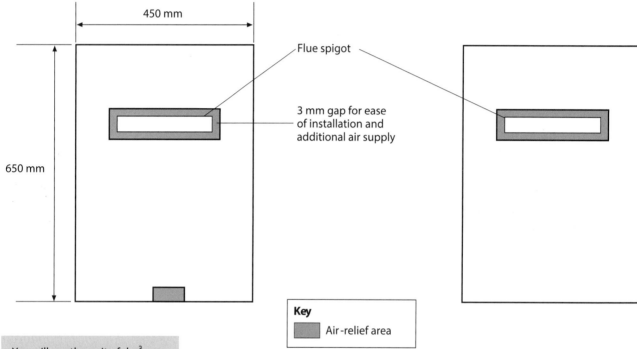

Figure 12.70 Example of a closure plate

You will see the unit of dm³ (decimetres cubed) has been used in Table 12.03. This may be new to you, so here is a simple example to explain it. There are 10 decimetres to every metre so there are 1,000 dm³ in a m³ (10 dm × 10 dm × 10 dm = 1,000 dm³).

A void below a flue spigot measures

$0.45 \times 0.37 \times 0.30 = 0.0499 \text{ m}^3$

To convert the void size from m³ to d m³, simply take 0.0499 m³ and multiply by 1,000 to give 49 dm³. The example void measured is 49 dm³ which, as you can see from Table 12.03, is satisfactory.

Debris catchment	Masonry chimneys			Block chimneys/flue systems to BS 715	
Min. void volume in dm³	Unlined	Lined clay cement or metal		New/ unused	Previously used
		New/ unused	Previously used		
2		✓		✓	
12	✓		✓		✓
Min depth in mm					
75		✓		✓	
250	✓		✓		✓

Table 12.03 Minimum void volumes and depths below gas-fire spigots

If the catchment area is not large enough, it can either be enlarged or an alternative catchment used in its place, such as a flue box.

Surround and hearth

Where the appliance is installed in a builder's opening, fireplace recess or flue box, the hearth should:

- extend throughout the whole base of the builder's opening, fireplace recess or flue box
- project at least 300 mm in front of the naked flame or incandescent part of the fire bed
- project at least 150 mm beyond each side of the naked flame or incandescent part of the fire bed.

Where the appliance is free-standing, the hearth should extend completely beneath the naked flame or incandescent part of the fire bed and project outwards at least 300 mm from all sides of the naked flame.

In all cases, the hearth thickness should be not less than 12 mm, and be such that the heat transmitted through it does not give rise to a temperature greater than 80 °C on its underside.

It is possible that a hearth thickness of greater than 12 mm may be needed, and this may be detailed in the manufacturer's instructions.

Gas supply

The gas supply can be taken from a 'restrictor elbow', which is normally fitted at the side of the chimney breast. Chrome-plated 8 mm tube can be fitted from the restrictor to the inlet of the fire, which has a union-type joint suitable for annual removal for servicing. Note that any gas supply pipe buried in the structure or running in the chimney recess must be suitably protected by a material such as PVC coating or wrapping. A suitable isolation valve must be included on each gas appliance for servicing purposes.

Figure 12.71 Restrictor elbow

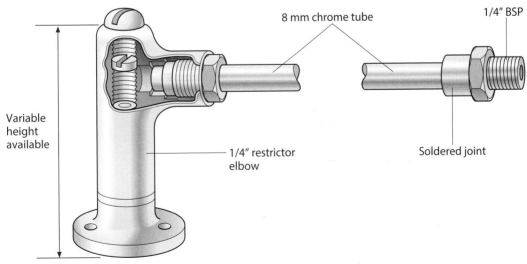

Figure 12.72 Restrictor elbow and connecting pipe

How to install an inset live fuel-effect gas fire

Checklist

PPE	Tools and equipment	Source information
• Overalls • Protective footwear • Barrier cream	• Electric drill • Spanners • Screwdriver • Leak detection equipment • Manometer • Smoke matches	• Manufacturer's instructions

1 Check before installing that the flue is fit for purpose and that the catchment area is sufficient to install the LFE. Follow the manufacturer's instructions to mark out the fixing positions and gas pipe position.

2 Fix the fire box into position using the purpose-made securing devices and check the flue outlet is sufficient distance away from the back of the chimney.

3 Fit the burner assembly into position and test for gas tightness.

4 Place the fire surround into position. – you will need to check manufacturer's instructions to see if coals should be fitted.

5 Place the coals into position following the manufacturer's instructions – each coal must be fitted into a given space to give the correct living flame effect. You must be gentle with the coals as they are easily damaged.

6 Turn on the gas fire and check for spillage. Check that the safety devices work and commission the appliance.

7 Hand over to the customer, explaining how to use and the need for regular maintenance.

Commissioning gas fires and water heaters

Before any gas fire is commissioned it must be inspected to ensure that all work has been carried out to the manufacturer's instructions, the British Standards and the relevant sections of the Gas Safety (Installation and Use) Regulations. The pre-commissioning inspection should include checking that:

- ventilation/combustion air is adequate for the number and type of appliances in the room, as sometimes extra appliances are added after the vent was installed
- the flue is correctly constructed – check the full length of the flue, especially in the loft space where changes may have been made that don't adhere to regulation
- the fire and hearth/surround are suitable
- installation work is satisfactory and complies with requirements
- gas installation and other works for the supply of gas are adequate.

Before lighting the fire make sure the flue has been inspected and a flue-flow test carried out. There are no exceptions to this. The test should comprise:

- Carrying out a tightness test.
- Purging the appliance and any other appliances on the installation.
- Lighting the fire and checking the flame pattern.
- Checking burner pressure.
- Checking the gas rate.
- Testing safety devices.
- Checking the user controls.
- Carrying out a spillage test (this is essential – no exceptions).
- Demonstrating the use of the appliance to the customer and emphasising the importance of regular servicing.
- Leaving the instruction manual and guidance with the customer.

Case study

A customer called out an engineer to service their gas fire. The engineer arrived to find the gas fire hanging off the wall in a precarious manner. He quickly isolated the gas supply and made it safe. It was an immediately dangerous situation and was reported to the Health and Safety Executive. The property belonged to a landlord and the tenant had continued to use the appliance in a dangerous condition.

What should have happened and who is responsible?

Discuss your answers with your tutor.

Servicing radiant gas fires

1 Gas fire.

2 Casing removed.

3 Radiants removed.

4 Cleaning the burner.

5 Fire removed.

6 Once fire and radiant replaced, set burner pressure.

7 Spillage test.

Servicing should include the following typical operations:

- Checking the ventilation requirements. Vents are often purposefully blocked by occupiers to reduce the draught they cause.
- Visually checking the flue, including termination. Bird nests and falling brickwork can cause blockages at any time.
- Checking for gas tightness.
- Lighting the fire and check operation.
- Disconnecting and removing the fire.
- Removing the closure plate.
- Checking for soot and rubble behind the closure plate and clean out.
- Carrying out a flue-flow test, remembering to pre-heat the flue to get the draught going in the correct direction.

- Resealing the closure plate, ensuring that the relief opening is clear.
- Cleaning dust and lint from the base of the fire where the cool air enters from the room.
- Cleaning the burner and the injectors.
- Checking the heat exchanger for corrosion damage and cleaning the flue ways.
- Reconnecting the fire and checking for gas tightness.
- Ensuring the fire is level and secure.
- Testing the ignition, cleaning or renewing any batteries or parts as necessary.
- Checking any electrical components are properly insulated. The heat from the gas fire can, over time, cause the insulation to go brittle, fall away and expose the wiring.
- Checking burner pressure and adjusting the regulator if required.
- Checking the gas rate.
- Checking the operation of the FSD.
- Checking user controls operate satisfactorily.
- Checking for spillage.
- Reporting back to the customer.

Fault finding and diagnosis

The fault list in Table 12.04 is not exhaustive and does not replace a manufacturer fault-finding chart for specific appliances.

Problem	Solution/checks
Space heater	
Heat exchanger cracked	Replace
Seals damaged between heat exchanger and flue spigot	Replace seals
Radiant support bar warped	Replace bar
Blocked heat exchanger	Unblock heat exchanger
Radiant, coals and logs	
Radiant cracked	Replace radiant
Discoloured radiant	Replace radiant
Poor located radiant or coals, logs	Clean and refit
Sooted or damaged radiant or coals/logs	Clean or replace and check function of gas appliance (burner pressure, flue seals, ventilation)
Outer casing	
Poorly attached to appliance	Check with manufacturer's instructions, refit or repair
Front glass panel or dress guard	
Damaged glass panel	Replace glass
Glass panel incorrectly fitted	Refit to manufacturer's instructions
Guard damaged or incorrectly fitted	Replace guard or refit correctly

Table 12.04 Fault finding and solutions – gas fires and water heaters

Progress check

1 What is the advantage of a duplex burner over a simplex burner?

2 What are the three types of flue that can be used for a stove?

3 In which room should an open-flued fire never be fitted?

4 What is the minimum vent required for a DFE?

5 Appliances over 12.7 kW net must not be fitted in which room?

DOMESTIC GAS COOKERS

Types of domestic gas cooker

There are a number of categories of cooking appliances available today:

- free-standing
- slide-in
- built-in.

These can be further categorised as:

- low-level grill/oven
- eye-level grill/oven
- hob, with/without fold-down lid.

The gas cooker is now a high-tech appliance with automatic ignition and cooking time control, together with forced-convection ovens, lights and easy-clean catalytic linings. Cookers must now carry a CE mark. Cookers manufactured before 1996 may not have this but will still comply by conforming to the Regulations, and some even carry a kite mark.

You must also check that the type of gas and the working pressure for which the appliance are suitable, as stated on the data plate. In most cases the inlet pressure to the cooker should be 20 mbar.

There are two main methods of controlling the operation of the oven section:

- gas mechanical controls – as discussed in Chapter 11
- electrical controls – many of these tend to be similar to the controls used in gas boilers.

Installing gas cookers

The first thing to do when installing a cooker is to check that the location is suitable and ensure that the manufacturer's instructions are adhered to. Some special requirements from the main standard covering gas cookers (BS 6172) include:

- cookers must not be installed in a bathroom or shower room
- cookers must not be installed in a bedsitting room of less than 20 m³ unless they have a single burner
- cookers for use on LPG must not be installed in basements/below ground level.

There are other considerations when siting a gas cooker:

- Avoid draughts – don't fix them too close to doors/windows.
- Be aware of curtains and other combustible materials close by.
- Follow manufacturer's instructions regarding clearances.
- Always fit a floor-standing cooker on a stable base and install a stability bracket or chain to prevent the cooker from rocking forward or being pulled over.
- Ensure the ventilation allowance complies with BS 5440 (reference here is to other appliances and/or use of extractor fans).

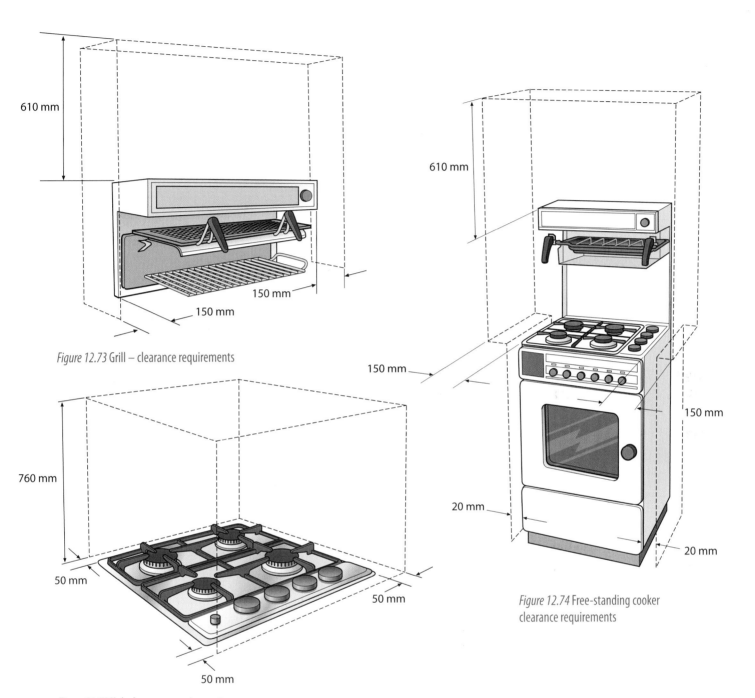

610 mm

150 mm

150 mm

Figure 12.73 Grill – clearance requirements

760 mm

50 mm

50 mm

50 mm

Figure 12.75 Hob clearance requirements

610 mm

150 mm

150 mm

20 mm

20 mm

Figure 12.74 Free-standing cooker clearance requirements

Figure 12.76 Backplate, elbow and bayonet

How to install a gas cooker

Checklist

PPE	Tools and equipment	Source information
• Overalls • Protective footwear • Barrier cream	• Electric drill • Level • Spanners • Screwdriver • Leak detection equipment • Manometer	• Manufacturer's instructions

1 Check with the manufacturer's instructions that the proposed position of the cooker is a safe location to install the appliance.

2 Measure the position of the bayonet back elbow as it has to fit correctly behind the cooker. The measurements are always in the manufacturer's instructions.

3 Fit the back plate elbow to the gas pipe and test for tightness and that it is secure. Using the correct cooker hose for the gas type, fit to the back of the cooker and connect to bayonet connection. Check bayonet connection with leak detection fluid, following connection to back elbow.

4 Ensure the stability bracket is fitted securely to the wall.

5 Level the cooker using the adjustable cooker feet.

6 Purge the cooker of air and light each burner in turn, checking their function. Check the operation of the cooker lid safety shut-off valve.

7 Check the door seals by pulling a piece of paper through the door seals to ensure it grips – if it does not grip, the door seals may be damaged or incorrectly fitted.

8 Hand over to the customer explaining how to use the cooker and leave them the instructions.

Gas connection

The gas installation pipe should be of an appropriate size to maintain a suitable working pressure at the appliance and to give the required heat input at the appliance. Fixed appliances should be connected by rigid pipework and an isolating valve, while slide-in appliances are normally connected by a flexible appliance connector, for use with a self-sealing plug-in device.

When using a flexible connector, take into account the following considerations:

■ Do not subject the connector to undue force.
■ Keep it away from excessive heat.

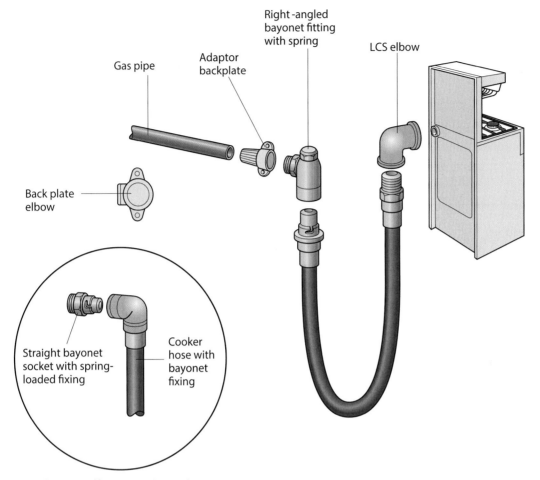

Gas pipe

Adaptor backplate

Right-angled bayonet fitting with spring

LCS elbow

Back plate elbow

Straight bayonet socket with spring-loaded fixing

Cooker hose with bayonet fixing

Figure 12.77 Bayonet and hose connection requirements

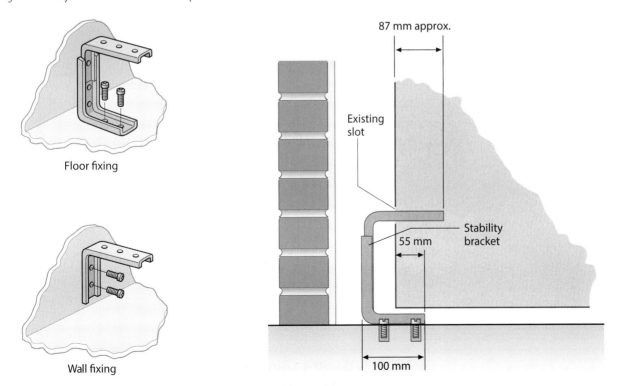

Floor fixing

Wall fixing

87 mm approx.

Existing slot

Stability bracket

55 mm

100 mm

Figure 12.78 Stability bracket

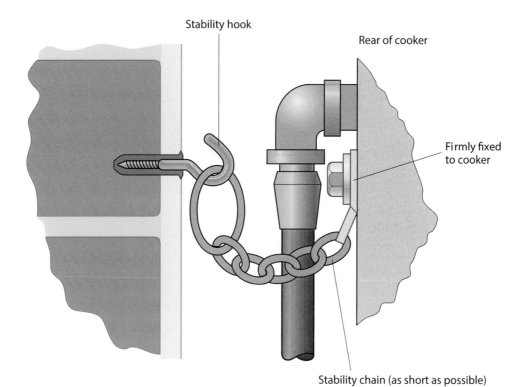

Stability hook

Rear of cooker

Firmly fixed
to cooker

Stability chain (as short as possible)

Figure 12.79 Stability chain

Key term

Bayonet fitting
The fitting is designed to enable quick release of the cooker hose from the pipework. It has a built-in spring to push shut as soon as the hose is removed.

- Ensure the **bayonet fitting** is secure to wall fitting.
- Make sure the hose hangs downwards; it should not be twisted.
- Fit the flexible hose in such a position that does not damage it.

Note that flexible connectors for use with LPG are to a higher specification, so care is needed in their selection. A free-standing cooker with a flexible connector must be fitted with a stability device firmly secured to the wall or floor.

Electrical connection

The electrical connection should be made as per the manufacturer's instructions, with particular regard to fuse rating, continuity, polarity and earth connection. The connection point should be no more than 1.5 m from the appliance, and the flexible cable should not be exposed to hot surfaces or POC.

Commissioning gas cookers

Before beginning to commission the appliance, check that:

- the appliance is suitable for the gas supplied
- the room/ventilation is suitable for the appliance
- the installation meets manufacturer's instructions
- the gas supply/connection meets requirements and is gas-tight (conduct the gas-tightness test with lid shut-off devices in the *open* position to ensure gas is being supplied to all parts of the cooker)
- proximity to combustible material has been checked and is acceptable
- the stability bracket has been fitted (where required).

Once you have conducted these checks and are sure the installation complies, you can install the cooker.

Gas cooker installation checklist

- Check the electricity is on or that batteries are fitted (as appropriate).
- Purge the appliance and check that the gas control taps are working smoothly.
- Check the igniters to all hob burners and light each burner.
- Check the flames for aeration/height.
- Check the burner pressure at the test point and the gas rate.
- Light the oven and check the operation of flame-supervision device.
- Check the oven thermostat operates correctly on bypass rate.
- Instruct the user on how the cooker works and advise them to have it serviced regularly.

Servicing and maintaining gas cookers

When servicing/maintaining a cooker it is essential to follow the exact procedure, as laid out in the manufacturer's instructions. The following procedure gives a general overview.

Preliminary checks

Before servicing or carrying out maintenance work on a gas cooker, check that:

- the appliance operates correctly when lit
- the appliance is suitable for the gas supplied
- the installation meets manufacturer's instructions
- the gas supply/connection meets requirements and is gas-tight
- proximity to combustible material has been checked and is acceptable
- the stability bracket has been fitted (where required).

Servicing procedure

- Isolate the gas/electricity supplies to the appliance.
- Check the condition of the flexible connector.
- Remove the burners and clean the air ports.
- Check and grease the gas taps.
- Check and clean the flue ways to the oven/grill.
- Check the condition of all the components and replace if necessary.
- Reassemble the components, checking any gas connections.
- Ensure the oven door seals are satisfactory by closing doors on a piece of paper and trying to pull it out – it should hold tight if the seal is good.

Servicing a cooker

Checklist

PPE	Tools and equipment	Source information
• Safety boots • Overalls	• Spanners • Screwdrivers • Manometer • Level	• Manufacturer's instructions

1 The cooker.

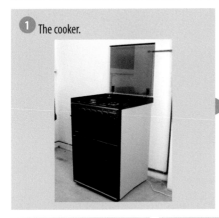

2 Burners and covers removed.

3 Checking grill burner/combustion.

4 Checking oven burner for incomplete combustion.

5 Checking FSD and flame picture after service.

6 Checking door seal.

7 Checking igniter and gas taps.

Fault finding and diagnosis

Problem	Solution/checks
Ignition	Check power supply, change batteries, test fuse Check ignition probes, clean or replace
Burners incorrect	Check flame picture, gas rate and incoming gas pressure Replace or clean injector Adjust incoming gas pressure
FSD does not operate	Replace FSD
Seals damaged	Replace damaged seals
Oven thermostat and bypass incorrect	Check and clean or replace
Hotplate lid cut-off functions incorrectly	Check mechanism and replace if necessary
Phial or sensing tube broken	Replace as per manufacturer's instructions
Stiff gas taps	Replace as per manufacturer's instructions

Table 12.05 Fault finding and solutions – domestic cookers

Re-commissioning gas cookers

- Re-light the cooker.
- Check the burner pressure is as stated on the data plate.
- Check the igniter function is okay.
- Check the flame-supervision device on the oven.
- Check the oven thermostat.
- Report back to the user and leave them with the manufacturer's instructions.
- Advise the customer of the on-going need for regular cleaning/ maintenance.

Progress check

1 What are the two main methods of controlling a gas cooker?

2 Why is a stability bracket used?

3 How do you check door seals?

4 If the cooker burners are functioning incorrectly, what checks should you carry out?

WARM AIR HEATERS

The vast majority of housing in the UK now uses a water central heating system. However warm air heating systems are still available and have their advantages.

A warm air system blows warm air around the house from a central heater through ducts and out into the room via vents and grills. Modern warm air units are condensing and electronically control room temperatures. This provides better fuel efficiency.

Modern warm air units have to be checked to ensure they are fully compliant with Part L of the building regulations before they are installed. The air in the unit is heated by the main burner through a metal heat exchanger and then this warm air is blown around the ducting.

The manufacturer's installation instructions describe any specific requirements for the condensing heater, e.g. a means for disposal of condensate, the air-gas valve setting and the water temperature operating range.

Replacement warm air units are available for existing units as they need replacing with age. The new units are quieter and more efficient and also scrub the air, preventing dust and allergies in the house.

New heaters

Any new heaters fall within the scope of the European Gas Appliances Directive, implemented in the UK by the Gas Appliances (Safety) Regulations, which require any new heaters to be CE marked.

You should check that the packaging and the heater's data plate are marked with at least the following information:

- the letters GB
- the type of gas and appliance inlet pressure as follows:
  — for a heater adjusted for natural gas, 'G20 and/or natural gas 20 mbar' plus the designation I2H
  — for a heater adjusted for butane, 'G30 and/or butane 29 mbar' plus the designation I3B
  — for a heater adjusted for propane, 'G31 and/or propane 37 mbar' plus the designation I3P
  — for a heater that is designed to burn either butane or propane gas at the correct pressure, 'G30/G31 and/or butane/propane 29/37 mbar' plus the designation I3+.

On a warm air unit data plate you will find the letters CAT I or CAT II followed by the gas type designations (i.e. 2H , 2P , 2B). This allows the unit to be adjusted for use on different types of gas by a qualified engineer.

Previously used heaters

Second-hand heaters should be checked to ensure they are suitable to be installed. If the heater is not suitable or in good condition it should not be installed. The data plate will contain information about the pressure and type of gas that the heater is suitable for.

Never install a second-hand heater without the manufacturer's instructions – these are usually available from the manufacturer if not supplied by the customer. Only go ahead with the installation if you have all the information you need and have no doubts about the unit's safety and suitability.

If the heater does not have a CE mark then you must make sure the appliance is fit for purpose with no damage and suitable for the type of gas supplied. Factors you will need to consider include surface temperature, heater stability and ventilation.

Information on the methods for designing, and the heat and fan requirements is given in the Warm Air Heating System Design Guide. Further information, particularly on heat losses and sizing, can also be found in CIBSE Guide B3: Ductwork.

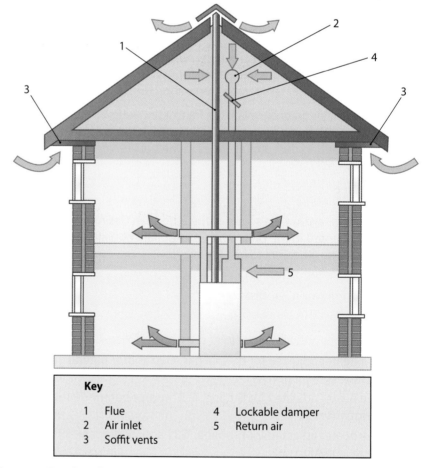

Key			
1	Flue	4	Lockable damper
2	Air inlet	5	Return air
3	Soffit vents		

Figure 12.80 Typical installation where combustion air is supplied from a ventilated roof space

Types of warm air heater

There are three basic designs for the warm air system: down flow, up flow and horizontal flow.

The air is drawn into the heat exchanger by the centrifugal fan. The air passes across hot plates which warm the air up and then is forced out through the other end into the duct system. The unit is fitted with a summer air circulation switch that turns the fan on (but not the burner) in the summer providing cool air around the house. The unit sits on a plinth or support that can withstand the weight. There should be sufficient access to all parts of the unit to allow for maintenance and removal.

Sufficient clearance should also be provided to enable the installation of the pipework and ductwork. The water connections should be made with compression fittings and the heat exchanger fitted with valves to enable easy removal.

It is important to note that draining and flushing the water system is recommended prior to the installation of the unit. A strainer should be fitted upstream of the unit.

Down flow

The down flow heater has an air intake at the top of the appliance and fires the heat downwards through the plenum to be distributed through the ducts to the rooms.

Up flow

The up flow heater has an air intake at the bottom of the appliance and fires the heat upwards through the plenum to be distributed through the ducts to each room.

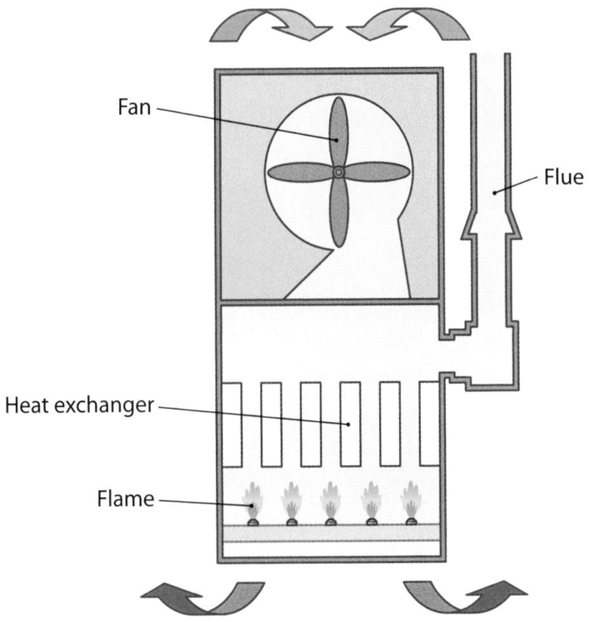

Figure 12.81 Down flow warm air system

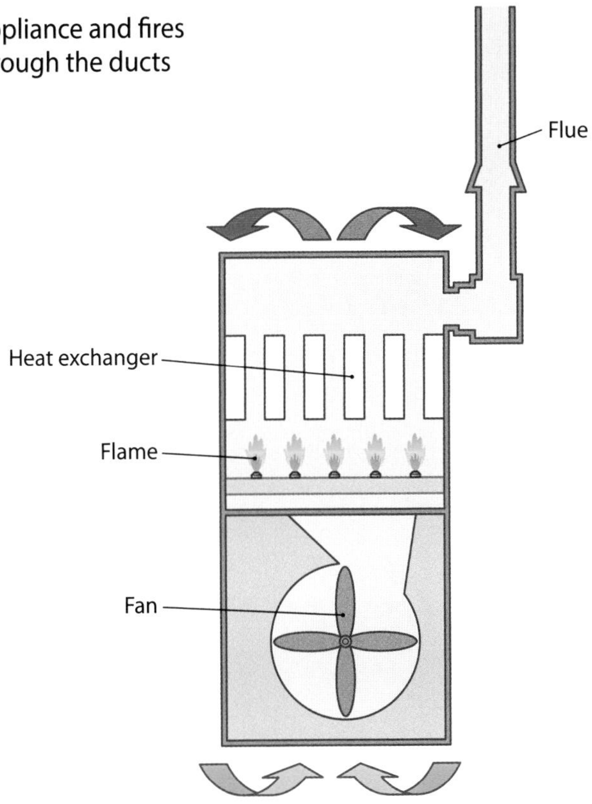

Figure 12.82 Up flow warm air system

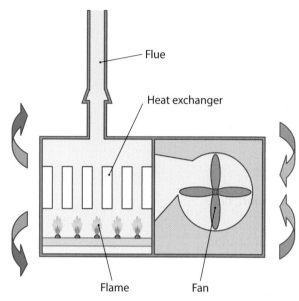

Figure 12.83 Horizontal flow warm air system

Horizontal flow

Air intake in horizontal flow systems is at the side and warm air is blown out from the opposite side of the unit into the duct system.

The unit normally sits on a plenum box as seen in Figure 12.86.

Warm air heating system components

Warm air ducting

When warm air ducting is thermally insulated, non-combustible insulation should be used. For insulated ducting within 2 m of the heater, the insulation shall, in addition, be thermally stable up to a temperature of 120 °C.

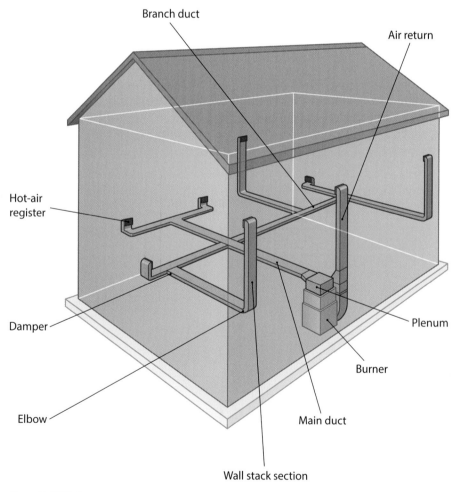

Figure 12.84 Air duct system

Return air ducting

The return air ducts can be made from flexible materials and therefore should be kept as short as possible and fixed according to manufacturer's instructions to prevent it collapsing.

The return air to the heater has to be free flowing with no obstructions from all heated rooms, except for the rooms that generate moisture such as kitchens and bathrooms where it is necessary to remove the moisture in the air before it passes back into the heater. Figure 12.85 shows a typical return air system installation.

The design of return air ducts should be made to avoid noise transmission from the fan unit. This requires that the air duct should not be too short, have at least one bend or be lined with noise-reducing material.

It is vital that the fan duct system does not affect the air used for combustion by that appliance or any other appliance as this will lead to incomplete combustion. It is therefore essential that the return air grills should be connected to the duct by air tight seals that prevent combustion air entering the duct.

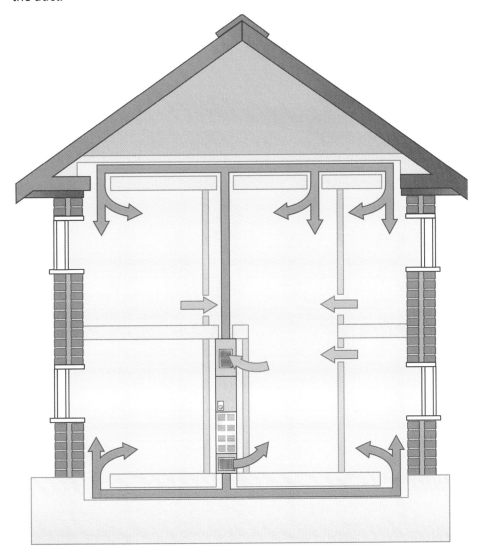

Figure 12.85 Warm air heaters

Plenums

Plenums are junction boxes that allow the air to be distributed in several directions and, as the unit sits on top of it, are very strongly constructed. The jointing of the ducting should be mechanical in the form of self-tapping screws or riveted and sealed with ducting tape. Only use the type of ducting tape advised by the manufacturer as this will ensure it is capable of withstanding the range of temperatures it will be subjected to during its operation.

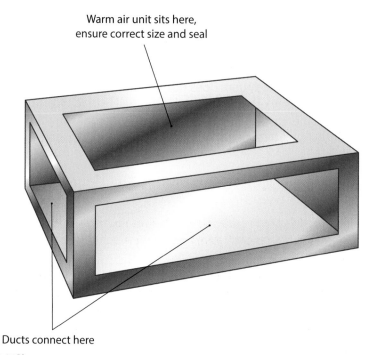

Warm air unit sits here, ensure correct size and seal

Ducts connect here

Figure 12.86 Plenum

Plenum adaptors

When an existing plenum is to be reused, any adaptor plates or fittings should be those either supplied or recommended by the heater manufacturer. As with the plenum itself, all joints to the plenum adaptor should also be fixed by mechanical means and sealed with a suitable self-adhesive ducting tape.

On completion of the installation, you should carry out a spillage test in accordance with the manufacturer's instructions. Take extra care when the rear of the appliance is not visible to ensure a complete seal has been achieved.

Additional information is given in the Heating and Ventilation Contractors' Association (HVCA) document, DW/144: Specification for Sheet Metal Ductwork.

Room temperature thermostats or thermistor thermostats

Only room temperature thermostats or thermistor thermostats which are compatible with the heater control system should be used.

The room temperature thermostat or thermistor thermostat has to control the whole warm air system from the temperature it senses, and therefore selection of the optimum location is most important to ensure maximum performance.

The preferred location is a room or space in which the temperature has the most significant effect on comfort. The room temperature thermostat or thermistor thermostat should be mounted:

- at a height of 1,200 mm to 1,500 mm above floor level
- on an inside wall
- away from local heat sources and direct sunlight.

For a building of more than two storeys, a room temperature thermostat or thermistor thermostat should be mounted between 1,200 mm and 1,500 mm above floor level.

Installing warm air heaters

For installations in new buildings of two or more storeys, refer to the applicable building regulations. Reference should also be made to BS 5588-1, which recommends that:

- transfer grilles in a protected stairway should not be fitted in any wall, floor or ceiling
- all ductwork passing through an enclosure to a protected stairway should be fitted so that all joints between the duct and the structure are fire-stopped
- where ductwork is used to carry warm air into the protected stairway through the enclosure of the protected stairway, the return air from the protected stairway should be ducted back to the heater
- warm air and transfer grilles or registers should be no more than 450 mm above floor level to prevent fire spread.

BS 5440-2 states that no air vents should penetrate a protected stairway or communicate with a bedroom or a room containing a bath or shower.

The above four recommendations do not apply to dwellings located within new buildings of two or more storeys which have their own access at ground or first floor level, or dwellings with a floor level not more than 4.5 m above ground or access level. Also exempt are existing dwellings in which an existing heater is to be replaced with a new heater of the same output and without significant alteration and/or extension to the duct system. In such instances, you should consult the manufacturer's instructions.

Location considerations

Compartments

If a heater is to be installed in a **compartment**, it needs to be installed according to manufacturer's instructions. The compartment has to:

- be fixed to a rigid structure – the internal surfaces have to be insulated from heat to the temperature specified in the manufacturer's installation instructions
- have air vents for compartment cooling, correct combustion and operation of the chimney, in accordance with BS 5440

Key term

Compartment
A purpose-built space designed specifically to house an appliance or equipment.

- have access for inspection and servicing of the heater and any ancillary equipment should be provided
- be fitted with a door that allows the withdrawal of the unit from the rest of the system
- have a notice fixed in a prominent position to warn against use as a storage cupboard
- have a return air grille(s) (for open-flued heaters) connected by the ductwork straight on to the return air inlet of the heater.

Where the compartment houses an open-flued heater, the door must not communicate with a room containing a bath or shower as the high moisture levels will affect the operation of the appliance.

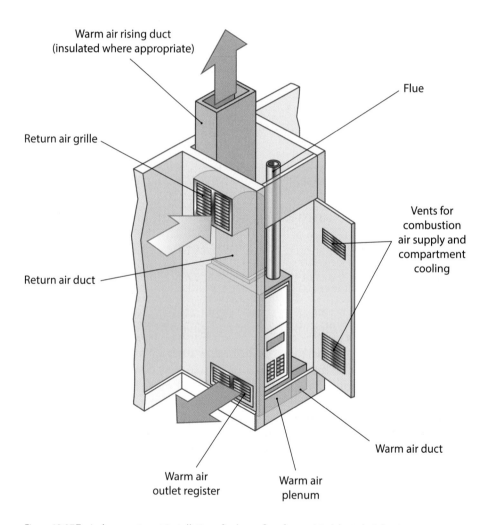

Figure 12.87 Typical compartment installation of a down-flow fan-assisted ducted-air heater

When a compartment has an open-flued heater fitted, the air vent or compartment door should not connect directly to a bedroom unless it is below 12.7 kW (14 kW gross) and the heater has a built-in safety device in case of POC build-up.

Its floor, ceiling and wall must be heat insulated as specified in the heater manufacturer's installation instructions.

When the air heater is an open-flued type, the draught diverter opening and air vents are located within the same heater space.

Where the heater manufacturer's installation requirements cannot be achieved, seek advice from the manufacturer. Where advice from the manufacturer is not available, any internal surface of the compartment which is of combustible material should be at least 75 mm from any part of the heater. Ways of determining whether a material is combustible or non-combustible are given in BS 476-22.

Airing cupboards

If an airing cupboard is to be adapted to house a heater, the heater should only be installed if the airing cupboard conforms to the same requirements as compartments, given above. When the flue pipe passes through the airing space, the contents of the space should be separated from the flue pipe by a guard comprising, for example, expanded metal or rigid wire mesh.

BS 5440-2 requires that any airing space is separated from a gas appliance by a fire-resistant partition and that, where such a partition is formed from perforated material, the major dimension of the apertures does not exceed 13 mm otherwise a fire risk could occur. For a double-walled flue pipe conforming to BS EN 1856-1, the level of insulation provided by the air gap between the internal and external walls may be sufficient to provide the necessary protection unless particularly combustible material is being stored.

For a single-walled flue pipe, a fire-resistant guard should be fitted around the flue pipe, with a minimum clearance of 25 mm between the pipe and the guard. The clearance between the flue guard and the compartment partition should be not less than 13 mm.

Under-stairs cupboards

If a heater has to be fitted under the stairs in a cupboard rather than a special compartment then different rules will have to be applied. Room-sealed heaters are the preferred type for this kind of installation.

If the building is no more than two storeys high the cupboard should follow the same rule as with compartments.

If the property is more than two storeys high then the cupboard will have to be fire resistant for a total of 30 minutes with a vent direct to the outside.

Any property greater than three storeys high will require approval from the local building control office and may well require intumescent type air vents to prevent the spread of fire.

Slot-fit

Only heaters specifically designed for slot-fit applications should be installed in this manner and they shall conform to the manufacturer's instructions. Figure 12.88 shows a typical slot-fit installation.

Roof space

Heaters can also be installed in the roof space to give extra room in the main part of the building or if there is no suitable space available.

The installation will require adequate support to be in place for the heater, ducts, walkways and working areas around the appliance to prevent engineers from falling through the ceiling when they are accessing it.

A 12 mm non-combustible sheet should be placed under the heater to reduce the risk of fire as, like most gas-burning appliances, it could possibly reach combustion temperatures. A guard should be fitted to prevent stored materials from being able to come in contact with the heater. Also be aware that if fitted in the loft any water heater attachments need a sufficient head of pressure to ensure the circulator can work.

To enable easy access for the engineer, a ladder access should be available (usually of the retractable roof type) and safety guard rails fitted to prevent accidents. Adequate lighting needs to be provided for the engineer so that they can see what they are doing (flash lights don't count).

A heater with integral circulator should only be installed when the circulation head of water from the circulator conforms with the manufacturer's instructions. This is in order to ensure adequate pressure levels in the system.

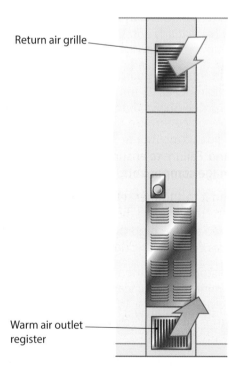

Return air grille

Warm air outlet register

Figure 12.88 Typical slot-fit installation

Commissioning warm air heaters

Pre-commissioning checks

Before the heater is commissioned, the installation should be inspected to ensure that the installation work has been carried out in accordance with the heater manufacturer's instructions (see table below).

Component	Check
Compartment construction	Heater 75mm away from combustible materials
Positive return	Must be positive ducted from outside
Return air grille position	Does the grill affect any other appliance?
Duct size	Is the duct correctly sized(use manufacturer's sizing charts)?
Ductwork	Ensure it is secure and correctly sealed
Draught diverter relief	Ensure draft diverter relief opening communicates directly with burner area
Heater flue pipes	Ensure flue is correctly installed and test to ensure it is working correctly
Heater alignment	Is heater correctly located to base and are no leaks visible between heater and plenum box?
Warm air outlets	Correctly installed and check direction vane adjustment
Temperature control	Position of room thermostat

Table 12.06 Installation checklist for warm air heater

The commissioning procedure

This commissioning procedure is as for a generic heater fitted with gas water circulator. Always ensure that you read and follow the manufacturer's instructions supplied with the product.

The wet part of the system must be filled and checked for leaks. To remove any air locks you will have to manually operate any motorised valves to the open position by a lever on the side of the motor. Then vent the system at any air vent positions you can find. Failure to ensure the heat exchanger and pipework are free of air will damage components.

Check the fan operates by setting it to summer setting so that the thermistor thermostat is not calling for heat. Then turn on the power to the unit – this should mean the fan operates without the main burner, so turns on. However, if it fails to operate, go through fan fault finding described in the section on 'Fault finding and diagnosis' (page 448).

The ducting system grilles and dampers should be adjusted to provide the required design velocity balance of the air flow. To do this you will need an air flow meter which is positioned in front of the grille and will give the velocity reading and can be compared with that in the design specifications.

Open any isolation valves on the system and the lock shield valve to allow water to flow into the heat exchanger. Set the summer air circulation switch to off position and ensure that the fan stops. If it fails to stop, check the fault finding section of the manufacturer's instructions.

Turn on the main burner and allow the water system to come up to temperature and then turn on the room thermostat and call for heat. This should activate the burner on the warm air unit and turn the fan on. The three-way valve should open to allow heating of the warm air.

Adjust the lockshield valve to give a 10 °C temperature differential between flow and the return. The heat exchanger is designed to have a flow temperature of 80 °C and a return temperature of 70 °C.

The system should warm up and then you should check that the return air temperature at the registers is around 40 °C cooler than the outlet temperature (i.e. the differential). To do this you will require a thermometer to measure the air temperature at the inlet and outlet of the unit.

Test the function of the thermistor thermostat and the programmer (if installed) by setting them to the off position. Check that the three-port valve moves over to the bypass position allowing the heat exchanger to cool down. The fan should switch off after approximately 3 minutes – the fan control unit turns the fan on when the water temperature reaches 70 °C and turns it off at 60 °C.

Turn the summer air circulation switch on with the system running, and ensure that the three-port valve changes over to the bypass position and that the fan continues to run. Set the thermistor thermostat and timer to the user's requirements and explain its operation to the customer.

Instructions for the user

Before leaving, ensure the user is aware of the following:

- the need for regular cleaning of the filter (once every month)
- how to adjust the heating system by opening and closing the system's air vents
- any warm air outlets in the rooms served by the heater must never be blocked
- that the return air inlet must never be blocked
- how to use the summer air circulation system.

Leave the instructions with the user for any future use by a service engineer.

Maintaining warm air heaters

Check that the installation complies with manufacturer's instructions. Turn on the appliance to ensure it is functioning correctly before carrying out any work on the appliance.

Maintenance should be carried out at least once a year. Check that the heat exchanger airways are free from obstructions. If necessary, clean with a vacuum cleaner from the air inlet end of the unit, taking care to not damage the airways.

Maintenance checks

- Check the overall condition of the strainer, cleaning as necessary.
- Check that the air filter is being regularly cleaned in accordance with the user's instructions using a vacuum cleaner.
- Check air passages within the heater are clean and unobstructed using a torch and visual inspection.

Safe working

Before working on the unit, always ensure that it is safely isolated from the electricity supply.

Safe working

Treat the elements of the heat exchanger very carefully – they are very fragile and can be easily broken if treated roughly.

- Check the burner bar and pilot assembly and remove from the unit to clean if required.
- Check the heat exchanger for any signs of damage and cracks, and replace anything as necessary.
- Check and clean the heat exchanger using purpose-made brushes and a vacuum cleaner.
- Turn on the main burner and check the gas pressure and gas rate.
- Check the flame picture against the manufacturer's instructions by looking through the site glass.
- Check the flue ways and look for signs of spillage – test if necessary using smoke pellets.
- Check that the FSD operates after around 60 secs of flame failure. Blow out the thermal couple and time how long before you hear the click of the thermo-electric valve closing down.
- Check the limit switch operation by creating a temporary overheat situation following the manufacturer's instructions.
- Check the operation of the fan and measure the temperature across the heater (450 °C to 550 °C) using a temperature probe.
- Check the thermostat operation by taking a set temperature and measuring the room temperature as it shuts off.
- Check the air speed to ensure no blockage is in the duct and the fan is not losing power.

Checking the heat exchanger of open-flued heaters

Visual inspection

To visually inspect a heater, you will need to first remove the casing. As always make sure the appliance is safely isolated from the electric supply before you begin work.

Remove access panelling, making sure it is not scratched, and carefully remove the fan assembly, putting the fixing screws in a box to prevent loss. This will reveal the heat exchanger. Look for holes and cracks/splits that are usually found around the welded seams.

Then look at the connection to the flue system. Look for cracks, identify if the primary flue is still in good condition and identify any signs of rust or damage. Place a torch into the heat exchanger and look from outside to see if you can see any light shining through gaps or cracks. If everything is satisfactory, replace all the panels in reverse order to how you removed them and replace the seals.

After reassembly, check the gas rate, burner pressure and adjust to manufacturer's specifications if necessary. Have a good look at the flame picture. Turn on the warm air circulatory fan and see if it disturbs the flame. If it does, check the heater and plenum for damaged seals. Have a good look at draught diverters for leaks at the back of open-flued appliances and also the primary flue. Put any faults found onto an inspection report, inform the customer and rectify as required.

Finally, let the appliance run for 10 to 15 minutes to reach its normal operating temperature and check again for any flame disturbance – as the metal expands it can open up seals that can be closed in colder conditions.

Smoke method

A good method for checking the seals and heat exchanger on a warm air unit is to use a smoke pellet. Turn on the appliance and let it reach operating temperature. After 10 minutes, turn off the appliance. Shut down all the warm air outlets except the one nearest to the heater. Put a lighted smoke pellet into the combustion chamber away from the burners and turn on the fan – see if there are any traces of smoke from the warm air outlet. If there are any signs of smoke it could indicate a possible fault. Check the heat exchanger again. If nothing is found then the appliance is safe to use.

Spillage test (open-flue heater)

It is necessary to carry out a spillage test on an open-flue heater to ensure the POC can be safely removed.

The first task is to check for any equipment in the room or adjacent to it could interfere with the appliance, such as extractor fans or chimneys. These can pull the air down an open flue causing spillage.

Close all windows and doors and, if installed, turn on any extractor fans as this will cause a negative pull, but if the ventilation is adequate it should have no effect on the appliance.

The spillage test on an open-flued heater is the same as for most open-flued appliances. Turn on the appliance and allow the flue to warm, then light the smoke pellet and place it under the draught diverter hood. Check for spillage. If spillage is identified you will need to check the flue for blockages. Once complete, then test again this time with the air circulation fan running to see if this has any effect. If no problem is found, the testing is complete and all extraction can be turned off and the heater returned to normal operation.

If for some reason the draught diverter is too difficult to reach, an alternative method is to use a smoke pellet in the heat exchanger and check if any smoke comes out of the draught diverter. You should then do it again, as described, with the air fan running.

Fault finding and diagnosis

Any electrical check should only be carried out by a competent person. The first checks to do are earth continuity, resistance to earth and polarity. Any de-assembling of components that involve the removal of electric connections should be re-tested as above before bringing the heater back on.

Before carrying out any fault finding, you must first check that the programmer works, as this can be a source of the fault. To do this, turn

Safe working

If a room next to the appliance being tested has a doorway directly into the room, then any extraction equipment in that room should be turned on and the door left open to check it does not interfere with the heater.

on the power to the appliance and check that the settings on the programmer and thermostat are calling for heat. If the appliance attempts to fire, it is okay, so continue with the rest of the fault finding. All electrical components should be replaced not repaired, otherwise you will void the manufacturer's guarantee.

Fault	Possible problem	Solution
Heater not operating	Fan voltage at terminal as per manufacturer's instructions Fan fails to start Check water temperature – below 70 °C, above 70 °C	No voltage present Set 'summer air' switch to 'I' If fan starts and no voltage is present, replace electrical fan Replace fan Check water supply Adjust lock shield control to restrict flow
No voltage present	Faulty wiring Socket connection between terminals in fan compartment	Replace fan
Water valve not operating	Check thermistor thermostat is calling for heat Check mains electrical supply Check supply to electrical panel at supply terminals. Check 500mA fuse. Check time control (if fitted) is at an 'on' setting Check for 230 V at terminal No voltage present	Replace electrical panel Check for 230 V at terminal, no voltage present Replace time control (if fitted) Replace valve actuator or check shorting link between motor terminal
Fan running when no heat required	Check 'summer air' switch is set to 'O', and check water temperature If water temperature is below 60 °C	Replace fan control
Burner flame disturbance	Cracked heat exchanger and fan is blowing air into the burner/combustion chamber Plenum not aligned with the heater so blowing air into the compartment and causing pressure fluctuations at the burner	Replace heat exchanger Re-set heater
Air blowing in the wrong direction	Unit changed from an up flow unit and a down flow being put in by mistake POC coming from an air grille is usually caused by lack of ventilation to the burner	The correct unit needs to be fitted for the system installed Increase ventilation

Table 12.07 Fault finding, possible causes and solutions – warm air heaters

Progress check

1 Where a compartment houses an open-flued heater, the air vents and/or compartment door you should not do what?
2 All ductwork passing through an enclosure to a protected stairway should be so fitted that all joints between the duct and the boundary are what?
3 Having serviced a warm air heater, what should you explain to the customer before leaving?
4 How long should you leave an appliance to reach normal operating temperature when checking the heat exchanger's integrity using a smoke pellet?

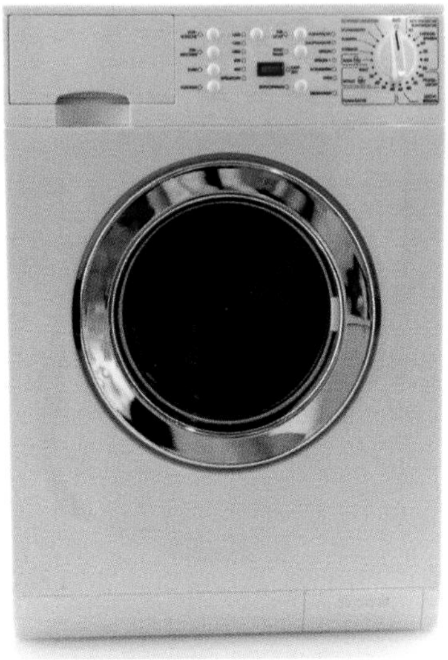

Figure 12.89 Gas tumble dryer

GAS-FIRED TUMBLE DRYERS

The gas-fired tumble dryer works in the same way as an electrical one except that it uses gas and has lower running costs.

A 240 V motor turns the drum using a belt system under a spring tensioner. The motor also powers a fan which drives the exhausted air through the dryer and pulls in induced airflow.

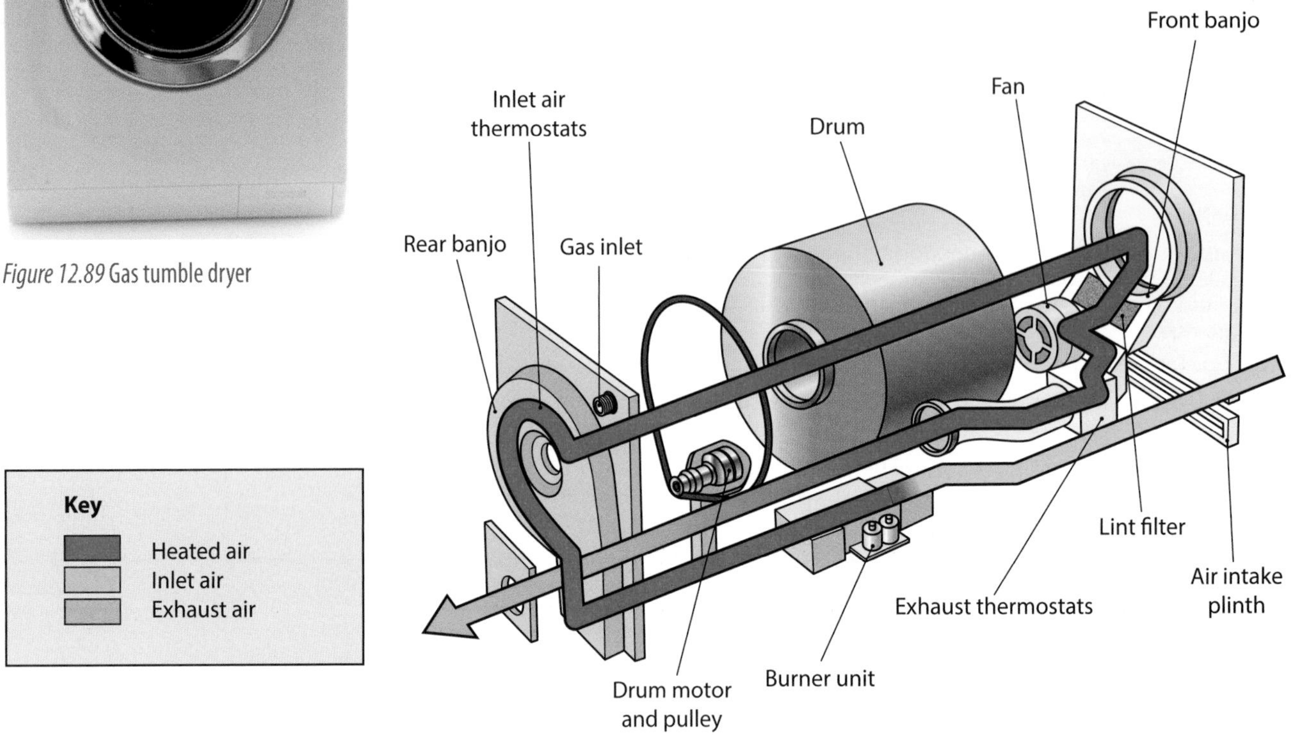
Key
Heated air
Inlet air
Exhaust air
Front banjo
Fan
Inlet air thermostats
Drum
Rear banjo
Gas inlet
Lint filter
Air intake plinth
Exhaust thermostats
Burner unit
Drum motor and pulley

Figure 12.90 Internal parts of a tumble dryer

The burner is controlled by 240 V twin solenoid arrangements that are controlled by two thermostats mounted at the rear banjo and two thermostats at the exhaust end ensuring full control.

The air is heated to 100 °C and enters the drum from the rear banjo into the perforated rear diaphragm on the drum. The exhaust air picks up the moisture from the clothing and passes out through the front banjo by the fan.

Installing a gas-fired tumble dryer

Pre-installation checks

Using the manufacturer's instructions check the following key points.

Position – the tumble dryer should be sited on an even floor able to take the weight of the appliance when loaded. Being sited near an outside wall to allow for venting to outside will also is useful. It should not be fitted in a bedroom or bedsit where the room volume is less than 7 m^3/kW. LPG appliances should not be installed in cellars because LPG is heavier than air.

Gas supply – the supply of gas is usually via a plug-in connector. The flexible connector should conform to BS 669 Part 1 1989 for natural gas and BS 3212 1991 for LPG. The gas point should be located 750 mm from the floor and 140–330 mm from the left side of the appliance.

Ventilation – although many tumble dryers have exhaust venting kits, they are still classed as flueless appliances and should conform to BS5440 Part 2 and communicate directly to the outside air.

Exhaust vents – all tumble dryers now need to be fitted with an exhaust kit: two versions are used. Those with max 3 kW (net) can use a vented hose through an openable window. Those above 3 kW can only be fitted with a permanent vent kit and the terminal must be at least 300 mm from the ground.

Case study

A gas tumble dryer was fitted with an outside vent and over time a plant had grown in the vent's pathway. This caused the exhaust gases to back up into the tumble dryer causing the appliance safety device to turn off the appliance. The engineer has to remove the plant to clear the path for the exhaust gases.

How have you resolved this type of problem in the long term in your own work?

Commissioning gas-fired tumble dryers

It is important that when commissioning you follow the manufacturer's instructions carefully.

Commissioning checklist

Be sure to take the following steps when commissioning tumble dryers:

- Check that the installation conforms to regulations.
- Test the system for gas tightness.
- Purge the system of air.
- Check that it is safe to connect to electrical equipment.
- Check that the ventilation is adequate to BS 5440-2.
- Check that the drum rotates/fan runs/burner lights correctly.
- Check that the gas pressure is to correct amount as stated in manufacturer's literature.
- Check that the fan does not affect other flues to BS 5440-2.
- Check that the outlet, if fitted, is clear of obstruction.
- Check when doors are opened, the tumble stops.
- Check the controls, timers, temperature operation.
- Explain how the filter needs cleaning to the customer.
- Explain the operation of the dryer to the customer.
- Leave the manufacturer's instructions with the customer.

Safe working

Children have been known to climb into tumble dryers if the door lock is faulty, which puts their safety at risk. This can be prevented if the lock is tested at each service.

Servicing a gas-fired tumble dryer

The servicing of tumble dryers need only be once a year but the filters will need cleaning at more regular intervals.

Servicing checklist

- Isolate the appliance from the electric supply.
- Isolate the appliance from the gas supply.
- Check the door seal is in good condition.
- Check the visible electrical connections for signs of damage.
- Check that the fan and inlet filter are clean and in working order.
- Check the burner for damage/corrosion and clean the assembly.
- Check that the vent hose is clear and fitted correctly.
- Check the drum belt for signs of wear and replace it if it is warn.
- Check that it is safe to reconnect the appliance to electrics.
- Reconnect the gas supply and test the installation/appliance for gas tightness.
- Check the appliance is level and stable.
- Check the drum rotation and gas ignites.
- Check and adjust the burner pressure at the gas valve.
- Check and operate the controls timer and thermostats.
- Check that all safety devices are functioning correctly.

Fault finding and diagnosis

Problem	Possible causes	Solutions/checks
No electrics to fan / drum motor	Check condition of electrical connections Check timer operation Check door switch Check 240 V to fan motor	Replace fan motor
Fan or drum failure	Check electric supply Check all switches and controls operate Check door properly shuts Check fan belt not broken or located wrongly	Replace as necessary
No ignition	Check electric supply Check gas supply	Reset appliance by turning the timer off and then on or by opening and shutting door Set timer for 30 min and heat selector low
Sparking but no ignition	Check gas supply Check solenoid valves are open Check inlet gas pressure Check condition and position of spark electrodes and probe Check polarity Check condition of electric connections	Replace control unit if necessary

Problem	Possible causes	Solutions/checks
Fan and drum turn but no ignition	Check the stat thermostats are operating Check high tension leads are correctly attached Check electrode position and gap Check probe is clean Check spark in correct place Check insulation in good condition Check wiring correct and connected Check centrifugal switch operation if installed	Replace as necessary
Gas supply	Check timer and thermostats are working correctly Check probe is clean and in correct position Check inlet gas pressure is not too high Check operation of door switch	Replace as necessary
Overheat	Check that the dryer has not been overloaded Check clearance dimensions Check timer and thermostats are working correctly Check drum filter is clear Check burner pressure and heat input is correct	Replace or adjust as necessary

Table 12.08 Fault finding, possible problems and solutions – tumble dryer

Progress check

1 A 240 V motor turns the drum using a belt system under a spring tensioner. What else does the motor power?

2 The air in the dryer is heated to what temperature?

3 Although many tumble dryers have exhaust venting kits, what are they still classed as?

4 Tumble dryers above 3 kW can only be fitted with a permanent vent kit and the terminal must be at least how far from the ground?

5 What should you check when a tumble dryer has overheated?

Knowledge check

What does the term 'range rated' mean?

a An appliance has a range of prices it can be purchased at

b An appliance can have its kW input adjusted

c How much your gas supplier your charges

d A range of appliances with different outputs

2 What is an air-pressure switch used for?

a To switch the air intake on in a boiler

b To blow up pressure vessels

c To stop the boiler firing if the fan has failed

d To identify holes in the flue

3 What is a duplex burner?

a A double-holed burner

b A double-holed injector

c Two separate burner units

d Two injectors separately operating two or more burner bars on a gas fire

4 Why is an ASD fitted to a fire?

a To prevent air build-up

b To turn off appliance if sufficient air supply is not available

c To ignite the main burner

d To prevent gas escape from the burner

5 What room should an open-flued fire never be fitted in?

a Living room

b Bedroom

c Dining room

d Bathroom

6 What is the common name for a self-sealing plug device?

a Bayonet fitting

b Safety lid

c Sprung gas valve

d Self-sealing valve

7 What are the two types of water heaters called?

a Duplex and simplex

b Boiler and heater

c Single point and multi-point

d Combi and boiler

8 What is a slow ignition device used for?

a To slowly open the gas to the burner.

b To ignite the gas slowly

c To heat up the water slowly

d To enable the water to boil

9 Before carrying out any fault finding on a warm air heater, what must you first check?

a That the customer is safely in another room

b That the programmer works

c That the electricity is on

d That the manufacturer is informed

10 The air in a gas tumble dryer is first heated to what temperature?

a 100 °C

b 110 °C

c 120 °C

d 80 °C

Liquefied petroleum gas

Safe working

In the event of a fire with an LPG installation, contact the fire brigade immediately. Warn them about the presence of LPG.

Safe working

Any LPG emergency control should have a label telling you who the supplier is and how to contact them in the event of an emergency.

INTRODUCTION

It is essential that anyone carrying out gas work is competent to do so and this includes working with liquefied petroleum gas (LPG), which is different from working with natural gas in some important respects, including its properties, characteristics, equipment and use.

Note: This chapter is an introduction to the basic principles of LPG. You will need a proper LPG qualification to carry out any work on these installations.

WHAT IS LPG?

Liquid Petroleum Gas (LPG) is a by-product of producing petrol/gasoline, as are many other liquids and gases. Fractional distillation is the process that separates the fossil fuels into all the liquids and gases that we are so familiar with in our daily lives. Some 40 per cent of LPG comes from the refining of crude oil and 60 per cent from field production.

There are two main sources of LPG:

- Oil/gas fields, where LPG is removed and condensed from natural gas – this process provides 60 per cent of the world's LPG supplies.
- Refineries where the crude oil is processed – this process provides 40 per cent of the world's LPG supplies. The quality and composition of the fuel varies depending on the processes involved and the type of crude oil used. Refineries use large quantities of LPG, especially butane, which is also added to gasoline to help the volatility of the explosion in the engine.

In domestic settings, the LPGs propane and butane are most commonly used.

Supplying LPG

Figure 13.01 shows the complete cycle of the LPG chain, from oil rig to end consumer. The multimillion pound cost of this production cycle determines the final cost to the end user. With worldwide, ever-increasing energy needs, LPG is constantly growing in demand, not only as a greener fossil fuel with fewer emissions, but also as an alternative to the more CO_2 heavy fossil fuels which may help to reduce energy poverty around the world.

LPG was first produced in 1910 by Dr Walter Snelling. Its first commercial use was in 1912 and it currently provides only about 3 per cent of the all energy consumed. It burns relatively cleanly, has low sulphur emissions and, because it's a gas, it does not pollute ground or water, but can cause air pollution.

THE PROPERTIES OF LPG

The first thing you must remember about LPG is just how dangerous the fuel can be, compared to natural gas. In order to work safely with LPG you must start with an understanding of the properties of LPG.

Generally we think of LPG as either propane or butane. In reality it is actually a blend of both with other fuels added in varying percentages. Many people use LPG without knowing the dangers of the fuel they are using. All gases have a different heat output, as shown in Figure 13.02, page 458. From this, you can see the LPG propane is over twice as hot as natural gas and butane even hotter. LPG is a much more volatile fuel than many people realise.

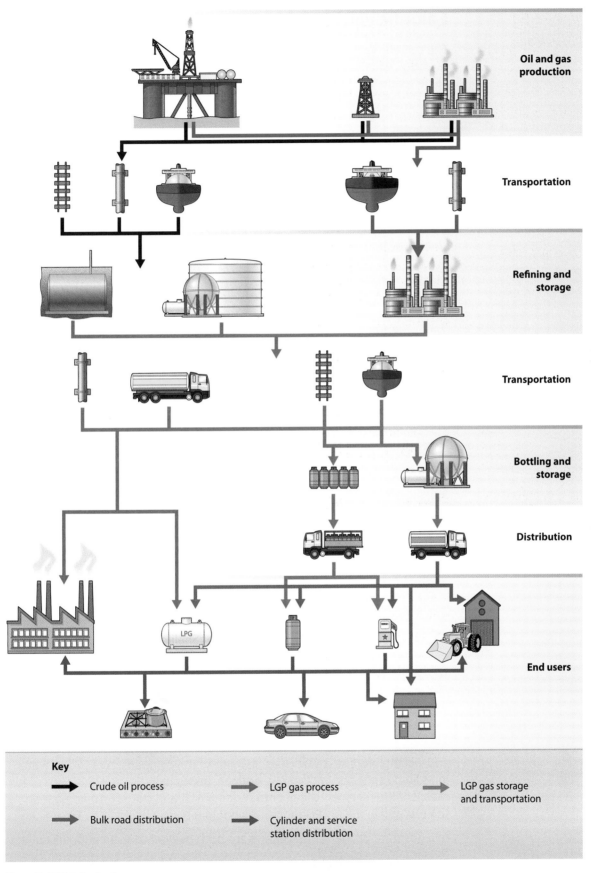

Figure 13.01 LPG distribution

Calorific values (heat energy)

The calorific value (CV) of a fuel is the amount of heat given when a unit quantity of fuel is burnt. For more information, see the section on 'Calorific value' (page 88).

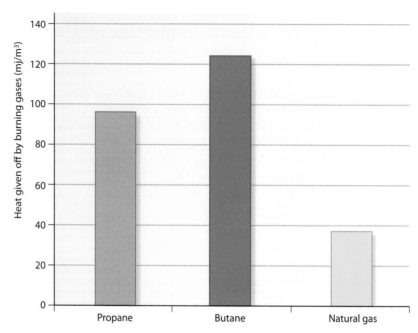

Figure 13.02 Heat output from gases

Property	Natural gas	Propane	Butane
Formula	CH_4	C_3H_8	C_4H_{10}
Calorific value gross (imperial)	1,040 Btu/ft^3	2,500 Btu/ft^3	3,200 Btu/ft^3
Calorific value gross (metric)	38.8 MJ/m^3	93.1 MJ/m^3	121.8 MJ/m^3

Table 13.01 Comparison of calorific value of each gas

Relative density

A fuel's relative density or specific gravity is a measure of its weight in relation to the weight of air or water.

As Table 13.02 shows, propane and butane are both much heavier than air and will therefore sink. The chart also shows the comparison of LPG to water in its liquid state. As you can see, both are about half the weight of water; this means, if the fuel was to leak onto water in its liquid state, it would float on the top, and in the right conditions it could remain a liquid.

In a gas state propane and butane are both heavier than air and will collect in voids. In buildings this will often mean they will collect in basements or ground floors. Propane is one and a half times heavier than air and butane is twice as heavy.

Example

If the relative density of air is 1 and natural gas is 0.6, then natural gas is nearly half the weight of air. This means that in the atmosphere natural gas will rise.

Property	Natural gas	Propane	Butane
Relative density as gas (air = 1)	0.6	1.5	2
Relative density as liquid (water = 1)	N/A	Approx. 0.512	Approx. 0.575

Table 13.02 Relative densities of gases

Boiling point

Boiling point is the temperature at which a fuel boils at atmospheric pressure. Below this temperature, the fuel will exist as a liquid and above, it will vaporise. Table 13.03 shows why only propane can be used for winter applications, not butane, because if the temperature drops below −2 °C, butane will simply not vaporise or 'gas'.

	Water	Natural gas	Propane	Butane
Boiling point	100 °C	−162 °C	−42 °C	−2 °C

Table 13.03 Boiling points

Storage pressure

Natural gas is stored at or just below its boiling point at atmospheric pressure in refrigerated storage. LPG is stored at a moderate pressure at ambient temperature (see Table 13.04).

	Natural gas	Propane	Butane
Storage pressure	Atmospheric	7 bar	2 bar

Table 13.04 Storage pressures

Flammability limits

A **BLEVE** is an explosion of a container of pressurised liquid as a result of the liquid's temperature having exceeded its boiling point. It is obviously crucial that BLEVE situations are avoided at all costs and understanding a gas's flammability limits will help you avoid such disasters.

Key term

BLEVE
Boiling **L**iquid **E**xpanding **V**apour **E**xplosion

	Natural gas	Propane	Butane
Flammability limits (% of gas in air)	5 – 15%	1.7 – 10.9%	1.4 – 9.3%

Table 13.05 Flammability limits

Table 13.05 represents the percentage of 'gas in air' needed for these fuels to burn. Imagine a room full of air and a gas is introduced, then we still have 'gas in air' whereas a room full of gas with air being introduced is now 'air in gas' so when the mixture reaches the flammability limits in Table 13.05 this is when an explosion or BLEVE could occur.

Any percentage below these limits and there is not enough fuel to create a combustible mixture; any percentage above these limits and there is too much fuel and not enough oxygen (from the air) to burn it. As you can see, the percentage for LPG is much lower than for natural gas, so the chances of there being explosive mixtures of LPG in air are much higher than with natural gas. Given that this fuel is often used in much smaller rooms and also sinks to the floor, making it harder to detect, you need to be extra careful when working with it and make users aware of the additional dangers of using it.

Ignition temperatures

From Figure 13.03 you can see that LPG is easier to ignite than natural gas as it ignites at lower temperatures. It also needs much less storage room (1.8 per cent), and it is over twice as hot if an explosion occurs. To put this into perspective, a 3,000 gallon LPG bulk tank BLEVE, would leave people standing 1 km away with second- or third-degree burns.

	Natural gas	Propane	Butane
Ignition temperature	650 – 704 °C	460 – 580 °C	410 – 550 °C

Table 13.06 Ignition temperatures

With lit cigarettes, the tip is roughly around 500 °C, so it would not light an explosive mixture of natural gas, but could easily ignite LPG within its flammability limits.

Expansion ratio

The main reason for storing these fuels in their liquid state is to take up less space than would be needed to store them as gas vapour. Table 13.07 shows the expansion ratios for all three fuels. This is based on liquid being stored in litres, to the equivalent in gas vapour. For every litre of liquid propane there are 274 litres of gas vapour. If a football pitch had a 13 kg cylinder on its base line and it exploded, the reaction would nearly reach the centre line.

	Natural gas	Propane	Butane
Expansion ratio (liquid to vapour)	Approx. x 600	x 274	x 230

Table 13.07 Expansion ratios

Figure 13.03 A gas explosion from a 47 kg cylinder of LPG can have fatal consequences

Combustion properties

These are the carbon and hydrogen molecules for these gases.

Figure 13.04 shows the chemical symbols for each fuel, containing the inner carbon atoms and the outer hydrogen atoms, simply expressed by their chemical symbols.

Just by seeing the difference here, we can appreciate the weight of the propane and butane molecules compared to the natural gas molecule. This is why they sink to the floor, due to their extra weight at the molecular level. In order to burn these fuels we need to add a source of ignition and the correct amount of oxygen.

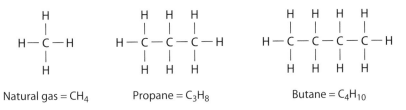

Natural gas = CH_4 Propane = C_3H_8 Butane = C_4H_{10}

Figure 13.04 Carbon and hydrogen molecules

Progress check

1 Who do you need to be registered with, to work on LPG?

2 Which fuel – propane, butane or natural gas – has the highest calorific value?

3 Which fuel – propane, butane or natural gas – is the heaviest?

4 What is the boiling point of propane?

5 What is the ignition point of propane?

Air:gas ratio

	Natural gas	Propane	Butane
Air: gas ratio	$10\,m^3 : 1\,m^3$	$24\,m^3 : 1\,m^3$	$30\,m^3 : 1\,m^3$
Oxygen: gas ratio	$2\,m^3 : 1\,m^3$	$5\,m^3 : 1\,m^3$	$6\,m^3 : 1\,m^3$

Table 13.08 Air:gas ratios

Considering the amount of oxygen in the atmosphere is approximately 21 per cent and nitrogen 79 per cent, we need more air than pure oxygen to burn these fuels correctly. Nitrogen is classed as an inert fuel which does not burn, so for most gas-burning appliances the nitrogen will just pass through the combustion process and be released into the atmosphere.

Due to the large volumes of air required to burn LPG, meeting the correct ventilation requirements is essential for safe combustion, especially in small areas such as caravans, boats and small rooms where cooking or heating via LPG. This is similar to the risk with natural gas.

Combustion equation for LPG

Propane
$$C_3H_8 + 5O_2 = 3CO_2 + 4H_2O + Heat$$

Butane
$$2C_4H_{10} + 13O_2 = 8CO_2 + 10H_2O + Heat$$

The equations show the ideal mixture for combustion; however as with natural gas, if the process is interfered with, you end up with products of combustion producing carbon monoxide due to the lack of oxygen. Due to the particular properties of LPG, the effects of CO poisoning will be increased with LPG than with natural gas. Both LPG and natural gas produce the same products of combustion – carbon dioxide and water vapour – but there are just more of them with LPG. Even if the combustion process is not interfered with, there is still a risk of asphyxiation due to the oxygen being used up quickly in a small area without the correct ventilation. See the section on 'Ventilation requirements' (page 469).

See Chapter 4 for more information on CO poisoning.

Vapour pressure and off-take rates

It is important to know how much gas is needed from a cylinder; this is calculated by the size of the cylinder and its **off-take rate**. The larger the cylinder or vessel, the more gas we can actually take from it – see the section on 'Off-take calculation and cylinder sizing' (page 464).

The difference between the vapour pressure of propane and butane is due to the boiling point of each gas; propane will exert more vapour pressure than butane because of its lower boiling point. The liquid inside the cylinder is already boiling at temperatures above −42 °C for propane and −2 °C for butane. The **ullage** space is left in all cylinders regardless of their size to allow for this expansion. Even the UK sun levels can cause considerable fluctuation in the ullage space in the height of the summer.

Off-take and vapour pressure affects the amount of evaporation – i.e. the process of molecules leaving a condensed liquid state and going into a gaseous state. The liquid state arises because molecules attract each other, making the condensed state lower in energy than the gaseous state. To form the condensed state from the gaseous state, the molecules need to give off energy in the form of heat. To form the gaseous state from the condensed state, the reverse must occur, that is, the system must absorb heat.

This is why evaporation has a cooling effect. The reason that different liquids evaporate differently is that the attractions between molecules are different for different substances.

The volatility of these fuels is measured by the vapour pressure. Inside the enclosed cylinder the fuel is in liquid state and, at its boiling point, this pressure will be the same as atmospheric pressure. However, as propane has a much lower boiling point than butane, it will exert more pressure under the same conditions. The trapped pressure inside a cylinder is referred to as 'saturated vapour pressure'. When no gas is being used the liquid will be still; it's not until you start to take gas off the cylinder that the liquid will start to boil.

Vapour pressure curve

When liquefied propane or butane turns to vapour it does so within a range of temperatures and pressures. Figure 13.05 shows this range of pressures. Designers of large underground LPG supply networks have to be aware of the possibility of vapour returning to liquid form inside the service supplies in winter; if the temperature is just right this is a possibility and could cause the release of liquid instead of vapour. In reality this rarely happens. Propane has a much high vapour curve than butane and can withstand more extreme temperatures; butane on the other hand has a wider vapour pressure curve and will continue to gas, but is very limited by temperature (see Table 13.03).

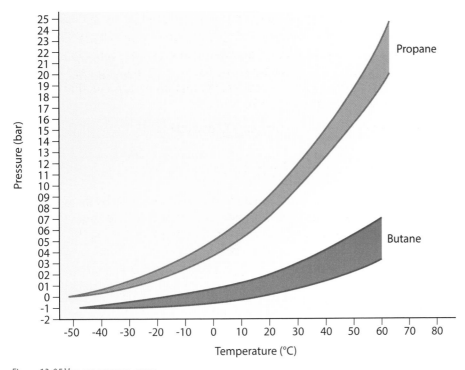

Figure 13.05 Vapour pressure curve

Regulators

As gas is continually drawn, the latent heat needed to convert the liquid to vapour is actually taken from the liquid itself, causing the temperature to fall even more. The more gas used, the more the liquid boils and creates further vapour, and the temperature falls even further. The temperature on the cylinder wall then falls below atmospheric temperature and, in the case of butane, condensation starts forming as a result. If this process carries on and you attempt to take more off the cylinder than it can supply, then auto-refrigeration takes place and, in the case of propane, frost forms on the cylinder. This temperature difference causes heat to be transferred from the atmosphere to the cylinder which then supplies the required latent heat.

Figure 13.06 LPG heat cylinder

Figure 13.07 Standard propane pre-set regulator

Installing regulators is a legal requirement for *all* LPG installations. The system design for propane will differ from butane, but due to the pressure differences, propane systems can often use butane and be multipurpose, but systems designed for butane must remain solely for that application.

There are many types of regulators for many different applications, domestic regulators have a 10-year recommended life span and the original date is stamped on the regulator. They should be replaced after this time period.

Off-take calculation and cylinder sizing

There are three main considerations when sizing gas cylinders:

▪ the maximum gas rate of the appliances
▪ if there is more than one appliance, the appropriate diversity factor
▪ the type of use/appliance.

Diversity

Diversity is a characteristic of the variety of gas loads whereby individual maximum demands usually occur at different times. Therefore, the maximum coincident load of a group of individual loads is less than the sum of the individual maximum loads. Diversity among customers' loads results in a diversity among the loads of distribution mains and regulators as well as between entire systems.

To determine the correct number and size of cylinders:

1 Calculate the total gross heat input rate of all the appliances in kW (see Chapter 4).
2 To allow for diversity and heating efficiency, multiply the total appliance gas rate by a 'coincidence' factor (typically 0.7).
3 Use Table 13.09 (on page 465) to determine the appropriate number and size of cylinders.
4 If more than one cylinder is required (for either off-take or total storage reasons), divide the total appliance input by the off-take rate (kW or kg/hr as long as units are consistent) to determine the number required.

It is advisable to provide an equal number of additional reserve cylinders, to ensure that adequate storage is provided.

Appliance	Gas rate range				Cylinders to supply demand						
	Butane		Propane		Butane			Propane			
	m³/h	kg/h	m³/h	kg/h	4.5 kg	7 kg	14.5 kg	3.9 kg	13 kg	19 kg	47 kg
Cooker (large domestic)	0.57–0.71	1.39–1.74	0.74–0.93	1.37–1.74	2.92	2.50	1.75	2.31	1.15	0.88	0.49
Cooker (normal domestic)	0.34–0.42	0.83–1.04	0.42–0.57	0.79–1.05	1.75	1.49	1.05	1.40	0.70	0.56	0.31
Cooker (caravan type)	0.23–0.25	0.56–0.63	0.28–0.31	0.53–0.58	1.00	0.91	0.63	0.77	0.38	0.31	0.17
Hotplate (2 burner)	0.11–0.14	0.28–0.35	0.14–0.20	0.26–0.37	0.83	0.72	0.50	0.70	0.35	0.28	0.15
Hotplate (3 burner)	0.14–0.20	0.35–0.49	0.20–0.25	0.37–0.47	1.16	1.01	0.70	0.90	0.45	0.36	0.20
Hotplate (2 burner and grill)	0.20–0.23	0.49–0.56	0.25–0.28	0.47–0.53	1.33	1.15	0.80	1.00	0.50	0.40	0.22
Boiling ring	0.06–0.08	0.14–0.21	0.08–0.11	0.16–0.21	0.50	0.43	0.30	0.40	0.20	0.16	0.09
Space heater, large (including boilers and air heaters)	0.14–0.23	0.35–0.56	0.20–0.28	0.37–0.53	1.33	1.15	0.80	1.00	0.50	0.40	0.22
Space heater, small	0.04–0.06	0.10–0.14	0.06–0.08	0.11–0.16	0.33	0.29	0.20	0.30	0.15	0.12	0.07
Lighting appliance	0.007–0.02	0.017–0.049	0.009–0.03	0.016–0.047	0.12	0.10	0.07	0.09	0.04	0.03	0.02
Sink storage water heater	0.06–0.08	0.14–0.21	0.08–0.11	0.16–0.21	0.5	0.43	0.30	0.40	0.20	0.16	0.09
Sink instantaneous water heater	0.31–0.37	0.77–0.90	0.40–0.48	0.74–0.90	2.00	1.85	1.30	1.70	0.85	0.68	0.38
Multipoint instantaneous water heater	0.68–0.91	1.67–2.23	0.88–1.19	1.64–2.22	5.35	4.58	3.20	4.20	2.10	1.68	0.93
Refrigerator	0.006–0.008	0.014–0.021	0.008–0.011	0.016–0.021	0.05	0.04	0.03	0.04	0.02	0.02	0.002
Iron	0.04–0.06	0.10–0.14	0.06–0.08	0.11–0.16	0.33	0.29	0.20	0.30	0.15	0.12	0.07

Note: This table is based upon the maximum gas rates and a 70 % diversity factor for cookers.

Table 13.09 Cylinder sizing

Cylinder size (kg)	Off-take when third full				
	(kg/hr)	**m³/hr**	**kw**	**ft³/hr**	**Btu/hr**
15	0.696	0.28	9.45	10	32 243
7	0.487	0.20	6.67	7	22 785
4.5	0.418	0.17	5.75	6	19 676

Table 13.10 Cylinder off-takes

Cylinder size (kg)	Off-take when third full				
	(kg/hr)	**m³/hr**	**kw**	**ft³/hr**	**Btu/hr**
47	2.373	1.27	33.51	44.84	114 336
19	1.139	0.71	18.74	25.07	63 941
13	1.054	0.57	15.04	20.10	51 316
6	0.777	0.42	11.08	14.80	37 804
3.9	0.527	0.28	7.39	9.90	25 215

Table 13.11 Propane cylinder off-take rates

Example

Suppose an installation consists of a caravan-style cooker, three lighting appliances, a refrigerator and a small space heater, then you could calculate the total kg/hr as set out in Table 13.12.

Appliance	Number of cylinders required			
	Butane (4.5 kg)	**Butane (14.5 kg)**	**Propane (3.9 kg)**	**Propane (13 kg)**
1 cooker, caravan type	1.00	0.63	0.77	0.38
3 lights	0.36	0.21	0.27	0.12
1 refrigerator	0.05	0.03	0.04	0.02
1 space heater, small	0.33	0.20	0.30	0.15
Total	1.74	1.07	1.38	0.67

Table 13.12 Total kg/hr required for example installation

Therefore, the number of cylinders required for normal use is either:

2 x 4.5 kg butane or 1 x 14.5 kg butane
2 x 3.9 kg propane or 1 x 13.0 kg propane

In choosing the type of cylinder you should take into consideration the likely usage and the time that the cylinder contents will last. For severe winter conditions, such as when caravanning in the winter, additional cylinders would be required. Alternatively, the number of cylinders required can be calculated by totalling the kg/h rates for each appliance and dividing this total by the recommended off-take for the size of cylinder chosen.

WORKING SAFELY WITH LPG

As you have seen, LPG can be dangerous if not treated correctly. There are a number of regulations which concern the way in which LPG is installed, the way the gas is used and how any accidents concerning LPG should be handled. As an engineer, you should be aware of these:

- The Highly Flammable Liquids and Liquefied Petroleum Gases Regulations 1972 (SI 1972/917)
- The Gas Safety (Installation and Use) Regulations
- Notification of Installations Handling Hazardous Substances Regulations 1982 (SI 1982/1357)
- The Control of Industrial Major Accident Hazard Regulations (CIMAH)1984 (SI 1984/1902)
- The Dangerous Substances (Notification and Marking of Sites) Regulations1990 (SI 1990/304)
- The Pressure Systems and Transportable Gas Containers Regulations1989 (SI 1989/2169)

You need to be familiar not only with the dangers of LPG, but also with the tightness testing procedure for domestic cylinder installations, the installation requirements, the commissioning and safe use/handling of LPG. Bulk tank installations are completely different to domestic and require a host of different qualifications in order to become competent.

LPG competency

According to the Gas Safety (Installation and Use) Regulations:

> No person shall carry out any LPG 'work' in relation to gas fitting, gas storage vessels or gas installations unless they are competent to do so.

You must also be registered with an approved Health and Safety Executive registration body i.e. the Gas Safe Register.

You must also hold a valid certificate of competence for *each* work activity you wish to undertake. The certificate must have been issued under the Nationally Accredited Certification Scheme (ACS) or Aligned Scottish/National Vocational Qualification. If you are thinking of becoming qualified in this area, you can qualify for LPG core and appliances only, if you need to, but the best way to approach this competency is to achieve the domestic natural gas core (CCN1) and appliances first, then convert to LPG. The benefit here is that any additional appliances you complete under the natural gas scheme will be automatically converted over to LPG.

Parts A and B of LPG core are covered by CONGLP1 (conversion from natural gas core) PD (Permanent Dwellings) + LAV/RPH/B (Leisure Accommodation Vehicles/Residential Park Homes/Boats) and these competencies are what a natural gas engineer requires to work on domestic LPG. You will need a special qualification to work on caravans only or residential park homes only.

Figure 13.08 Gas safe register symbol

There are also several codes of practice (COP) covering LPG activities. The list of relevant COP available occasionally changes, but the most common are:

- COP22 LPG Piping System Design and Installation 2002
- COP24 The Use of LPG Cylinders
- Part 1: The Use of Propane in Cylinders at Residential Premises 2006

The LPG industry covers a lot of legislative ground, but for installers who need to focus on domestic installations only, the COP and the British Standards you need to be fully aware of comprise a much shorter list, but it's still a considerable learning curve. LPG is a specialised trade and, due to the dangers of the fuel, should be treated with as much underpinning knowledge as possible to ensure the safety of all concerned.

Ventilation requirements

Ventilation is an important part of any fuel-burning appliance, but with LPG there are additional problems of where this fuel is used. As mentioned earlier the size of room is of great importance and LPG is used widely in small spaces such as in caravans and boats.

With the exception of permanent dwellings, before any assessment of ventilation requirements can be made you will need to establish the year of manufacture for the home/vehicle.

Caravans and residential park homes do have permanent ventilation, but they require particular attention due to the small room size and the need for a gas:air ratio of only 2 per cent to be within the explosive limits. The problem with a boat of course, is there cannot be low level ventilation as the vents would be below water and would compromise the buoyancy of the vessel.

Case study

In a North Wales caravan park an elderly person attempted to use a blow torch to defrost a gas cylinder to make a cup of tea, because he thought the supply had frozen. However, this caused the cylinder to explode and he ended up in hospital with burns and smoke inhalation. Luckily he survived the ordeal.

This was a butane cylinder so in winter it will not gas below –2 °C, therefore the occupier was attempting to heat the cylinder to make it gas. Due to the excess heat, this small cylinder of butane would still create a big enough vapour cloud to do considerable damage to the caravan. The occupier was not in the direct vicinity of the explosion; otherwise the outcome could have been fatal.

What would the correct procedure have been for using LPG in wintry conditions?

Ventilation calculations for PD/LAV/RPH/B

Due to the updates in the British Standards, gas engineers need to first establish the date of manufacture in order to apply the correct ventilation calculation or chart reference for any high/low level ventilation.

From 1977 to February 1999 the ventilation requirements were taken from BS 5482-2, therefore all caravan manufacturers complied with the relevant ventilation standards. After this date the ventilation requirements for manufacturers changed to BS EN 721:1999. A certification scheme run by the National Caravan Council (NCC) covers most caravans sold in the UK and there is a data badge usually displayed in a prominent position near the entrance to the caravan which should include the date of manufacture on it. For caravans compliant with the 1977–99 BS 4989, an orange on white background NCC badge should be displayed. For caravans compliant with current standards (i.e. BS EN 721, from 1999 to current day) a green on white background NCC badge should be displayed.

Prior to February 1999, the ventilation calculation was based on the maximum gross heat input rating of all appliances. The following formula was applied:

V= mm^2 (2200 x U) + (440 x F) + (650 x P)

V = the total ventilation requirements to be divided equally between high and low level; it must never be lower than 4,000 mm^2

U = the total kW input (gross) of all flueless appliances including cookers

F = the total kW input (gross) of all open flued appliances

P = the number of people the caravan is designed for

Example

Take the following values and apply them to the equation above to work out the ventilation requirements.

U = open-flued water heater with a maximum heat input of 12.5 kW (gross), plus a caravan-type space heater with a maximum heat input of 3.0 kW (gross)
F = a flueless cooker with a maximum heat input of 10.5 kW (gross)
P = the caravan is designed for four people

V = (2,200 x 10.5) + (440 x [12.5 + 3.0]) + (650 x 4)
V = 23,100 + 6,820 + 2,600
V = 32, 520 mm^2 or 325 cm^2

For an equal distribution between high and low

$\dfrac{325 \text{ cm}^2}{2}$ = 162.6 cm^2 at each level.

After February 1999 BS EN 721 applies. Table 13.13 can be used to determine the minimum ventilation requirement in relation to principle habitation compartments. BS EN 721 states that the high-level ventilation can be provided through a roof vent which has been designed for that purpose. The location of any high-level ventilation should be at least 1.8 m above floor level and not less than 300 mm above the upper surface of the highest sleeping berth. Low-level ventilation should be located not more than 100 mm above floor level and, as with any ventilation, it must be free from obstruction and not be the closable hit and miss type (i.e. the ability to close off the air supply fully, instead of it being fixed permanently open).

Table 13.13 can be used to simply size any LAV according to its floor area in m^2, and then simply apply the size requirements in mm^2 for high and low level.

Total plan area of leisure accommodation vehicles (m²)	Minimum size of high-level ventilation (roof vents) (mm²)	Minimum size of low-level ventilation (mm²)
Up to 5	7,500	1,000
Over 5 and up to 10	10,000	1,500
Over 10 and up to 15	12,500	2,000
Over 15 and up to 20	15,000	3,000
Over 20	20,000	5,000

Table 13.13 Minimum ventilation requirements

LPG in contact with other materials

Commercial LPG has no effect on metal, but in contact with aluminium the metal corrodes and oxidises, so any use of aluminium is kept to the vapour system equipment and components only on domestic cylinders. Any traces of caustic soda, chlorides or moisture can also have a strong chemical reaction. Natural rubber can have a strong effect when in contact with LPG as the gas causes the rubber to become spongy. This is why the rubber tubing used on an LPG manometer for tightness testing has to be made from neoprene not natural rubber. All other rubber products used in LPG fittings, such as valve seals, hoses, gaskets, do not use natural rubber for this same reason.

The solvent action from traditional gas jointing compounds can also have an effect, so it is important that only compounds which are impervious to LPG or PTFE tape are used. Some plastics can become either soft or brittle when in contact with LPG, this can create some plasticiser to be introduced into the LPG system and cause problems with regulators or control valves.

Safe storage

LPG is always thought of as being stored in cylinders. However, there are many other storage options available from bulk vessels to motor vehicles and commercial usage.

Bulk vessel storage

If you are thinking of working on or installing bulk tanks, then not only will you need domestic LPG competencies, but you will need a host of commercial qualifications too, so only a basic understanding of these installations is required here. Bulk storage vessels also appear in some domestic installations, particularly in Wales where gas is not mains supplied.

Chapter 13: LIQUIFIED PETROLEUM GAS (LPG)

Maximum water capacity of any single vessel (litres)	Nominal LPG capacity (tonnes)	Maximum separation distances of above ground vessels (metres)		
		A	B	C
150 – 250	0.05 – 0.25	2.5	0.3	1.0
500 – 2500	0.25 – 1.1	3.0	1.5	1.0
2500 – 9000	1.1 – 1.40	7.5	4.0	1.0

A = From buildings, boundary, property line or any fixed source of ignition
B = As A but with a fire wall
C = Between vessels

Table 13.14 Bulk storage vessels

Bulk storage vessels can contain many thousands of kilograms of LPG, and are predominately white or green. They are used to store LPG for heating, cooking, and numerous commercial and industrial applications.

If the standard four 47 kg cylinders are not enough to meet the needs of a domestic property, then a bulk tank installation may be required. Customer tanks vary in size from small domestic to commercial (typically holding between 200 kg and 2,000 kg) to even larger industrial tanks (with capacities of 100 tonnes or more).

Bulk storage vessels should be sited:

- in the open and well ventilated
- to allow access for refill
- so the delivery driver has sight their vehicle while standing at the vessel during filling
- the minimum distance from buildings, boundaries, property lines and trees, according to the size of the vessel (see Table 13.14)
- a minimum of 1 m from a hedge used as a screen (on one side only).

One of the requirements of the Gas Safety Regulations is that a bulk storage vessel installation must be protected against both over- and under-pressure situations, for example the regulator fails or the vessel is allowed to become empty. As such, an Over Pressure Shut Off device (OPSO) and a Under Pressure Shut Off device (UPSO) must be fitted.

Engineers wanting to carry out this type of work have to adhere to these standards as well as be qualified for the installation of these vessels. The relevant standards and codes are:

- The storage of LPG at fixed installations
- Health and Safety Executive HS/G34
- LP Gas Association Code of Practice 1
- Part 1 – Design, installation and maintenance of bulk LPG storage at fixed installations
- Part 2 – Small bulk LPG installations for domestic purposes.

Some bulk tank installations will be metered for either single user or multiple users.

Cylinders

There are many different types, colours and varieties of cylinders. In the UK, you will be more familiar with red cylinders for propane and blue for butane, but do not take this for granted. The colour may not correspond and the language written on the cylinder may not be English, so unless the cylinder has been painted over after it has left the gas supplier, there should be a UN number painted onto each cylinder to help identify them. These are the United Nations numbers and are UN1978 for propane and UN1011 for butane.

The two most common LPG gases are known as Commercial Propane and Commercial Butane as defined in BS 4250.

Butane is predominately stored in blue cylinders of up to 15 kg and is generally used for leisure applications and mobile heaters. **Propane** is predominately stored in red cylinders of up to 47 kg.

Some general safety rules on cylinders are:

- Never store cylinders below ground or near to drains.
- Always store cylinders upright in a cool well ventilated outdoor area away from other flammable materials and naked flames, and secure from vandalism.
- Always handle cylinders with care.
- Only ever store and use propane cylinders outdoors.

Propane cylinders

Propane cylinders *must not* be:

- stored in any cellar, basement or sunken area
- installed less than 1 m measured in the horizontal plane from the nearest cylinder valve, or less than 300 mm measured vertically above the cylinder valve(s), from:
  — fixed sources of ignition
  — openable windows
  — unprotected electrical equipment
  — excessive heat sources
  — ignitable/combustible materials etc.
  — openings in houses, e.g. vent ducts, airbricks, flue terminals etc.
- installed within 3 m of any corrosive, toxic or oxidising materials, unless a fire-resistant barrier is placed between.

The Republic of Ireland (ROI) and Northern Ireland (NI) have slightly different requirements for propane cylinders. They must not be installed within 3 m (ROI) or 2 m (NI) measured in the horizontal plane, from untrapped drains or unsealed gullies, or openings to cellars, unless a gas dispersion wall not less than 250 mm high is provided in between.

Figure 13.09 Safe storage of gas cylinders

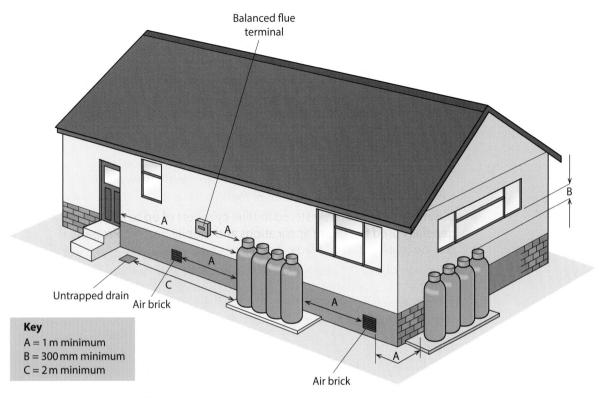

Balanced flue terminal

Untrapped drain
Air brick

Key
A = 1 m minimum
B = 300 mm minimum
C = 2 m minimum

Air brick

Figure 13.10 Multi-cylinder installations

LPG INSTALLATIONS IN DOMESTIC DWELLINGS

Multi-cylinder installations

Multi-cylinder installations are an ideal alternative means of supply to bulk tank installations where the site may be restricted in terms of the available space for gas storage or tanker access.

The cylinders are generally arranged in two sets – 'service' and 'reserve' – and linked by a pressure regulator or changeover valve. This valve automatically switches the supply from the service to the reserve cylinders when the gas cylinders are nearly empty without any noticeable interruption of the gas supply. This is a good system if the customer remembers to keep an eye on the changeover valve indicator to see if the service cylinders have emptied and are now using the reserve set. Once the service cylinders are empty, the user should contact their gas supplier to replace the empty service cylinders. The empty cylinders then become the reserve cylinders and so the cycle continues.

If the cylinder supply comprises four or more cylinders connected to an automatic changeover device, the Gas Safety (Installation and Use) Regulations state that an Over Pressure Protection Shut Off device (OPSO) must also be provided.

The Gas Safety Regulations also requires that, if there are two or more refillable cylinders connected by an automatic changeover valve, an Emergency Control Valve (ECV) must be fitted.

Figure 13.10 shows a common second stage regulator, complete with internal under pressure shut-off valve (UPSO), OPSO and a limited capacity pressure relief valve to comply with BS 3016. In the event of a supply failure, or an increase in pressure failure above the set value, the regulators will ensure the supply is safely shut down. The user is permitted to reset the UPSO, for example if the tank runs out and is refilled. The user may not re-set the system in the event of an over pressure situation as this would require a competent engineer.

Automatic changeover valves

These devices would normally be connected to two or four cylinders and have a regulator attached. The high-pressure hoses (also known as 'pigtails') must be checked for damage and date of manufacture. It is a recommendation only that these flexible hoses are changed every five years, not a necessity, but due to the dangers of LPG it would be good practice.

If the cylinder has emptied, the indicator on the top section will be red and pointing to the left. Once the consumer notices this they should then manually turn the indicator to the right cylinder and it will then change colour (or clear) to indicate the spare cylinder has fuel. They should then contact their gas supplier, who then changes the empty cylinder, and the whole process starts again.

Bulk tank installations

The position of the UPSO/OPSO will determine the service supply pressure to the premises. The most common, favoured by the larger gas suppliers, is to have the first stage regulator and the UPSO/OPSO on the bulk tank to provide a low pressure (37 mbar) service to the property. In other installations you may find the UPSO/OPSO at the property, which means

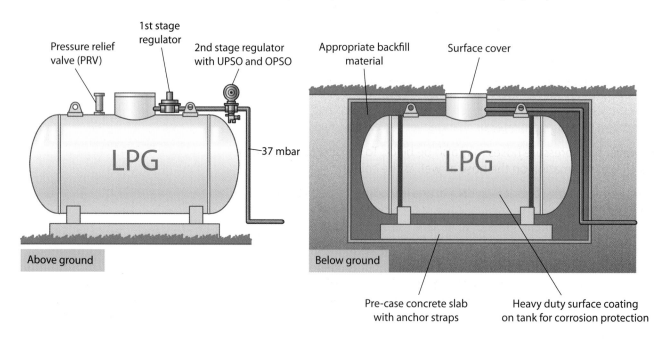

Figure 13.11 Installation of LPG units

that the service supply will be fed from the first stage regulator only and will be at around 1.5 bar. This is high pressure and can cause some major gas escapes should it fail. Built into these tanks are safety devices to help prevent BLEVEs and over filling; pressure relief valves are set to 14.5 bar on white tanks and 17.5 bar on green tanks, due to the extra heat absorption. Liquid level indicators (a manual device set at 87 per cent full) allow the delivery driver to ensure they do not overfill the tank, non-return valves for filling and emptying these tanks as well as a float valve to indicate how much fuel has been used, also the first and second stage regulators including a customer emergency control valve (ECV) for emergencies.

From mid-1994 these LPG installations must be installed and used in accordance with the Gas Safety (Installation and Use) Regulations, and the Pressurised Systems and Transportable Gas Containers Regulations 1989.

Typical settings of a second stage regulator are:

37 mbar, operating tolerance +/− 5 mbar

OPSO: set to close at 75 mbar, tolerance +/− 5 mbar

UPSO: set to close between 25 and 32 mbar

Relief valve: set to vent at 55 mbar, tolerance +/− 7 mbar.

You can find information and guidance on the safe use of LPG at small commercial and industrial bulk installations at www.hse.gov.uk/gas/lpg/safeuse.pdf

Hoses and connections

Having the correct connections on LPG hoses is essential for safe use. It is important that jubilee clips are of the type which does not cut through the hose and are designed for LPG. The crimping method is far superior and should also be designed for use with LPG.

Always check you have the correct colour of hose for the job at hand:

- black = low pressure
- orange = high pressure.

However high pressure hoses may also be black. To be sure, always double check the information on the hose as well as the colour.

The LPG hose looks identical to the natural gas cooker hose in Figure 13.15; the only difference is the red stripe along the side which indicates it's only suitable for LPG.

To identify which hose can be used for either high pressure or low pressure, check the numbers written on its side. This information will look something like: BS3212:1991/2/8.0, which indicates the BS number (BS3212:1991) followed by the hose type (/2) and the internal bore diameter (/8.0). This is then followed by the date of manufacture, to indicate the recommended changing interval.

Flexible connections must conform to:

- BS 3212 Type **1** (Low pressure)
- BS 3212 Type **2** (High pressure)

Figure 13.12 LPG hoses and connections

Figure 13.13 Gas micropoint and cooker hose – a red stripe will be needed on the side to indicate it can be used for LPG

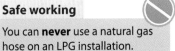

Safe working

You can **never** use a natural gas hose on an LPG installation.

Push-fit fittings

With the exception of fittings conforming to BS 669 installed in accordance with BS 6172, BS 6173 or BS 7624, quick-release fittings should not be used for gas installations. With the exception of fittings conforming to BS 7838 for use with corrugated stainless steel semi-rigid pipe, push-type fittings should not be used for gas installations.

Pipework materials

The following materials *are recommended* for general use for installation pipework:

- solid drawn copper tube used with copper or copper alloy fittings
- steel tube used with wrought iron, low-carbon steel or malleable cast iron fittings, or steel compression couplings and protected against corrosion, e.g. by galvanising and painting
- stainless steel tube and appropriate compression or screwed fittings.

The following materials are *not recommended* for general use for installation pipework for the reasons shown:

- aluminium (due to corrosion)
- composition and lead (due to creep – it is not to be used as per the Gas Safety Regulations)
- brass tubing (of a kind that is subject to season cracking)
- plastics (due to low melting point and turning brittle in low temperatures).

Pipe sizing

As with any supply of gas/liquid the supply pipe has to be of the correct size to accommodate the requirements of demand. LPG requirements are similar to those for natural gas, but their specific requirements need to be understood before you attempt any installations.

Length of tubing	Metric copper tube outside diameter									
(m)	6 mm		10 mm		15 mm		22 mm		28 mm	
	kW	m³/h	kW	m³/h	kW	m³/h	kW	m³/h	kW	m³/h
3	2.93	0.12	22.60	0.88	38.4	1.49	207	8.01	412	15.92
6	2.05	0.085	14.65	0.57	26.1	1.01	135	5.21	230	8.86
9	1.76	0.071	12.31	0.48	20.5	0.79	108	4.19	215	8.53
12	1.47	0.059	10.84	0.42	17.9	0.70	94	3.62	187	7.25
15	1.17	0.048	9.67	0.38	15.5	0.60	82	3.20	168	6.51
18	1.17	0.048	8.79	0.35	13.5	0.53	74	2.86	145	5.61
21	0.88	0.040	8.21	0.32	12.9	0.50	67	2.58	135	5.24
24	0.88	0.040	7.62	0.29	12.0	0.47	61	2.38	126	4.87
This chart is for copper tube only and is in accordance with BS EN 1057:1996.										

Table 13.15 Capacity of copper tube outside diameter

The heat input is based on propane at a low pressure of 37 mbar with 2.5 mbar maximum pressure drop over the length of the pipe (compared to 1 mbar for natural gas). Use Table 13.15 to ensure that when all appliances are running there will be enough LPG to supply them (provided, of course, that you also have the correct size cylinder to begin with).

Connecting cylinders

When connecting cylinders:

- ideally always stand cylinders upright on a firm level base
- never use a cylinder when it's tilted or on its side
- never change or connect cylinders near a source of ignition – i.e. cigarette, open fire, electric fire etc.
- always check that the male 'bullnose' connection on propane regulators or pipework is not damaged or dirty before making connection
- clip on butane connections, check that the sealing washer is in position and in good condition
- always use the correct spanner to tighten screwed connections (left-hand thread)
- **never** rely on finger-tight joints
- **never** check for leaks with a naked flame.

Where regulators are connected to cylinder valves using flexible hoses for high pressure regulation, the regulator inlet should be sited above the level of the cylinder valve outlet. This ensures that the hose rises continuously between the cylinder and the regulator connection.

Flexible hoses should not extend from one room to another or pass through walls, partitions, ceilings or floors.

For flexible hoses for low pressure application, these should be secured to nozzles using either the correct size metal crimped clips or worm drive hose clamps (at regular intervals).

As with natural gas, a harmless sulphur-based chemical (mercaptan) is added to LPG. But unlike natural gas, this isn't because LPG is odourless and difficult to detect. In its natural state, LPG is odourless, colourless and tasteless. Mercaptan, the same harmless chemical that is added to natural gas, contains sulphur, which makes it smell a bit like rotten eggs.

The general area of an LPG leak can often be detected by smell and its exact position determined by brushing leak detection fluid (LDF) over the suspected area. Always use special LPG-quality flexible hose, which conforms to British Standards, remembering:

- natural rubber and most plastics are not suitable for use with LPG
- to always check hoses for wear before making a connection
- to always use a hose clip for securing hoses to regulators and appliances (on ridged nozzles position the hose clip on the plain section).

You should never use gas appliances in locations without adequate ventilation. Gas appliances must have good ventilation to ensure that they are operate correctly. This will also prevent CO being produced by a lack of combustion air. Remember:

- always read and observe equipment manufacturers' instructions
- always use the correct type of pressure regulator

- never tamper with the pressure regulator
- always have a means of ignition ready before you turn on the gas
- if the flame goes out accidentally, never attempt to relight it until the escaped gas has been dispersed
- never apply external heat to a cylinder even if it is has frosted
- always keep the number of cylinders in a work area to the absolute minimum
- remove any empty cylinders to a secure storage area after use.

TIGHTNESS TESTING LPG SYSTEMS

Safe working

The importance of tightness testing on LPG cannot be taken lightly; this is a dangerous substance and must be 'gas tight' according to these test procedures. Do not use a match or lighter to look for a leak on *any* gas installation. Only the correct equipment and the correct test procedure for each type of LPG installation should be implemented.

You will be familiar with some of the following procedures as there are common themes with natural gas. However, there are some additional key procedures that are unique to LPG and these are explained below.

LPG is a dangerous substance and must be 'gas tight' in accordance with these test procedures. **Do not** use a naked flame to look for a leak on *any* gas installation. Only the correct equipment and the correct test procedure for each type of LPG installation must be implemented.

The Institution of Gas Engineer's Utilisation Procedures (IGE/UP) produce the tightness testing and purging procedures and these are covered in IGE/UP1/B edition 3. It applies to LPG/air mixture, natural gas and LPG installations, including meters and installation pipework. It also covers the air testing procedure when natural gas or LPG is not used or available to carry out the tightness test. The following limits apply:

- Maximum operating pressure (MOP) at the outlet of the ECV not exceeding 2 bar for natural gas and LPG
- a pipe's nominal bore no greater than 35 mm (DN32, R1¼)
- maximum meter capacity not exceeding 16 m^3/hr
- maximum installation volume (IV) of 0.035 m^3
- LPG/air mixture installations not to exceed 21 mbar at the outlet of the primary meter at any point in the section
- LPG installations not to exceed 37 mbar at the outlet of the primary meter or any point in the section.

Installations larger than 0.035 m^3 are rare in domestic premises, but need to be checked prior to testing. Propane and butane typically operate at 37 mbar and 28 mbar. LPG service pipework is not covered by IGE/UP/1B 3.

If the above limits are exceeded then, IGE/UP/1 Edition 2, IGE/UP/1A Edition 2, or IGEM/UP/1C will apply and these procedures are not covered in this introduction to LPG. As a responsible engineer you will be called upon to make professional judgments. You should not deviate or undertake tasks that are beyond your competency and expertise. Where MOPu exceeds 75 mbar the requirements for competency to do work are set at a higher level than for installations that do not exceed 75 mbar.

For any part of an installation upstream of either an MIV (if meter fitted) or final stage regulator (no meter) and downstream of an ECV with a maximum operating procedure (MOP) exceeding 75 mbar, the components should be:

- pre-assembled and strength tested by the manufacturer
- strength tested in accordance with IGE/UP/1.

LPG service pipework is not covered by IGE/UP/1B edition 3.

Strength testing is unlikely to be needed in domestic premises, and in general tightness testing is carried out to ensure that any installation has a leak rate below the level which could be considered a hazard caused by an escape, assuming there is adequate ventilation. On new installations the test is to verify that within the tolerances of the test equipment and the time to carry out the test there are no leaks, i.e. nominally zero leakage.

On existing installations where appliances are connected the test is to verify that the pipework is nominally gas tight and the appliances connected to that installation are within the acceptable limits (due to the age and wear). This is providing there is no smell of gas.

After any test, if you can smell gas or discover an indication of gas on a detector, this is unacceptable. Upon completion of any work carried out on an installation that may affect its gas tightness, you must test the installation for tightness again. Fuel gas or air should be used as a test medium, if a section of pipe contains fuel gas then it should be pressurised with fuel gas.

Where a new installation is connected, it is recommended that the installation is pressurised via the fuel gas supply.

Prior to testing, check to ensure there are no open ends on the installation and, if found, you must seal them with the appropriate fitting.

Perceptible Movement

A movement of 0.25 mbar or less on a fluid (water) gauge is considered to be 'not perceptible'. If the gauge moves, it can be inferred that the pressure within the installation has altered by more than 0.25 mbar. When using an electronic gauge that can register movement of less than 0.25 mbar the pass criteria of 'no perceptible movement' is considered to be a maximum of 0.25 mbar, except for gauges that read to one decimal place when it is considered a maximum of 0.2 mbar.

Testing for let by

The let by test determines that the main emergency control is not passing gas whilst in the off position. If the valve is letting by gas, then this could give a false reading, such as no pressure loss or a smaller pressure loss than actually exists.

The current standard set by the IGM for let by states the test is to be carried out at a pressure of between 7-10 mbar for a period of one minute. If, after this time, the gauge has risen by more than 0.25 mm the engineer must confirm 'let by'. If there is no perceptible movement of the gauge reading, then the valve has passed the test.

Before you begin the let by test procedure:

- For cylinder installations using high pressure hoses, a rise in pressure can occur. This could also be due to hose relaxation. Should this be the case, additional time must be allowed for temperature settlement before starting the let by test.

- An UPSO valve installed downstream of the supply control valve, will automatically close as the pressure downstream is reduced. To ensure correct pressure testing, this should be manually reset and allowed to re-shut, which will release any trapped upstream pressure.

Safe working

No-one other than an authorised operative working on behalf of the ESP should attempt to remove, repair or dismantle the valve.

Confirming let by

Let by can be confirmed by disconnecting the outlet union of the valve and applying leak detection fluid (LDF) to the valve barrel or ball of the emergency control valve (ECV).

If the ECV is letting by then you must make the installation safe by isolating and sealing the installation with an appropriate fitting. Immediately contact the relevant gas supplier or Emergency Service Provider (ESP). The installation must not be put into use until the ESP has rectified the fault.

Let by of a refillable cylinder valve

If let by is confirmed on a cylinder valve, and if it is safe to do so, the cylinder must be changed. The faulty cylinder valve must be sealed off with an appropriate fitting to prevent further escape. The gas supplier must be immediately notified of the fault so arrangements can be made for the collection of the cylinder and rectification of the fault.

Lock up of the regulator

Lock up means that the regulator is in a fully closed position. If this occurs, it can prevent the installation from being fully tested or from proven as gas tight from the outlet of the ECV to the inlet of the regulator. If lock up has occurred, it is possible for a small gas escape to exist on these connections which will not register as a loss of pressure on the manometer.

It is important to be aware that even when the correct procedure is followed, it is not always possible to be sure that lock up has not occurred. For this reason you must spray these connections with LDF (outlet of ECV and inlet connection to the regulator) even if the installation is deemed to have passed the test.

After completing the let by test (between 7-10 mbar), raise the pressure in the installation to the appropriate Tightness Test Pressure (TTP). The actual test pressure is dependent on the type of gas, the operating pressure being used and whether a regulator is fitted. Tables 13.16 and 13.17 show the operating pressures for this test.

Nominal operating pressure	Tightness test pressure
14 mbar	13 – 14 mbar
21 mbar	20 – 21 mbar

Table 13.16 Operating pressures for air test

Type of installation	Operating pressure	TTP	
		Propane	Butane
Installations with a regulator in the section to be tested	28 mbar		20 – 21 mbar
	30 mbar	28 – 29 mbar	28 – 29 mbar
	37 mbar	30 – 31 mbar	
Installations without a regulator in the section to be tested	28 mbar		27 – 28 mbar
	30 mbar	29 – 30 mbar	29 – 30 mbar
	37 mbar	36 – 37 mbar	

Table 13.17 Operating pressures for propane/butane test

To ensure the regulator is not locked up during the tightness test period then the following guidance should be followed:

- Avoid higher pressures to prevent regulator lock up.

- Operating pressures are found on touring caravans, designed for use on the road, where the gas installation has been installed in accordance with BS EN 1949 and the regulator complies with BS EN 12864 or BS EN 13786.

- For butane installations with NOP 28 mbar, if while raising the pressure to the required tightness test pressure it exceeds 21 mbar but not 23 mbar, re-adjust the pressure to between 20 and 21 mbar. If the pressure exceeds 23 mbar, drop the pressure back to between 7 and 10 mbar before raising the pressure to between 20 and 21 mbar.

- For butane/propane installations with NOP 30 mbar, if when raising the pressure to the required tightness test pressure it exceeds 29 mbar but not 31 mbar, re-adjust the pressure to between 28 and 29 mbar. If the pressure exceeds 31 mbar, drop the pressure back to between 7 and 10 mbar before raising the pressure to between 28 and 29 mbar.

- For propane installations with NOP 37 mbar, if while raising the pressure to the required tightness test pressure it exceeds 31 mbar but not 33 mbar, re-adjust the pressure to between 30 and 31 mbar. If the pressure exceeds 33 mbar, it is necessary to drop the pressure back to between 7 and 10 mbar before raising the pressure to between 30 and 31 mbar.

- For installations with NOP 14 mbar if while raising the pressure to the required tightness test pressure it exceeds 14 mbar but not 16 mbar, re-adjust the pressure to between 13 and 14 mbar. If the pressure exceeds 16 mbar, drop the pressure back to between 7 and 10 mbar before raising the pressure to between 13 and 14 mbar.

- For installations with NOP 21 mbar, if while raising the pressure to the required tightness test pressure it exceeds 21 mbar but not 23 mbar, re-adjust the pressure to between 20 and 21 mbar. If the pressure exceeds 23 mbar, drop the pressure back to between 7 and 10 mbar before raising the pressure to between 20 and 21 mbar.

Installations tested with air

This tightness test procedure is for an installation that is connected to a live LPG gas supply. However where the installation is not connected to a live LPG gas supply, the installation can be tested with air.

Where the LPG installation is new or an existing installation has been exposed to air, providing the initial tightness test of the installation has passed and after purging the installation of all the air, it is essential that a second tightness test is carried out to ensure the installation is gas tight.

This is because of the differences in the viscosity of LPG and air. LPG is more searching and certain levels of escape will not be detected on the initial tightness test where the installation contains air.

As with natural gas, providing there is no smell of gas, existing LPG installations with appliances connected are permitted a small leakage. Higher leakage rates (above 0.25 mbar) may be allowed for existing installations, where appliances are connected. This will require the installation volume to be calculated to determine the allowed maximum permissible pressure drop (see Table 13.18).

Installation volume	Maximum permissible pressure drop
≤ 0.0025 m3	2 mbar
>0.0025 m^3 – ≤ 0.005 m^3	1 mbar
>0.005 m^3 – ≤ 0.01 m^3	0.5 mbar
>0.01 m^3 – ≤ 0.035 m^3	No perceptible movement (fall)

Table 13.18 Maximum permissible pressure drops on existing LPG installations with appliances connected

Test Equipment

The test equipment required is:

- A 600 mm manometer (water U gauge) or an electronic gauge. The water U gauge used for LPG is twice the size of a natural gas manometer and reads up to 60 mbar.
- Flexible rubber tubing.
- A stopwatch or other timing device.

Figure 13.14 LPG water gauge

Pre-test inspection for LPG and LPG/air mixture

Regardless of whether the test will be carried out using LPG or LPG/air mixture it is necessary to do the following safety checks.

- Ensure that all sections of pipe-work are connected.
- Inspect joints to ensure they are correctly fabricated.
- Ensure exposed gas ways are plugged or capped.
- If appliances are installed, ensure the isolation valves are open and burner control taps are turned off.
- Ensure any cooker fold down lid is raised.
- Ensure that the LPG is isolated at the cylinder tank or tank valve(s).
- Ensure the pressure gauge is correctly assembled, zeroed and connected to a suitable point on the installation in the section of pipe-work. This must be on the outlet of the supply control valve and final stage regulator (where applicable).

TIGHTNESS TESTING PROCEDURE (AS PER IGEM/UP/1B EDITION 3)

LPG (Propane and Butane)

Begin by carrying out the pre-test inspection, as above. Then carry out the let by test.

- This test procedure covers let by testing of the cylinder(s) tank valve(s) or supply control valve.
- Slowly open and close the supply control. Adjust pressure to between 7–10 mbar.
- If there is an UPSO valve on the supply, operate the UPSO lever. This may cause an increase in pressure. If necessary adjust the pressure to the required test pressure of between 7–10 mbar.
- Observe the gauge for one minute.
- Where no perceptible movement is observed, the valve has passed the test. Any rise greater than 0.25 mbar indicates the valve is letting by.
- Confirm let by, disconnect the union and apply LDF to the valve barrel or ball. If a cylinder valve is found to be leaking, the cylinder should be exchanged. If a bulk tank valve is found to be leaking, the gas supplier should be notified.
- If the valve is letting by then you must make safe the installation, and contact the Gas Supplier/ESP.
- Slowly open the supply control valve and raise pressure to the appropriate tightness test pressure. (See Table 13.17).
- Close supply control valve. Remember to check for locking up.

Remember, LPG by nature may require longer stabilisation periods than the figures quoted. After completing the let by test continue with the tightness test.

- Allow one minute for temperature stabilisation (wait longer if necessary).
- Observe the gauge for two minutes.
- For all installations (with or without appliances connected) if there is no perceptible movement of the gauge reading and no smell of gas, the installation is deemed to have passed the test.
- If the installation fails the test either trace and repair the escape(s) and re-test the installation make the installation safe by sealing it with an appropriate fitting.
- Remember, where the LPG installation is new or if an existing installation has been exposed to air, purge the installation and repeat the tightness test procedure.
- Remove the pressure gauge and re-seal the test point.
- Slowly turn on the gas supply and test the pressure test point, supply control valve outlet connection and regulator connections with LDF.
- Record the results and where appropriate, inform the responsible person.

Purging procedures for LPG

Within the vicinity of the purging activity, ensure the following safety precautions are taken throughout the purging process.

- Avoid any accumulation of gas within confined spaces.
- Prevent inadvertent operation of any electrical switches or appliances.
- Extinguish all potential sources of ignition.
- Ensure there is no smoking or naked lights.
- Ensure good ventilation by opening doors and windows.
- Give advice to the responsible person and those in the area of the intent to purge and that there will be a smell of gas.

These precautions are applicable even if a source of ignition is held adjacent to the purged gas, as a mixture of un-ignited gas/air may be released until a suitable mixture is achieved.

- Ensure all appliances are turned off.
- Slowly turn on the gas supply and, if applicable, take a note of the meter reading.
- From a suitable purge point on the installation turn on a burner control tap on an appliance with an open burner.
- Ignite the purge mixture at an open burner as soon as possible, by holding a source of ignition adjacent to the burner head or by continually operating the appliance ignition system. It may be necessary to connect a temporary burner onto the installation.
- Confirm the presence of gas by igniting the burner and turn off the control tap.
- Upon establishing the presence of gas, return to the meter where applicable and note the volume of gas that has passed. Continue to purge until the correct purge volume has been passed (see Table 13.19).
- Regardless of the installation type (meter or no meter), ensure that every branch of pipe-work is purged.
- Establish a stable flame picture at each appliance.

Where appliance(s) are identified as not being commissioned, either the appliance(s) must be commissioned or disconnected from the supply and sealed with an appropriate fitting. Attach a label indicating the appliance(s) are not commissioned.

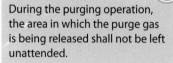

Safe working

During the purging operation, the area in which the purge gas is being released shall not be left unattended.

Type of installation		Purge volume (PV)
Meter designation	Pipework diameter	
U6, G4, E6	< 28 mm	0.01 m³ (0.35 ft³)
U6, G4, E6, U16, G10	28 mm ≤ 35 mm	1.5 IV

Table 13.19 Purge volume

Tightness testing for LPG/Air Mixture

This applies to installations that are supplied with LPG/air mixture, such as parts of the Isle of Man. It does not apply to installations where air is used for the tightness test procedure. Begin the test by carrying out the pre-test inspection then the let by test, as with LPG (see above). If let by is detected, confirm it and apply LDF to the valve barrel or ball. You must always make the installation has been made safe.

- Slowly open the supply control valve and raise pressure to the appropriate tightness test pressure (see Table 13.16).
- Close supply control valve and allow a minimum of one minute for stabilisation.
- Observe the gauge for two minutes.
- For all installations (with or without appliances connected) if there is no perceptible movement of the gauge reading and no smell of gas the installation shall be deemed to have passed the test.
- If the installation fails, either trace and repair the escape(s) and re-test the installation or make the installation safe by sealing it with an appropriate fitting.
- If the installation passes the test remove the pressure gauge and re-seal the test point. Slowly turn on the gas supply and test the pressure test point, supply control valve outlet connection and regulator connections with LDF.
- Record the results and where appropriate, inform the responsible person.

Type of Installation		Maximum permissible pressure drop
Meter Designation	Pipe-work Diameter	
No Meter ECV/AECV only (e.g. flat)	<35mm	1.5 mbar
Diaphragm meter <6 m³hr (e.g. U6 G4)	<35mm	1.5 mbar
Diaphragm meter > 6 m³hr < 16 m³hr (e.g. U16, G10)	<35mm	0.5 mbar

Table 13.20 Maximum permissible pressure drop for an individual dwelling with existing appliances connected (LPG/Air)

Purging procedures LPG/air mixture

Before taking on a purging activity, you will need to follow the same safety steps as you followed for LPG (see page 484). These precautions are applicable even if a source of ignition is held adjacent to the purged gas, as a mixture of un-ignited gas/air may be released until a suitable mixture is achieved.

The purging procedure you will follow is very similar to the one for LPG (see page 484). You will need to carry out the same process here, taking note of the correct purge volumes, as shown in Table 13.19.

TIGHTNESS TESTING INSTALLATIONS WITH AIR

This test came into force 1st July 2012. This test below may be undertaken before the installation is connected to a live gas supply. After the installation is connected to a live gas supply it must be tightness tested again and purged with fuel gas.

Tightness test process

Begin by carrying out the pre-test inspection (as above). Then follow the procedure below.

For OP's of 30 mbar found on caravans designed for use on the road, where the gas installation has been installed in accordance with BS EN 1949, the installation will have been tightness tested with air by the manufacturer.

- Connect the pressure gauge to the installation via a branch of a test T piece, which is valved on the other branch for air to be pumped into the installation.
- Slowly raise the pressure in the installation with air to the appropriate value given in Table 13.21 then turn off the pressure source. If a regulator is fitted to the section to be tested, avoid higher pressures to prevent regulator lock up.
- Allow one minute for the pressure and temperature within the installation to stabilise. At the end of the stabilisation period re-adjust the pressure to the TTP then turn off the pressure source. A slight increase or decrease in pressure readings during this time may be due to pressure and/or temperature within the installation stabilising. A major pressure decrease is probably due to an escape in the installation – this will need to be rectified before recommencing the test.
- Over the next two minutes check for any perceptible movement of the gauge reading. If there is none the installation is deemed to have passed the test.
- If the installation fails, either trace and repair the escape(s) and re-test the installation or make the installation safe by disconnecting appliance(s) or the relevant section, as appropriate, and sealing all open ends with an appropriate fitting.
- Upon completion of the test, remove the pressure gauge and re-seal the test point/test T connection.
- Record the results and, where appropriate, inform the responsible person.
- Once the installation has been connected to a live gas supply the installation shall be tightness tested following the appropriate procedure, before the installation may be purged with fuel gas.

Fuel	Nominal operating pressure (mbar)	Tightness test pressure (mbar)
LPG/air	14	13 – 14
LPG/air	21	20 – 21
Natural gas	21	20 – 21
Butane	28	45 – 46
Propane	37	45 – 46

Table 13.21 Tightness test pressures for air

Knowledge check

1 What is the United Nations number for propane?

a 1012
b 1011
c 1978
d 1976

2 Butane cylinders normally go up to what size?

a 19 kg
b 13 kg
c 15 kg
d 6 kg

3 Which is heavier than air?

a Propane
b Butane
c Methane
d Ethane

4 To be competent on LPG, who do you have to be registered with?

a Corgi
b Gas Safe Register
c Health & Safety Executive
d Building Regulations

5 As a liquid is LPG lighter or heavier than water?

a Lighter
b Heavier
c Same
d It's not a liquid

6 The lowest temperature propane will still boil at is:

a –44 °C
b –45 °C
c 42 °C
d –42 °C

7 What does BLEVE stand for?

a Boiling Liquid Exciting Vapor Explosion
b Boiling Liquid Expanding Vapor Explosion
c Boiling Liquid Extreme Vapor Explosion
d Boiling Liquid Example Vapor Explosion

8 The flammability limit of butane is:

a 3–9 per cent
b 2–10 per cent
c 1–8 per cent
d 1.8–9 per cent

9 What is the expansion ratio of propane?

a 1 ltr = 220 ltrs
b 1 ltr = 250 ltrs
c 1 ltr = 230 ltrs
d 1 ltr = 274 ltrs

10 Which of the following is easier to ignite than natural gas?

a Propane
b Butane
c Methane
d A and B

11 How much oxygen does propane require to burn 1 m^3 of fuel?

a 5 m^3
b 6 m^3
c 10 m^3
d 2 m^3

12 How much faster is CO absorbed into the bloodstream than oxygen?

a 200 times
b 300 times
c 400 times
d Same as oxygen

13 What type of vent cannot be used to provide ventilation?

a Hit and miss
b Non-closeable
c Permanently open
d Correct size

14 What is the maximum off-take for a 34 kg cylinder in m^3/hr?

a 0.83
b 0.74
c 0.93
d 1.03

15 What distance should a propane cylinder be from an untrapped drain?

a 2 m
b 3 m
c 4 m
d 5 m

INDEX